MECHANICS
OF
MACHINERY

MECHANICS OF MACHINERY

Dr. Mahmoud A. Mostafa
University of Alexandria, Egypt

CRC Press
Taylor & Francis Group
Boca Raton London New York

CRC Press is an imprint of the
Taylor & Francis Group, an **informa** business

CRC Press
Taylor & Francis Group
6000 Broken Sound Parkway NW, Suite 300
Boca Raton, FL 33487-2742

© 2013 by Taylor & Francis Group, LLC
CRC Press is an imprint of Taylor & Francis Group, an Informa business

First published in paperback 2017

No claim to original U.S. Government works
Version Date: 2012912

ISBN 13: 978-1-138-07223-7 (pbk)
ISBN 13: 978-1-4665-5946-2 (hbk)

Library of Congress Cataloging-in-Publication Data

Mostafa, Mahmoud A.
  Mechanics of machinery / Mahmoud A. Mostafa.
    p. cm.
  Includes bibliographical references and index.
  ISBN 978-1-4665-5946-2 (hardback)
  1. Mechanical movements. 2. Machinery--Equipment and supplies. I. Title.

TJ181.M67 2012
621.8--dc23   2012030992

Visit the Taylor & Francis Web site at
http://www.taylorandfrancis.com

and the CRC Press Web site at
http://www.crcpress.com

Dedication

This text is dedicated to my recently born grand child Yahya Omar Mostafa

Contents

Foreword

It is not an easy task to write a foreword for a book authored by Professor Mostafa, who taught me the principles of machine dynamics and vibrations. In the course of his teaching and research career, Professor Mostafa has been the teacher and mentor to his students. He kept updating his methods of instruction according to the continuous developments in his field of expertise.

This book is the pinnacle of Professor Mostafa's contributions over the years. The chapters are based on his lectures at several universities in Egypt and the Middle East. The book covers both the graphical and analytical methods of the kinematics and dynamics of different types of mechanisms with low and high pairs. It presents new analytical approaches, which are helpful in the programming, and the kinematic and dynamic analysis of mechanisms and cams.

The book also presents new topics such as the analytical plot of cam contour, minimum cam size, and in-place balancing.

I am sure that both academia and the industry will benefit much from this book. The new topics, lucid language, and step-by-step examples are all assets to its success.

Professor Sohair F. Rezeka
Mechanical Engineering Department
Alexandria University, Egypt

Preface

This book is intended to serve as a reference for students in the mechanical field and practicing engineers, and is concerned with the analysis of machines.

This book discusses the kinematics and dynamics of mechanisms. It is intended as an informative guide to a more complete understanding of kinematics and its applications. It is hoped that the fundamental procedures covered here will transfer to problems the reader may encounter later.

This book was developed from an earlier version published in 1973. That early version was based on graphical analysis, which did not meet the requirements of modern developments. This version includes analytical analysis for all the topics. These analytical analyses makes it possible to use math software for fast, precise, and complete analysis.

Chapter 1 introduces several mechanisms to familiarize the reader with different motions and functions they can perform. Analytical analysis for the performance of the mechanism is also presented, which is adapted to use math software to facilitate the study of the performance of mechanisms.

Chapter 2 deals with the study of velocities and accelerations in the mechanism. This is a necessary step for the design of machines. The graphical method, which is based on vector equations, is presented and is applied to different mechanisms. The graphical method gives insight to the velocities and accelerations for members in the mechanism. Also, analytical analysis is presented and adapted for use with math software for an overall study of the mechanisms. One distinct feature of this book is the analysis of sliding links using a theory developed by the author. It is a replacement for Coriolis components, which are generally difficult to apply in most cases.

The subject of cams is presented in Chapter 3. For specified motion cams, the profile is obtained by graphical method. To obtain the contour analytically, equations in Cartesian coordinates, which was developed by the author, is presented. Special emphasis is directed toward the factors affecting the cam design, such as the pressure angle and the radius of curvature.

Chapters 4 through 6 are devoted to giving a realistic study of the geometry and kinematics of all types of gears. The study of gear reduction units is very important for machine application.

Chapter 7 is concerned with the study of force analysis in mechanisms. Force analysis is divided into three parts—static force analysis, friction force analysis, and dynamic force analysis. In this book, the traditional graphical method is used in addition to the analytical method. The analytical method lays down the foundation for using math software to perform the analysis. Programs using MathCAD are presented for complete analysis of all kinds of mechanisms, which include position analysis, velocities and acceleration analysis, and force analysis. This chapter also includes the study of the torque variation and the use of flywheels to reduce the speed variation.

Chapter 8 covers the study of balancing of machines. It explains how to balance rotating parts and reciprocating parts. In-place balancing of machines using vibration measurements is also presented.

Professor Mahmoud Mostafa
Mechanical Engineering Department
Faculty of Engineering
University of Alexandria
Alexandria, Egypt

Acknowledgments

The author wishes to express his gratitude to the coauthors, Professor Nomaan Moharem, late Professor Elssayyed Elbadawy, and late Professor Hasan Elhares, of the previous version who encouraged the idea of making a complete version, with updated materials, in the field of "Mechanics of Machinery."

I would like to express my deep appreciation to my wife Suzy, my sons Hosam and Omar, and my daughter Gina for their support and their help and encouragement to finish this book. So, this book is also dedicated to them.

Also, I would like to thank my students of the Faculty of Engineering, University of Alexandria, for their help to rectify some materials in the book.

Finally, I would like to express my appreciation to the editorial staff of Taylor & Francis Group for their encouragement. Their comment that my book will be a very good addition to their list made me proud.

Author

Dr. Mahmoud Mostafa is currently a professor in the mechanical engineering department, Faculty of Engineering, University of Alexandria. He received his BSc degree in mechanical engineering from the University of Cairo. He also received his MSc and PhD degrees from Oregon State University. He held several positions such as research engineer in E.I. Du Pont Company, USA, June 1964; visiting professor in University of Riyadh, Saudi Arabia; and visiting professor in Beirut Arab University. Dr. Mostafa was the head of the mechatronics department at Alexandria High Institute of Engineering and Technology and is still supervising the department. His area of research is in the fields of mechanical vibrations, kinematics of machinery, and synthesis of mechanisms. He was a consultant for several corps in Egypt. He attended international conferences in the United States, Canada, Pakistan, and Egypt. He has several inventions, which incorporate certain mechanisms.

1 Mechanisms

1.1 DEFINITIONS

Machine: A machine, according to Reuleaux, is a combination of resistant bodies (rigid, elastic, or fluid) so arranged that by their means the mechanical forces in nature can be compelled to produce some effect or work accompanied by certain determinate motions. Figure 1.1 shows a cross section of a single-cylinder engine (or compressor). For an engine, a mixture of air and vapor of flammable fluid enters the cylinder, is ignited, pushes the piston, and the connecting rod, which in turn causes the crank to rotate. Thus, the engine transmits the gas force to be a torque on the crank. Also, it converts the reciprocating motion of the piston to a rotary motion for the crank. The function of the compressor is opposite to the engine.

Mechanism: A combination of bodies meant for transmitting, controlling, or constraining the relative motion between the bodies. If we look at a machine only from the point of view of motion, then it is a mechanism. Figure 1.2 shows the skeleton outline of an engine and is considered to be a mechanism.

Planar and spatial mechanisms: Mechanisms can be divided into planar mechanisms and spatial mechanisms according to the relative motion of rigid bodies. In planar mechanisms, all the relative motions of rigid bodies are in one plane or in parallel planes. If there is any relative motion between the bodies that is not in the same plane or in parallel planes, the mechanism is called a spatial mechanism. In other words, planar mechanisms are essentially two dimensional, whereas spatial mechanisms are three dimensional. This chapter covers only planar mechanisms.

Kinematics: Kinematics of mechanisms is concerned with the motion of the parts without considering the actual shape of the bodies or the forces in a machine. In other words, kinematics deals with the motion, velocity, and acceleration of the parts.

Kinetics: Kinetics deals with all the forces in a machine, including the forces resulting from the masses and the acceleration.

Dynamics: Dynamics is a combination of kinematics and kinetics.

Links: A link is defined as any part of a machine having motion relative to some other part. It must be capable of transmitting a force. There are three types of links:

1. Rigid links, which may transmit tension or compressive forces such as crank, connecting rod, and piston
2. Tension links, which transmit only tensile forces such as belts, ropes, and chains
3. Compression links, which transmit only compressive force, for example, the fluid in hydraulic jacks or the automobile braking system

1

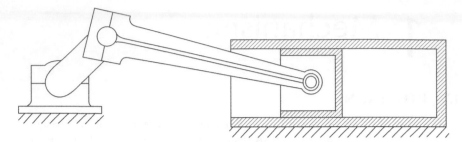

FIGURE 1.1   A cross section of a single-cylinder engine.

FIGURE 1.2   The skeleton outline of an engine.

Frame: The frame in a mechanism is considered as a fixed link. In a machine, the frame is considered as all the fixed bodies connected together by welding or bolting.

Pairs: A pair is a joint between the surfaces of two rigid bodies which keeps them in contact and to have relative motion. Pairs are divided into two types, that is, lower pairs and higher pairs.

Lower pairs: A joint between two bodies is defined as a lower pair when the contact between them is on a surface. There are two types of lower pairs in plane mechanisms: One is the revolute joint as the case of doors (Figure 1.3a). In the study of mechanisms, the revolute joint is represented as skeletons (Figure 1.3b and c). The other type of lower pairs is the prismatic (sliding) joints as in the case of drawers and the ram of the shaping machine. The skeleton outline of these joints is shown in Figure 1.4a and b.

Higher pairs: In higher pairs, the contact between two bodies is through a point as in the case of ball bearings or through a line as in roller bearings, gears, cams, and cam followers. The skeleton outline of this type of pairs is shown in Figure 1.5a. In fact, a higher pair can be considered as two lower pairs, that is, sliding and revolute pairs (Figure 1.5b).

Other types of joints:

Spherical joint: A ball and a socket represent a spherical joint (Figure 1.6). The shift stick in an automobile is an example for the spherical joint. The handle can move in all directions.

Screw joint: Bolts and nuts are examples of screw joints. Power screws and screw jacks are other examples (Figure 1.7).

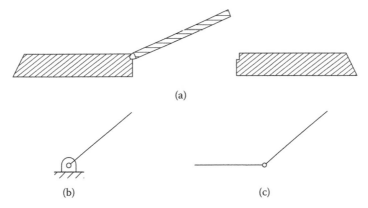

FIGURE 1.3 (a) Represents a door connection, (b) and (c) represent the revolute joints.

FIGURE 1.4 Representation of prismatic joint. (a) Link slides inside another (b) link slide on the surface of another.

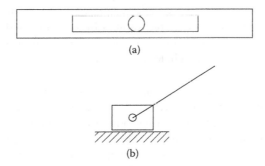

FIGURE 1.5 (a) Represents a higher pair joint, (b) represents the equivalent higher pair joint.

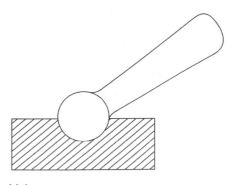

FIGURE 1.6 Spherical joint.

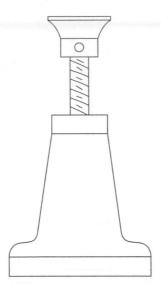

FIGURE 1.7 Screw joint.

Complete, incomplete, and successful constraints:
- Complete constraint determines in a definite direction the relative displacements between two links independently of the line of action of the impressed force, for example, a square bar sliding in a square hole.
- In the case of incomplete constraint, a little change in direction of the impressed force may alter the direction of the relative displacement, for example, a cylinder in a hole. The cylinder may rotate and slide inside the hole.
- In the case of successful constraint, an external force, for example the force of gravitation or a force applied to a spring or fluid, is impressed on an element to prevent motions other than the desired relative motion within the limits of the displacement. For example, the relative motion between the piston and the cylinder of an engine is not completely constrained. But the connecting rod between the piston and the crank prevents the piston from rotating inside the cylinder.

Kinematics chain: Kinematic chains are combinations of links and pairs without a fixed link. If one of the links is fixed, we get a mechanism. All links have at least two pairs. The relative motion between the links is completely constrained.

Kinematics analysis: Kinematics analysis is the investigation of an existing mechanism regarding its performance and motion and estimating the velocity and acceleration of its links.

Kinematics synthesis: It is the process of designing a mechanism to accomplish a desired task. It is involved with choosing the type and dimensions of the mechanism to achieve the required performance.

Degrees of freedom (DOFs): The number of DOFs of a system is defined as the number of independent relative motions among the rigid bodies of the system.

For example, for a revolute pair (Figure 1.3b and c), relative motion between the links is a rotational motion about the joint. So, the revolute pair has only one DOF. This applies to prismatic pairs also (Figure 1.4) for which the relative motion is sliding motion. For higher pairs (Figure 1.5a), the number of DOFs is two. This is because the joint allows both rotational and translational motions as demonstrated in Figure 1.5b. For a spherical joint (Figure 1.6), the motion is not restricted to certain directions. Thus, it has infinite DOFs.

If the number of DOFs of a chain is zero or negative, then it forms a structure, that is, there is no relative motion between the links.

The number of DOFs of a mechanism is also called the mobility of the device. Mobility of a device is the number of input parameters (usually pair variables) that must be independently controlled to bring the device into a particular position.

Important: The number of pairs at a joint is equal to the number of links connected to the joint subtracted by one.

1.2   DEGREES OF FREEDOM OF PLANAR MECHANISMS

The number of DOFs of a mechanism can be estimated by using Gruebler's equation, which is written in the following form:

$$DOFs = 3(n - 1) - 2l - h$$

Where

DOF is the number of degrees of freedom in the mechanism.
n is the number of links including the fixed link.
l is number of lower pairs.
h is the number of higher pairs.

EXAMPLE 1.1

Find the number of DOFs for each of the following chains:
Figure (a)

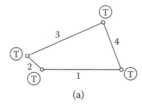

(a)

$n = 4$
$l = 4$
$h = 0$

$$DOFs = 3 \times (4 - 1) - 2 \times 4 - 1 \times 0$$
$$= 1$$

Figure (b)

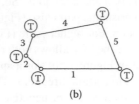

(b)

$n = 5$
$l = 5$
$h = 0$

$$DOF = 3 \times (5 - 1) - 2 \times 5 - 1 \times 0$$
$$= 2$$

Figure (c)

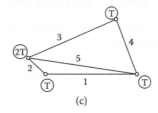

(c)

$n = 5$
$l = 6$
$h = 0$

$$DOF = 3 \times (5 - 1) - 2 \times 6 - 1 \times 0$$
$$= 0$$

Figure (d)

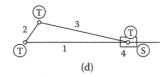

(d)

$n = 4$
$l = 4$
$h = 0$

$$DOF = 3 \times (4 - 1) - 2 \times 4 - 1 \times 0$$
$$= 1$$

Figure (e)

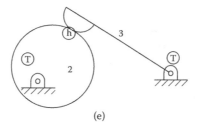

(e)

$n = 3$
$l = 2$
$h = 1$

$$DOF = 3 \times (3-1) - 2 \times 2 - 1 \times 1$$
$$= 1$$

EXAMPLE 1.2

Find the number of DOFs for the mechanism shown in Figure 1.8.

$n = 8$
$l = 10$
$h = 0$

$$DOF = 3 \times (8-1) - 2 \times 10 - 1 \times 0$$
$$= 1$$

Note that,
   T denotes a turning joint.
   S denotes a sliding joint.

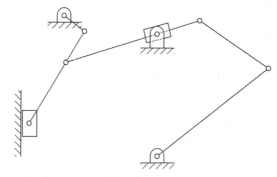

FIGURE 1.8   Examples for degrees of freedom.

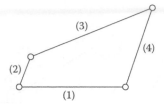

FIGURE 1.9   Four-link chain.

1.3   FOUR-REVOLUTE-PAIRS CHAIN

The four-bar linkage is the simplest of all chains. It is often used to construct very useful mechanisms. Consider a four-link chain with four revolute pairs (Figure 1.9). Consider that the links have different lengths, r_1, r_2, r_3, and r_4. Grashof's theorem states that a four-bar mechanism has at least one revolving link if

$$r_1 + r_2 < r_3 + r_4$$

Also, the three moving links rock if

$$r_1 + r_2 > r_3 + r_4$$

For this chain, if we fix one link at a time we obtain, in a general sense, several mechanisms that may be different in appearance and in the purposes for which they are used. Each mechanism is termed an inversion of the original kinematics chain.

1.3.1   FOUR-BAR MECHANISM

A four-bar mechanism (Figure 1.10) is obtained by fixing link (1) in the four revolute chains shown in Figure 1.9. This mechanism transfers the rotary motion of one link to an oscillatory motion for another link or vice versa. The links of the four-bar mechanism are denoted as follows:

Link (1) is called the frame.
Link (2) is called the crank.
Link (3) is called the coupler.
Link (4) is called the rocker.

1.3.1.1   Performance of the Four-Bar Mechanism

Referring to Figure 1.10,
- Link (2) makes a complete revolution
- Link (4) oscillates through an angle β, called the rocking angle. These motions are assured if the following conditions are applied:
- According to the extreme right position,

$$r_2 + r_3 < r_1 + r_4$$

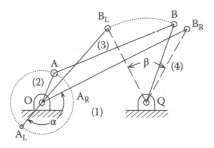

FIGURE 1.10 The four-bar mechanism.

- According to the extreme left position,

$$r_3 - r_2 < r_1 + r_4$$

From the previous two conditions, we can deduce that

$$r_3 < r_1 + r_4$$

- When the crank coincides with link OQ to the right,

$$r_1 - r_2 < r_3 + r_4$$

- When the crank is along link OQ to the left,

$$r_1 + r_2 < r_3 + r_4$$

From these two conditions we can deduce that

$$r_1 < r_3 + r_4$$

- The extreme right position, point B_R of the rocker, is when the coupler is along the crank and the crank is at point A_R.
- The extreme left position, point B_L of the rocker, is when the coupler coincides with the crank at point A_L.
- The crank rotates through an angle α when the rocker moves from the extreme left position to the extreme right position, assuming that the crank rotates counterclockwise.
- The crank rotates through an angle $2\pi - \alpha$ when the rocker moves from the extreme right position to the extreme left position.
- If α is not equal to π, the motion of the rocker is described as quick return motion. That is, the rocker moves faster when going from left to right than when going back. The ratio of the two angles, assuming the crank rotates with uniform speed, is called the time ratio, λ:

$$\lambda = \frac{2\pi - \alpha}{\alpha}$$

- The static driving force is transmitted from the crank to the rocker through the coupler. This force is either tension or compression. The angle between the rocker and the coupler (angle ABQ) is called the transmission angle. The torque transmitted to the rocker has a maximum value when this angle is $\pi/2$. When this angle deviates from $\pi/2$, the torque transmitted to the rocker decreases. It is advisable to keep the value of this angle as close to $\pi/2$ as possible.
- The motion of the mechanism is traced by the following steps (Figure 1.11):
  - Draw a circle with radius equal to the length of the crank and center at point O.
  - Divide the circle to an equal number of divisions. The more divisions the more accurate results.
  - Draw an arc of a circle with radius equal to the length of the rocker and center at point Q.
  - At each point A on the circle, line OA makes an angle θ with the horizontal position of the crank, draw an arc with radius equal to the length of the rocker to intersect the arc of the rocker at point B.
  - Measure the angle of line QB, that is, angle φ in Figure 1.11.

The relation between the output angle φ and the input angle θ is shown in Figure 1.12. The angles are measured from the horizontal datum.

1.3.1.2 Coupler Curves

A point on a coupler link traces a curve (Figure 1.13). Tracing is carried out by using the steps described as in Section 1.3.1.1 and then locating the position of point C at different locations. By changing the position of this point, we can obtain a vast number of curves that may be helpful in several mechanical engineering applications.

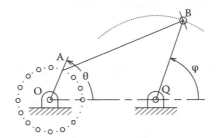

FIGURE 1.11   Tracing the four-bar mechanism.

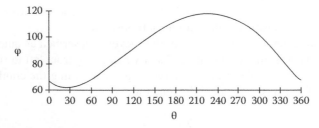

FIGURE 1.12   Relation between the rocker angle and the crank angle.

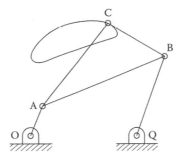

FIGURE 1.13 The coupler curve.

Important: The process of tracing a mechanism is tedious. However, the analytical method which will be explained in this chapter is much simpler and more powerful than tracing.

1.3.1.3 Synthesis of a Four-Bar Mechanism

It is interesting to design a four-bar mechanism to give a certain performance. This is explained in the following examples.

EXAMPLE 1.3

Design a four-bar mechanism such that the length of the crank (r_2) is 30 mm, the length of the fixed link (r_1) is 100 mm, the rocking angle (β) is 60°, and the time ratio (λ) is 1.

SOLUTION

In Figure 1.10, at the extreme right position, line $A_R B_R$ is along line OB_R. Also, line $A_L B_L$ is along line OB_L. Since the time ratio (λ) is 1,

$$\lambda = \frac{2\pi - \alpha}{\alpha} = 1$$

$$\alpha = \pi = 180°$$

Since $\alpha = 180°$, lines OB_L and OB_R coincide. Also, the distance $B_L B_R$ is twice the length of the crank. We use this data and proceed with the following steps:

PROCEDURE

1. Draw line $B_L B_R$ with length = 60 mm.
2. Draw lines QB_L and QB_R, which are equal and make an angle 60°. Line QB_R is equal in length to the rocker.
3. Draw an arc of a circle of radius 100 mm (the length of the fixed link) to intersect line $B_R B_L$ extended at point O. Point C is in the middle of line $B_L B_R$. Line OC represents the coupler.

Therefore, the lengths of links of the four-bar mechanism in Figure 1.14 are as follows:

$$r_1 = 100\,mm, r_2 = 30\,mm, r_3 = 85.4\,mm, r_4 = 60\,mm.$$

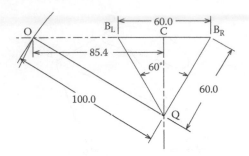

FIGURE 1.14 Graphical solution of Example 1.3.

EXAMPLE 1.4

Design a four-bar mechanism such that the length of the fixed link (r_1) is 80 mm, the length of the rocker (r_4) is 60 mm, the rocking angle (β) is 90°, and the time ratio (λ) is 1.4.

SOLUTION

Since the time ratio (λ) is 1.4,

$$\lambda = \frac{2\pi - \alpha}{\alpha} = 1.4$$

$$\alpha = 150°$$

In Figure 1.10, the angle between lines OB_R and OB_L is equal to $180° - \alpha = 30°$.

PROCEDURE

1. Draw lines QB_L and QB_R each of length 60 mm. Angle B_LQB_R is equal to 90°. These two lines represent the rocker at the two extreme positions as in Figure 1.15.
2. Draw lines $Q'B_L$ and $Q'B_R$ such that the angle $B_LQ'B_R$ is equal to 30°.
3. Draw a circle passing through points B_L, Q', and B_R. We should bear in mind that lines from points B_L and B_R to any point on this circle make an angle equal to 30°.
4. From point Q, draw an arc of a circle of radius 80 mm (the length of the fixed link) to intersect the circle at point O.

Line OB_R represents $r_3 + r_2 = 130.4$ mm
Line OB_L represents $r_3 - r_2 = 58.7$ mm
Therefore, the lengths of links are as follows:

$$r_1 = 80\,\text{mm},\ r_2 = 40.850\,\text{mm},\ r_3 = 89.550\,\text{mm, and}\ r_4 = 60\,\text{mm.}$$

Synthesis can also be performed for coupler curves by satisfying certain precision points [19] or for specific outputs [59], which is not within the scope of this chapter.

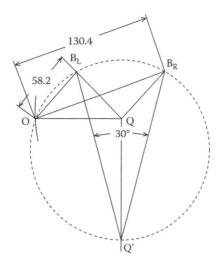

FIGURE 1.15 Graphical solution of Example 1.4.

1.3.2 DRAG (DOUBLE ROTATING) LINK MECHANISM

If the shortest link in a chain, link (2) in the chain shown in Figure 1.9, is fixed, we obtain a mechanism in which two links rotate continuously (Figure 1.16). This condition is ensured by satisfying the following conditions:

• When link (4) is horizontal to the right,

$$r_1 + r_4 < r_2 + r_3$$

• When link (2) is horizontal to the left,

$$r_1 + r_2 < r_3 + r_4$$

This mechanism is called double-crank mechanism or commonly named as drag link mechanism. It is usually used as a part of compound mechanisms to obtain certain performance, as will be explained later in Section 1.8.1.1.

1.3.3 DOUBLE-ROCKER MECHANISM

If link (2) in the chain shown in Figure 1.9, which is the shortest link, is used as a coupler and link (3) is fixed, we obtain a mechanism in which the other two links oscillate, as shown in Figure 1.17. This condition is ensured by satisfying the following conditions:

• When link (4) is at the extreme right position,

$$r_1 + r_4 < r_2 + r_3$$

• When link (2) is at the extreme left position,

$$r_1 + r_2 < r_3 + r_4$$

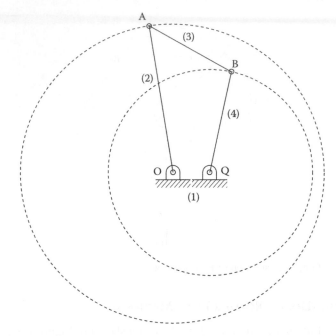

FIGURE 1.16   Drag link mechanism.

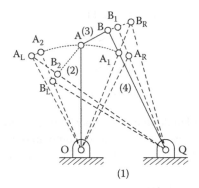

FIGURE 1.17   The double rocker mechanism.

1.3.3.1   Performance of the Double-Rocker Mechanism

- We start the trace by moving link (4) to the right. It reaches its extreme right position when link (3) is along link (2). Point B becomes point B_R and point A becomes point A_1.
- From this position, link (2) keeps moving to the right while link (4) starts to move to the left. Link (2) reaches its extreme right position when link (3) coincides with link (4). At this position, point B becomes B_1 and point A becomes point A_R.

- From this position, link (4) keeps moving to the left dragging link (2) behind it. Link (4) reaches its extreme left position when link (3) coincides with link (2). At this position, point B becomes point B_L and point A becomes point A_2.
- From this position, link (4) moves to the right while link (2) keeps on moving to the left. Link (2) reaches it extreme left position when link (3) is along link (4). At this position, point B becomes point B_2 and point A becomes point A_L.

The motion is repeated as described.

1.3.4 APPLICATIONS BASED ON FOUR-BAR LINKAGES

There are many practical applications that are based on four-bar linkages. Some of them are listed in Sections 1.3.4.1 through 1.3.4.3 and some others are listed later.

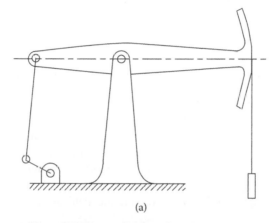

(a)

(b)

FIGURE 1.18 The beam engine. (a) Skeleton outline of a beam engine used in deep oil wells (b) photograph of the engine.

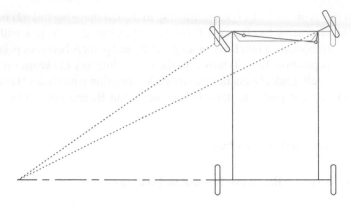

FIGURE 1.19   Ackermann steering mechanism.

1.3.4.1   Windshield Wiper of Automobiles

The oscillating motion of automobile wipers is achieved by using a four-bar mechanism.

1.3.4.2   Beam Engine

The skeleton outline of the beam engine used in deep oil wells is shown in Figure 1.18a. A photograph of the engine is shown in Figure 1.18b.

1.3.4.3   Automobile Steering Mechanism

This mechanism is essential for vehicles. During turns, if the steered wheels, usually the front wheels, of a vehicle are kept parallel, each wheel will have a different center of rotation. This will cause slip in the wheels, accelerating their damage. The correct situation is to make the whole vehicle rotate around one center only (Figure 1.19). This is accomplished by adjusting the angles of rotation of the front wheels. This is accomplished by using the Ackermann steering mechanism as shown in Figure 1.19.

1.4   SINGLE-SLIDER CHAIN

A single-slider chain is obtained by replacing one of the revolute pairs in the four-revolute-pairs, discussed in Section 1.3, by a prismatic pair, as shown in Figure 1.20. The prismatic pair is between links (1) and (4). We obtain different mechanisms if we fix one link at a time. Each mechanism is an inversion of the original single-slider chain.

1.4.1   ENGINE MECHANISM

The engine mechanism (Figure 1.21) is obtained by fixing link (1) in the single-slider chain shown in Figure 1.20. This mechanism transfers the rotary motion of one link to a reciprocating motion for another link or vice versa.

Link (1) is called the frame.
Link (2) is called the crank; it has a length R.
Link (3) is called the connecting rod; it has a length L.
Link (4) is called the piston.

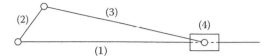

FIGURE 1.20   Single-slider chain.

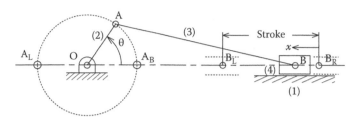

FIGURE 1.21   The engine mechanism.

1.4.1.1   Performance of the Engine Mechanism

- The crank, link (2), makes a complete revolution.
- The piston, link (4), has a reciprocating motion.
- The extreme right position of the piston, called top dead center, occurs when the connecting rod is along the crank. At this position, point B becomes point B_R and point A becomes point A_n.
- The extreme left position of the piston, called bottom dead center, occurs when the connecting rod coincides with the crank. At this position, point D becomes point B_L and point A becomes point A_L.
- The distance between the top dead center and the bottom dead center is called the stroke.
- In this configuration, the centerline of the piston passes through the center of rotation of the crank. In some designs, this line is shifted away from the center of rotation of the crank. In this case, the time taken by the piston to move from right to left is not the same as the time taken when it moves from left to right.
- The motion of the piston from the top dead center as the crank rotates through an angle θ is given as follows:

$$x = R(1 - \cos\theta) + L\left[1 - \sqrt{1 - \left(\frac{R}{L}\sin\theta\right)^2}\right]$$

1.4.1.2   Radial Engine

The radial engine mechanism is used in automobile engines. It is available in multibanks either in line or radial (called V engine). Radial engines are used in aircraft engines where a group of cylinders are arranged in radial positions with the crank shaft (Figure 1.22a). The number of cylinders is usually an odd number. In some engines, the cylinder bank is fixed while the crank rotates. Some engines are available in which the crank is fixed while the cylinder bank rotates as in the motorcycle

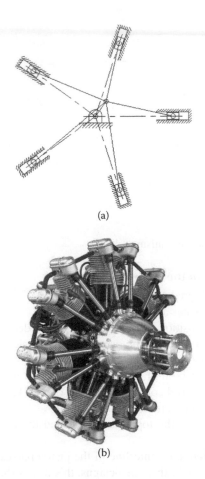

(a)

(b)

FIGURE 1.22 The radial engine. (a) Schematic view (b) photo for the radial engine.

engine shown in Figure 1.22b. This gives more mass moment of inertia for the rotating parts. The photograph is of Rotec's R2800-7 Cylinder 110 HP, courtesy of Rotec Engineering, Houston, Texas.

1.4.2 QUICK RETURN OSCILLATING LINK MECHANISM

This mechanism is an inversion of the single-slider chain of Figure 1.21. Link (3) in the slider chain is fixed and is denoted as link (1) in the quick return oscillating mechanism. The sliding joint between links (3) and (4) in the chain is placed at the end of the crank, link (2), as shown in Figure 1.23. It is also the same as fixing link (2) and making it longer than link (3).

The extreme positions of the oscillating link (4) occur when it is tangent to the crank circle. The extreme positions of the crank are located at points A_R and A_L. Link (4) moves through an angle β between these two positions. When the crank moves from the extreme right position to the extreme left position, assuming clockwise rotation, it rotates through an angle α and returns back through an angle $2\pi - \alpha$:

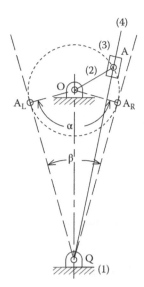

FIGURE 1.23 The quick return oscillating link mechanism.

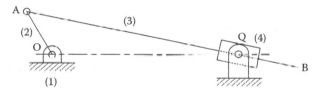

FIGURE 1.24 The tilting block mechanism.

$$\alpha = 2\cos^{-1}\frac{OA}{OQ}$$

Link (4) has a quick return motion with a time ratio, λ, given by

$$\lambda = \frac{2\pi - \alpha}{\alpha}$$

1.4.3 OSCILLATING (TILTING) BLOCK MECHANISM

This mechanism (Figure 1.24) is practically similar to the oscillating link mechanism described in Section 1.4.2. In this case, block (3) in Figure 1.23 is replaced by a link and link (4) is replaced by a block.

1.4.4 DOUBLE ROTATING LINK MECHANISM

In this mechanism (Figure 1.25), the shortest link is fixed and the other links rotate continuously. It is usually used as a part of compound mechanisms to obtain certain performance as will be explained in Section 1.8.1.

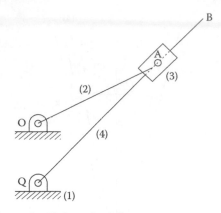

FIGURE 1.25   Double rotating link mechanism.

1.5   DOUBLE-SLIDER MECHANISMS

Some of the mechanisms using two sliders are presented in Sections 1.5.1 through 1.5.3.

1.5.1   Scotch Yoke Mechanism

This mechanism (Figure 1.26) transfers the rotary motion of the crank to a reciprocating motion for the yoke, link (4). The displacement of the yoke from the extreme right position is given by

$$x = R(1 - \cos\theta)$$

The function of this mechanism is similar to the engine mechanism presented in Section 1.4.1. The difference is that the motion of the yoke is pure harmonic motion, which is useful in many applications.

1.5.2   Ellipse Trammel

This mechanism (Figure 1.27) is used to trace an exact ellipse. It consists of a board, that is, link (1) in the figure, which has two perpendicular slots. Each slot has a slider, links (2) and (4), which slides along it. The two sliders are connected to link (3) by revolute pairs. Point P on link (3) traces an exact ellipse. Line AP represents the major axis and line BP represents the minor axis of the ellipse. For an angle θ, the coordinates of point P are given as follows:

$$x = AP\cos\theta$$
$$y = BP\sin\theta$$

This is the parametric equation of the ellipse.

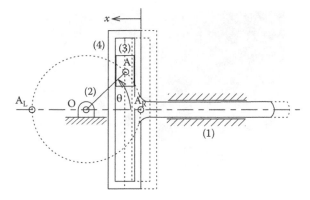

FIGURE 1.26   The Scotch yoke mechanism

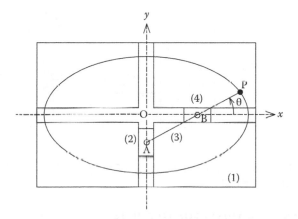

FIGURE 1.27   Ellipse trammel.

1.5.3 OLDHAM COUPLING

Oldham coupling transmits uniform angular speed between two parallel shafts whose axes do not coincide and who are a radial distance apart.

The parts of the mechanism are shown in Figure 1.28a. The assembled mechanism is shown in Figure 1.28b. It consists of a driving disk (2), intermediate disk (3), and a driven disk (4).

Both driving and driven disks have rectangular recesses, which are positioned perpendicular to each other. The intermediate disk has two perpendicular rectangular slots, one at each side, which engage the recesses of the driving and driven disks. The center of the intermediate disk is located at the intersection of the centerlines of the driving and driven recesses. When the recess of the driving disk rotates through an angle θ, the center of the intermediate disk is located at point C, which is the intersection of the two recesses. Therefore, this center rotates on a circle with diameter, O_1O_2, equal to a (notice that angle O_1CO_2 is 90°; Figure 1.28c). When the driving disk rotates through 90°, the center of the intermediate disks describes 180° on its path circle. Thus, this center rotates with twice the angular speed of the driving and driven shafts.

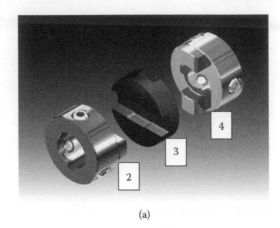

(a)

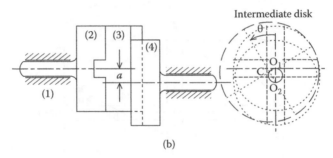

(b)

FIGURE 1.28   The Oldham coupling. (a) Photo of the parts (b) assembled drawing (the photograph is courtesy of Knoll).

1.6   MECHANISMS WITH HIGHER PAIRS

Higher pairs, as explained in Section 1.1, are pairs in which the contact between two bodies is through a point or a line. Some mechanisms with higher pairs are listed in Sections 1.6.1 through 1.6.3.

1.6.1   CAM MECHANISMS

Cam mechanisms (Figure 1.29) are used to transmit motion from a machine element (cam) to another machine element (follower) through direct contact. The nature of the contact depends on the type of the follower tip.

1.6.2   GEARS

Gears (Figure 1.30) are used to transmit positive motion between shafts, change the direction of motion, and change the speed of rotation.

1.6.3   GENEVA WHEEL

Geneva wheels (Figure 1.31) are used to transfer the rotary motion of a shaft to an intermittent motion for another shaft.

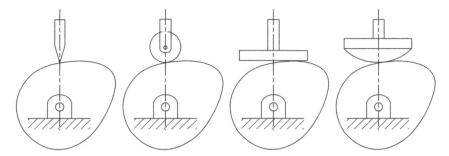

FIGURE 1.29 Cam mechanisms.

FIGURE 1.30 A pair of spur gears.

FIGURE 1.31 Geneva wheel.

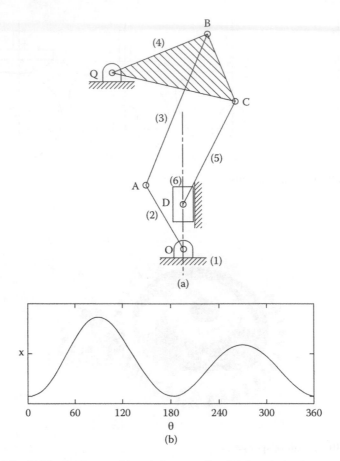

FIGURE 1.32 (a) The sewing machine skeleton outline (b) the relation between the needle motion "x" and the crank angle θ.

1.7 COMPOUND MECHANISMS

Usually, plane mechanisms have more than four links to give more complicated motions than those provided by simple mechanisms. To demonstrate how a compound mechanism may generate special motions, consider the sewing machine mechanism shown in Figure 1.32a. It consists of six links. If we trace the motion of the needle represented by point D, we find that it makes two strokes for one crank rotation (Figure 1.32b).

1.8 SPECIAL MECHANISMS

1.8.1 QUICK RETURN MOTION MECHANISMS

Quick return motion mechanisms are quite useful, especially for shaping machines. In these machines, the motion of the ram during the cutting stroke is slow. To save time, we need the ram to return faster. Some of the mechanisms used in these machines are listed in Sections 1.8.1.1 through 1.8.1.3.

1.8.1.1 Drag Link Mechanism

This mechanism (Figure 1.33) consists of the four-bar drag linkage described in Section 1.3.2 [links (1), (2), (3), and (4)] and the engine mechanism described in Section 1.4.1 [links (5) and (6)]. The crank is link (2), the coupler is link (3), the drag link is link (4), the connecting rod is link (5), and the ram is link (6). The cutting tool is fixed to the ram, that is, link (6). The extreme right position of the oscillating ram occurs when the tip of the crank is at point A_R. The extreme left position of the ram occurs when the tip of the crank is at point A_L. Assuming a clockwise rotation, the crank rotates through an angle $2\pi - \alpha$ when the ram moves from the extreme right position to the extreme left position. It returns back when the crank rotates through an angle α. The time ratio, λ, is given by

$$\lambda = \frac{2\pi - \alpha}{\alpha}$$

1.8.1.2 Shaper Mechanism

A shaper mechanism consists of two single-slider mechanisms, that is, the quick return oscillating link mechanism and the engine mechanism (Figure 1.34). The extreme positions are indicated in Figure 1.34.

The time ratio, λ, is given by

$$\lambda = \frac{2\pi - \alpha}{\alpha}$$

$$\alpha = 2\cos^{-1}\left(\frac{OA}{OQ}\right)$$

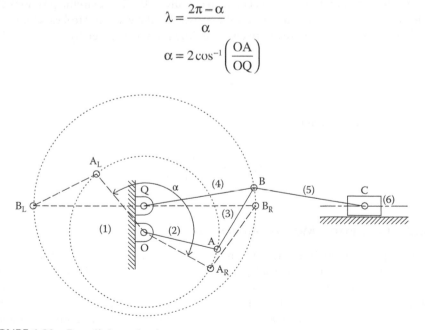

FIGURE 1.33 Drag link mechanism.

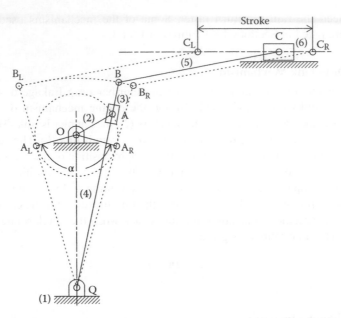

FIGURE 1.34   The shaper mechanism.

1.8.1.3   Whitworth Mechanism

The mechanism consists of two single-slider mechanisms, that is, the double rotating link mechanism and the engine mechanism (Figure 1.35). The extreme positions are indicated in the figure. The angles for the forward and backward strokes (clockwise rotation) are α and $2\pi - \alpha$, respectively. The time ratio, λ, is given by

$$\lambda = \frac{2\pi - \alpha}{\alpha}$$

The angle, α, is given by

$$\alpha = 2\cos^{-1}\left(\frac{OQ}{QA}\right)$$

1.8.2   INTERMITTENT MOTION MECHANISMS

In many practical applications, certain parts of machines are required to stop for a finite interval of time, especially automatic processing machines, for example, turrets, slide projectors, indicators, counters, and so forth. Of course, intermittent motion can be obtained now by stepper motors and by programming. But mechanical systems are still cheaper and more reliable. Some of the intermittent motion mechanisms are listed in Sections 1.8.2.1 through 1.8.2.3.

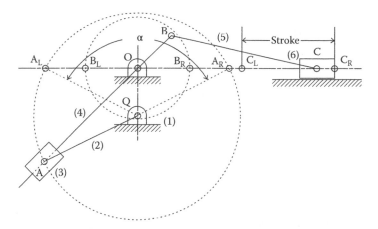

FIGURE 1.35 Whitworth quick-return mechanism.

1.8.2.1 Geneva Wheel

Geneva wheel (Figure 1.36) is perhaps the best known intermittent motion mechanism. This is due to its very smooth action. The main elements of the mechanism in Figure 1.36 are as follows:

- The driving disk (1)
- The locking disk (2)
- The driving pin (3)
- The driven wheel (4)

The locking disk and the driving pin are fixed on the driving disk. The driven disk has slots and is placed on a level above the driving disk. The driven disk is truncated with circular arcs having radii equal to the radius of the locking disk to prevent its motion during the stopping interval. At the same time, the locking disk is truncated with some curve to allow the wheel to move during the action interval. The depth of truncated part is determined when one of the slots of the wheel is along the line of centers as shown in the Figure 1.36. The truncated part of the locking disk must start exactly at the middle of the truncated arc of the wheel as shown in Figure 1.37.

The angle between the slots of the wheel, α, is determined from the moving period and, consequently, the number of slots (which must be a whole number) is determined.

At the start of the movement interval, the driving pin engages the wheel slot as shown in Figure 1.37. Line OA is perpendicular to line QA. If the distance between the driving shaft and the driven shaft; the number of slots, n; and the diameter of the driving pin are given, the procedure for designing the system is outlined as follows: The angle between the slots is

$$\alpha = \frac{360}{n}$$

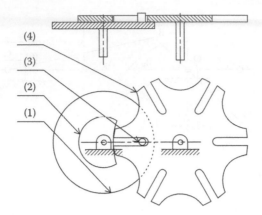

FIGURE 1.36   The Geneva wheel.

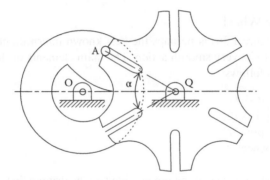

FIGURE 1.37   The Geneva wheel at the star of engagement.

Consider Figure 1.38. The steps for constructing the Geneva wheel are outlined as follows:

1. Draw line OQ equal to the center distance.
2. From point Q, draw a line that makes an angle $\alpha/2$ with OQ. This represents the centerline of the slot.
3. From point O, draw a line perpendicular to the aforementioned centerline to intersect it at point A.
4. From point Q, draw a circle with radius equal to QA, which represents the wheel.
5. From point A, draw a circle representing the path of the driving pin to intersect line OQ at point A'.
6. Draw two lines parallel to the centerline of the slot and tangent to the pin circle. The depth of the slot is determined when the driving pin is on line OQ with a reasonable clearance (the depth is equal to QA − OQ + OA + radius of the pin + clearance).

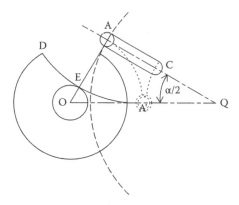

FIGURE 1.38 Constructing the Geneva wheel.

7. From point O, draw a circle representing the locking disk with reasonable radius.
8. From point O, draw a circle with radius equal to OQ – QA – clearance. It intersects line OA at point E.
9. Draw arc BED; D is symmetrical to B about line OA.

1.8.2.2 Locking-Slide Geneva

The locking-slide Geneva is basically similar to the Geneva wheel. The difference between the two is in the way the wheel is locked during the stopping interval. In this mechanism (Figure 1.39), one pin locks and unlocks the wheel, whereas the second pin drives the wheel during the action interval. In the position shown in Figure 1.39, the driving pin is about to engage the slot of the wheel, whereas the locking pin is just clearing the slot (the crank is rotating clockwise).

1.8.2.3 Ratchet Wheel

The ratchet wheel shown in Figure 1.40 is a simple mechanism that produces an intermittent motion. It consists of a driving pawl (3) (sometimes called a detent or catch), which is hinged to an oscillating link (2), a ratchet wheel (1), and a holding pawl (4). The ratchet wheel and the oscillating link rotate independently about the same axis. The angle of oscillation of the link is equivalent to one tooth pitch. The holding pawl is used to hold the ratchet in place and prevent it from rotating backward during the idle stroke.

1.8.3 STRAIGHT-LINE MOTION MECHANISMS

It is frequently necessary to constrain a point to move along a straight line. Of course, sliding pairs can give this motion satisfactorily, but we must not forget that they are relatively bulky and are subjected to comparatively rapid wear. So, in some circumstances, it is desirable to obtain the necessary motion by using turning joints.

Straight-line motion mechanisms can be classified into three categories: exact generated, exact copied, and approximate.

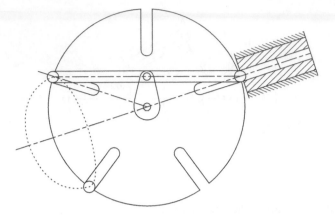

FIGURE 1.39   The locking-slide Geneva.

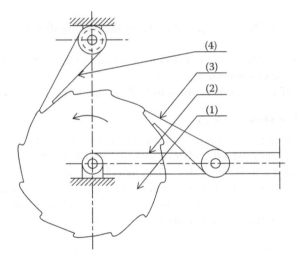

FIGURE 1.40   The ratchet wheel.

1.8.3.1   Exact Generated Straight-Line Motion

Theory

Consider Figure 1.41. If line OQP rotates about a point O, point Q is constrained to move on a circle; if the product OQ × OP is constant, then point P moves on a straight line perpendicular to a diameter passing by point O.

Proof

Point A lies on the circle such that OA is a diameter. Join points A and Q. Draw line PB perpendicular to line OA extended. The two triangles, OBP and OQA, are similar since they have a common angle and both have a right angle. Then

$$\frac{OQ}{OB} = \frac{OA}{OP}$$

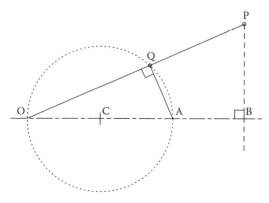

FIGURE 1.41 Proof of the straight line theory.

Thus

$$OQ \times OP = OA \times OB$$

But OA and OB are constants.
Therefore,

$$OQ \times OP = \text{constant}$$

This is true for any line passing through point O as long as point Q lies on the circle. There are two mechanisms that fulfill this condition.

1.8.3.1.1 Peaucellier Mechanism

In Peaucellier mechanism (Figure 1.42), link CQ is equal to link OC. Thus, point Q moves on a circle passing through point O. Links OA and OB are equal. Links QA, QB, AP, and BP are equal and form a lozenge. Point D is at the intersection of the two diagonals. Thus, QD = DP. From Figure 1.42,

$$\overline{OA}^2 = \overline{OD}^2 + \overline{DA}^2$$
$$\overline{QA}^2 = \overline{QD}^2 + \overline{DA}^2$$

On subtraction, we get

$$\overline{OA}^2 - \overline{QA}^2 = \overline{OD}^2 - \overline{QD}^2$$
$$= (OD + QD) \times (OD - QD) = \text{constant}$$
$$OD - QD = OQ$$
$$OD + QD = OP$$

Then,

$$OP \times OQ = \text{constant}$$

Therefore, point P traces a straight line normal to OC.

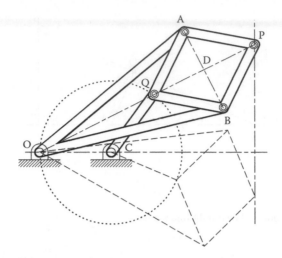

FIGURE 1.42   Peaucellier mechanism.

1.8.3.1.2   Hart Mechanism

In Hart mechanism (Figure 1.43), the four links AB, BE, ED, and EA form a crossed parallelogram in which AB = DE and AE = BD. Point Q traces a circle having its center at C and passing through O. Due to the geometry of the mechanism, lines AD, BE, and OQP are always parallel. For the triangle DEB,

$$\overline{BD}^2 = \overline{DE}^2 + \overline{BE}^2 - 2DE \times BE \cos D\hat{E}B$$

Also,

$$\cos D\hat{E}B = \frac{BE - AD}{2DE}$$

Then

$$\overline{BD}^2 = \overline{DE}^2 + \overline{BE}^2 - BE \times (BE - AD)$$
$$= \overline{DE}^2 + BE \times AD$$

Since BD and DE are fixed links,

$$BE \times AD = \text{constant} \tag{1.1}$$

The triangles AOQ and ABE are similar, thus

$$\frac{OQ}{BE} = \frac{OA}{AB}$$
$$OQ = BE \times \frac{OA}{AB} \tag{1.2}$$

The triangles BOP and BAD are similar, thus

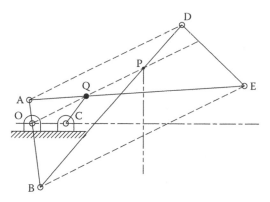

FIGURE 1.43   Hart mechanism.

$$\frac{OP}{AD} = \frac{OB}{AB}$$

$$OP = AD \times \frac{OB}{AB} \tag{1.3}$$

Multiply Equations 1.2 and 1.3:

$$OQ \times OP = BE \times AD \times \frac{OA \times OB}{\overline{AB}^2}$$

The lengths OA, OB, and AB are constants. According to Equation 1.1,

$$OQ \times OP = constant$$

Therefore, point P traces a straight line normal to OC.

1.8.3.2   Exact Copied Straight-Line Motion

In these mechanisms, the straight line is not generated. Rather, it is copied from an existing straight line. The Scott–Russell mechanism (Figure 1.44) is an example of such mechanisms.

The motion of point B is copied by point P. The conditions required are OA = BA = AP and the line of action of point B passes through point O.

1.8.3.3   Approximate Straight-Line Mechanisms

A large number of four-bar linkages are designed such that a point may trace an approximate straight line. They are useful in many applications. Some of these mechanisms are listed in Sections 1.8.3.3.1 through 1.8.3.3.3.

1.8.3.3.1   Watt Mechanism

The best known motion among the approximately correct straight-line motions is the one introduced by James Watt (the inventor of the steam engine) to guide the piston

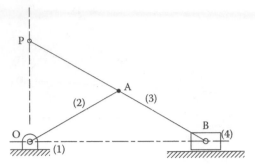

FIGURE 1.44   Exact copied straight line motion mechanism.

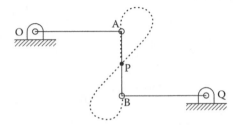

FIGURE 1.45   Watt mechanism.

rod of the early steam engine. It consists of two oscillating links, OA and QB, and a coupler, AB (Figure 1.45). Point P is located on link AB such that

$$\frac{OA}{QB} = \frac{BP}{AP}$$

Point P traces a loop if we consider all the possible locations of the mechanism. Part of this loop is a good approximate straight line as shown in the figure.

1.8.3.3.2   Grasshopper Mechanism

The grasshopper approximate line motion mechanism (Figure 1.46) is in fact a modification of the Scott–Russell mechanism. The slider is replaced by the rocker QB, which should be long. In this case, the motion of point B is along an arc of a circle. The coupler is lengthened to compensate the circular motion of the tip of the rocker. Point P describes an approximate straight line along the path PP″. To construct the mechanism, we perform the following steps:

- Draw line PP″; locate point P′ in the middle of this line.
- Draw line P′B′ with appropriate length.
- Draw line PB with length equal to P′B′; PB represents the coupler.
- Bisect line BB′ and draw a perpendicular line. Locate point Q on this line. It is noted that QB represents the rocker.
- Locate point A on BP and locate A′ on B′P′ such that BA is equal to B′A′.
- Draw line AA′ and draw a perpendicular line to intersect line P′B′ at point O. It is noted that OA represents the crank.

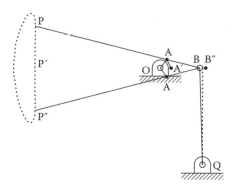

FIGURE 1.46 Constructing the grasshopper mechanism.

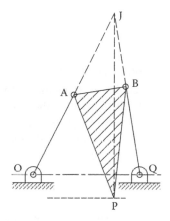

FIGURE 1.47 Robert mechanism.

The complete path of point P has the shape of letter D.

1.8.3.3.3 Robert Mechanism

The outline skeleton of Robert mechanism is shown in Figure 1.47. Point P traces an approximate straight line over a certain region when satisfying the following conditions:

$$OA = QB,$$
$$AP = BP,$$
$$AB = \tfrac{1}{2}OQ.$$

1.8.4 PARALLEL LINKS MECHANISMS

This mechanism comprises a combination of links so constructed that if one point in the mechanism moves in a certain path, another point moves in a similar path that may be equal to, larger than, or smaller than the copied path. Some applications of such mechanism are described in Sections 1.8.4.1 through 1.8.4.3.

FIGURE 1.48   Locomotive drive.

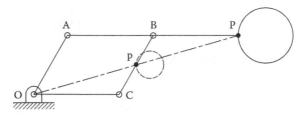

FIGURE 1.49   Pantograph.

1.8.4.1   Coupling Rod in Locomotive Drive

The motion of the locomotive engine is transmitted to its driving wheels. This motion is transmitted to other wheels to increase the number of driving wheels. The simplest method is to use the four-bar linkage shown in Figure 1.48. The length of the coupler link (AB) is equal to the distance between the centers of the wheels. Also, the driving link (OA) on the driving wheel is equal and parallel to the link of the driven wheel, link QB.

1.8.4.2   Pantographs

A pantograph (Figure 1.49) is a four-bar linkage that is used to copy the motion of some point to another point with a certain scale.

The conditions required are as follows:

$$OA = BC \text{ and } AB = OC$$

The magnification factor, X, between points P (on link AP) and point p (on link BC) is as follows:

$$X = \frac{OP}{Op} = \frac{AP}{AB}$$

It is obvious that demagnification is obtained by copying the motion of point p from point P. The amount of magnification is controlled by adjusting the lengths of linkages. Other configurations are shown in Figure 1.50.

Pantographs are used for redrawing maps to a different scale; in engraving machines where the operator is guiding a stylus at p to follow an accurate master, which may be 20 times the required size; in guiding cutting tools or the flame of a cutting torch in accordance with a given pattern; and so on.

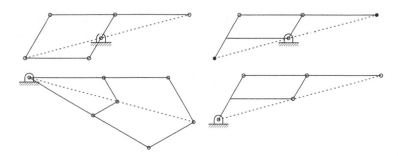

FIGURE 1.50 Different configurations for the pantograph.

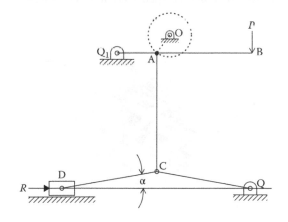

FIGURE 1.51 Some parallel link mechanisms. (a) Wind shield wiper (b) extending door mechanism (c) drafting board.

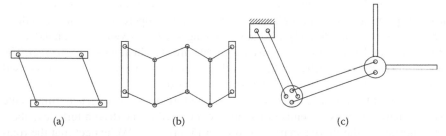

FIGURE 1.52 A mechanism with large mechanical advantage.

1.8.4.3 Other Applications

Some interesting applications of parallel link mechanisms are shown in Figure 1.51, such as the parallel ruler (Figure 1.51a and b) and the drafting machine (Figure 1.51c).

1.8.5 MECHANISMS WITH LARGE MECHANICAL ADVANTAGE

The purpose of mechanisms with large mechanical advantage is to produce a large force by applying a small force. An example of such mechanisms is the toggle link mechanism shown in Figure 1.52.

The mechanism can be operated either manually by lever Q_1B or automatically by crank OA. For the manual operation, we apply a force, P, to overcome the resisting force, R. Usually, QC = CD. The resisting force, R, is given by

$$R = \frac{Q_1B}{2 \times Q_1A \times \tan \alpha} P$$

For small values of α, $\tan \alpha$ tends to 0 and R may be very large.

1.8.6 UNIVERSAL JOINT

The universal joint, simply called U joint, is shown in Figure 1.53. It is also known as Cardan joint or Hooke's joint. It is used for connecting two rotating shafts whose axes lie in one plane and make an angle with each other during operation. It consists of three parts, that is, a driving fork (1) fixed to the driving shaft, a driven fork (2) fixed to the driven shaft, and a cross (3). The arms of the cross are at right angles to each other. The ends of the arms are attached to the forks through revolute joints. The U joint is used in many applications, but it is widely used in the transmission system of automobiles.

The ends of the cross connected to the driving fork describe a circle normal to the driving shaft, whereas the ends of the cross connected to the driven fork describe a circle normal to the driven shaft, as shown in Figure 1.54. We expect that the rotational angles of the driving shaft and the driven shaft are not equal due to the inclination angle α. As a result, their angular speeds are not the same.

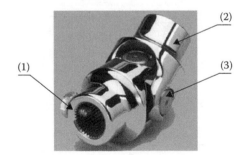

FIGURE 1.53   Universal joint.

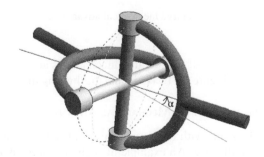

FIGURE 1.54   Universal joint showing the angle of intersection.

It is important to determine the relation between the angles of rotation of the two shafts. Consider Figure 1.55; it represents the plan view when the two shafts are in a horizontal plane. The axes of the two shafts are represented by lines OS_A and OS_B, respectively, and they make an angle α. Looking at the driving shaft, we see that the fork of the driving shaft describes a circle, A_oAA_oA, whereas the fork of the driven shaft describes an ellipse, $A_oB_oA_oB_o$.

Consider that at the start, the arms of the driving shaft are vertical at A_oA_o and the arms of the driven shaft are horizontal at B_oB_o. When the driving shaft rotates through an angle θ, its arm is located at point A. The arm of the driven shaft is located at point B. Line OA is normal to line OB. The actual position of point B is obtained by looking at the joint in the direction of the driven shaft or by projecting point B to point B' on the circle. The actual angle of rotation of the driven shaft is φ. The value of φ is obtained as follows:

$$\tan \varphi = \frac{B'D}{OD} \tag{1.4}$$

$$\tan \theta = \frac{BC}{OC} \tag{1.5}$$

But

$$B'D = BC,$$
$$OD = OB_1,$$
$$OC = OB_1 \cos \alpha = OD \cos \alpha.$$

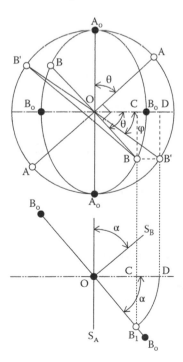

FIGURE 1.55 Geometry of a universal joint.

Substituting in Equations 1.4 and 1.5, we get

$$\tan \varphi = \cos \alpha \times \tan \theta \qquad (1.6)$$

The angular velocity and angular acceleration of the driven shaft are obtained by differentiating Equation 1.6,

$$\dot{\varphi} = \frac{\cos \alpha}{1 - \sin^2 \alpha \sin^2 \theta} \dot{\theta}$$

$$\ddot{\varphi} = \frac{\cos \alpha \sin^2 \alpha \sin 2\theta}{(1 - \sin^2 \alpha \sin^2 \theta)^2} \dot{\theta}^2$$

Figure 1.56 shows the variation in angular velocity (to scale $\dot{\theta}$) and angular acceleration (to scale $\dot{\theta}^2$) for the driven shaft. The variation in angular acceleration is harmful for the components attached to the driven shaft.

1.8.6.1 Double Universal Joint

To rectify the harmful effect of variation in angular velocity of the driven shaft of a universal joint, we use double universal joints, which are shown in Figure 1.57. In this case, we have a driving shaft, an intermediate shaft, and a driven shaft. Special attention should be paid to ensure that the forks at the ends of the intermediate shaft are in one plane, and the angle between the driving shaft and the intermediate shaft is equal to that between the driven shaft and the intermediate shaft.

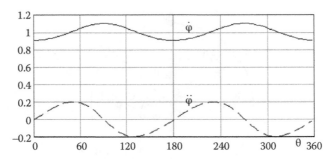

FIGURE 1.56   Relations between the angular velocity and the angular acceleration of drive shaft with the rotational angle of the driving shaft.

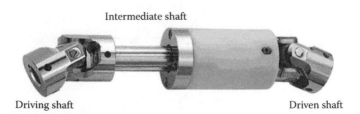

FIGURE 1.57   Double universal joint.

1.9   ANALYTICAL POSITION ANALYSIS OF MECHANISMS

It is necessary to configure a mechanism during the working cycle. This is possible by tracing the mechanism at different positions as explained in Section 1.3.1.1 for the four-bar mechanism. For compound mechanisms, this procedure is tedious. Analytical analysis could be a powerful tool for studying the work space of mechanisms. Vector algebra is used for this analysis.

1.9.1   Vectors

Vectors can be described in polar or Cartesian coordinates. Polar coordinates are most suitable for kinematics analysis, whereas Cartesian coordinates are suitable for force analysis. However, there is a relation between the two coordinate systems.

A plane vector $\mathbf{V}$ is represented in the Cartesian coordinate system in terms of its components along the x and y axes, V_x and V_y, respectively. Thus,

$$\mathbf{V} = V_x \mathbf{i} + V_y \mathbf{j} \tag{1.7}$$

where $\mathbf{i}$ and $\mathbf{j}$ are the unit vectors along the x and y axes, respectively. In terms of the polar coordinate system, a vector can be represented using complex numbers as follows:

$$\begin{aligned} \mathbf{V} &= V\,e^{i\theta} \\ &= V\cos\theta + iV\sin 0 \end{aligned} \tag{1.8}$$

where

- V is the magnitude of $\mathbf{V}$.
- The angle θ is the angle of inclination of $\mathbf{V}$ with the x axis; it is positive in the counterclockwise direction.
- It is noted that i is the complex number $\sqrt{-1}$.

The first part on the right-hand side of Equation 1.8 is called the real part, whereas the second part is the imaginary part.

Comparing Equations 1.7 and 1.8, we get

$$V_x = V\cos\theta$$
$$V_y = V\sin\theta$$

The presence of i in the second term on the right-hand side of Equation 1.8 indicates that the component is in the direction of the polar coordinate system.

1.9.1.1 Unit Complex Vector

The unit complex vector, $\mathbf{u}$, is a vector whose magnitude is unity and who makes an angle θ with the x axis. Thus,

$$\begin{aligned} \mathbf{u} &= e^{i\theta} \\ &= \cos\theta + i\sin\theta \end{aligned} \tag{1.9}$$

1.9.1.2 Complex Conjugate Vector

The unit complex conjugate vector, $\mathbf{u}_c$, has unit length and makes an angle θ. Thus,

$$\mathbf{u}_c = e^{-i\theta}$$
$$= \cos\theta - i\sin\theta \tag{1.10}$$

This means that in order to obtain the conjugate of any complex vector, it is enough to replace θ by $-\theta$ or, simply, i by $-i$.

1.9.1.3 Multiplication of Complex Vectors

Let

$$\mathbf{V}_1 = V_1\,e^{i\theta_1}$$
$$\mathbf{V}_2 = V_2\,e^{i\theta_2}$$

Then

$$\mathbf{V}_1 \times \mathbf{V}_2 = V_1\,e^{i\theta_1} \times V_2\,e^{i\theta_2}$$
$$= V_1 V_2\,e^{i(\theta_1+\theta_2)}$$

This means that the multiplication of two complex vectors is a complex vector whose magnitude is equal to the product of the magnitudes of the two original vectors and angle is equal to the sum of the angles of the original vectors.

Conclusions:

1. Multiplication of a vector with its conjugate:
   Let

$$\mathbf{V} = V\,e^{i\theta}$$

   Then

$$\mathbf{V}_c = V\,e^{-i\theta}$$

   Therefore,

$$\mathbf{V} \times \mathbf{V}_c = V^2$$

   That is, the multiplication of a vector by its conjugate is a real quantity, which is equal to the square of the magnitude of the original vector. This operation eliminates the imaginary part of the original vector.

2. Rotation of a vector:
   Let

$$\mathbf{V} = V\,e^{i\theta}$$

   If vector $\mathbf{V}$ is rotated through an angle φ, its angle becomes $\theta + \varphi$. The new vector is represented as follows:

$$\mathbf{V} = V\,e^{i(\theta+\varphi)}$$
$$= V\,e^{i\theta}\,e^{i\varphi}$$

If φ is equal to π/2, then

$$\mathbf{V} = V\,e^{i\theta}\left(\cos\pi/2 + i\sin\pi/2\right)$$
$$= iV\,e^{i\theta}$$

Therefore, to rotate a vector through an angle of π/2 in the counterclockwise direction, simply multiply it by i. Similarly, to rotate a vector through an angle of π/2 in the clockwise direction, simply multiply it by −i.

1.9.2 PLANE MECHANISMS

In Sections 1.3 and 1.4, we presented a group of simple mechanisms. The basic group includes the four-bar, the engine, the tilting block, and the shaper mechanisms (Figure 1.58).

If we remove the crank from these mechanisms (Figure 1.59), we get a group of chains, which can form any compound mechanism. As an example, consider the compound mechanism shown in Figure 1.60. It consists of eight links. It is clear that the mechanism is formed from the following chains:

- Line OA, link (2), is the crank.
- Links (3) and (4) form an engine chain. Point A is the input point to this chain.
- Links (5) and (6) form a tilting block chain. Point C is the input point to this chain.
- Links (7) and (8) form a four-bar chain. Point D is the input point to this chain.

1.9.3 ANALYSIS OF CHAINS

If each chain is analyzed separately, the analysis of any complex mechanism is an easy task. In fact, simple software can be designed to perform this analysis.

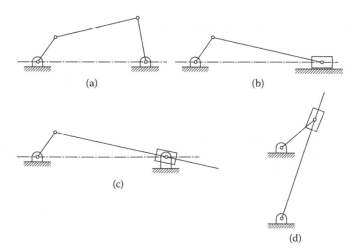

FIGURE 1.58 The simple known plane mechanisms. (a) Four-bar mechanism (b) engine mechanism (c) tilting block mechanism (d) shaper mechanism.

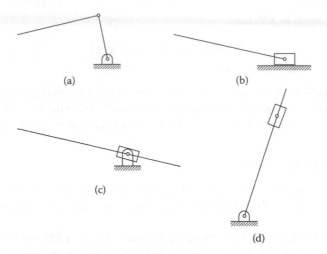

FIGURE 1.59 Chains used in complex mechanisms. (a) Four-bar chain (b) engine chain (c) tilting block chain (d) shaper chain.

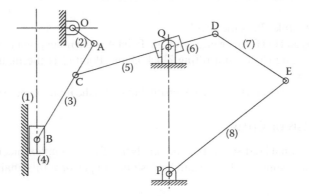

FIGURE 1.60 Example of a complex mechanism.

1.9.3.1 Crank

Usually, the crank is treated as the source of motion of a mechanism. So, its position is usually specified (Figure 1.61).
Given:

1. The location of point O, (x_O, y_O)
2. The crank length, r_2
3. The crank angle, θ_2

Find:

The position of point A, (x_A, y_A)
Analysis:

The position vector of point A is given by

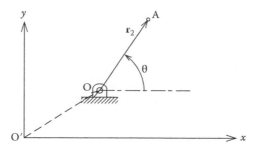

FIGURE 1.61   The crank.

$$\mathbf{r}_A = \mathbf{r}_O + \mathbf{r}_2$$
$$= \left(x_O + iy_O\right) + r_2\, e^{i\theta_2}$$
$$= \left(x_O + r_2 \cos\theta_2\right) + i\left(y_O + r_2 \sin\theta_2\right)$$

Thus,

$$x_A = x_O + r_2 \cos\theta_2 \tag{1.11}$$

$$y_A = y_O + r_2 \sin\theta_2 \tag{1.12}$$

Note: If the crank pivot is located at the origin of the coordinate system, then $x_O = y_O = 0$; therefore,

$$x_A = r_2 \cos\theta_2 \tag{1.13}$$

$$y_A = r_2 \sin\theta_2 \tag{1.14}$$

1.9.3.2   Four-Bar Chain

A four-bar chain is shown in Figure 1.62. The coordinate system is located at O'. The coupler AB is defined as link (3), whereas the rocker QB is link (4). The support of the rocker is at point Q. The input motion of the chain is delivered at point A. Given:

1. The length of the coupler, r_3
2. The length of the rocker, r_4
3. The location of point Q, (x_Q, y_Q)
4. The position of point A, (x_A, y_A)

Find:

The positions of links (3) and (4), which are represented by angles θ_3 and θ_4

Analysis:

First, join AQ to form the vector **d**. Consider the vector of point Q:

$$\mathbf{r}_Q = \mathbf{r}_A + \mathbf{d}$$

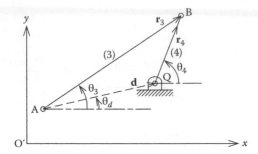

FIGURE 1.62   The four-bar chain.

Or

$$\mathbf{d} = \mathbf{r}_Q - \mathbf{r}_A$$

Writing each vector in its complex form, we get

$$d\,e^{i\theta_d} = (x_Q + iy_Q) - (x_A + iy_A)$$
$$= (x_Q - x_A) + i(y_Q - y_A) \tag{1.15}$$

Multiplying the vectors on both sides of the equation by their conjugates, we get

$$d\,e^{i\theta_d} \times d\,e^{-i\theta_d} = [(x_Q - x_A) + (y_Q - iy_A)] \times [(x_Q - x_A) - i(y_Q - y_A)]$$

This leads to

$$d^2 = (x_Q - x_A)^2 + (y_Q - y_A)^2$$

Or

$$d = \sqrt{(x_Q - x_A)^2 + (y_Q - y_A)^2} \tag{1.16}$$

Equating the real parts and the imaginary parts of Equation 1.15, we get

$$\sin\theta_d = \frac{y_Q - y_A}{d} \tag{1.17a}$$

$$\cos\theta_d = \frac{x_Q - x_A}{d} \tag{1.17b}$$

The exact value of θ_d is obtained from Equations 1.17a and b. In order to obtain θ_3, consider the loop for point B:

$$\mathbf{r}_4 = \mathbf{r}_3 - \mathbf{d}$$

Or

$$r_4\,e^{i\theta_4} = r_3\,e^{i\theta_3} - d\,e^{i\theta_d} \tag{1.18}$$

Multiplying by the conjugate, we get

$$(r_4\,e^{i\theta_4}) \times (r_4\,e^{-i\theta_4}) = (r_3\,e^{i\theta_3} - d\,e^{i\theta_d}) \times (r_3\,e^{-i\theta_3} - d\,e^{-i\theta_d})$$

This leads to

$$r_4^2 = r_3^2 + d^2 - r_3 d(e^{i\beta} + e^{-i\beta}) \tag{1.19}$$

where

$$\beta = \theta_3 - \theta_d$$

It is clear that

$$e^{i\beta} + e^{-i\beta} = 2\cos\beta$$

Thus, according to Equation 1.19,

$$\cos\beta = \frac{r_3^2 + d^2 - r_4^2}{2r_3 d}$$

It should be noted that β is an angle in the triangle QAB (Figure 1.62). Therefore, if $\cos\beta$ is positive, then $0 \le \beta \le \frac{\pi}{2}$. If $\cos\beta$ is negative, then $\frac{\pi}{2} \le \beta \le \pi$.

According to the configuration of Figure 1.62, vector $\mathbf{r}_3$ is leading vector $\mathbf{d}$. Thus,

$$\theta_3 = \theta_d + \beta$$

However, in some situations the four-bar chain is located such that vector $\mathbf{d}$ is leading vector $\mathbf{r}_3$. This depends on the orientation of points A, B, and Q. In this case,

$$\theta_3 = \theta_d - \beta \tag{1.20}$$

Generally speaking, θ_3 is obtained using $\theta_3 = \theta_d \pm \beta$ according to the chain configuration. Now, θ_4 can be obtained from Equation 1.18:

$$\sin\theta_4 = \frac{r_3\sin\theta_3 - d\sin\theta_d}{r_4} \tag{1.21a}$$

$$\cos\theta_4 = \frac{r_3\cos\theta_3 - d\cos\theta_d}{r_4} \tag{1.21b}$$

1.9.3.3 Engine chain

An engine chain, in a general sense, can be located as shown in Figure 1.63. The line of action of the slider is along the unit vector $\mathbf{u}$, which makes an angle, α, with the x axis, and is at a distance h from the origin of the coordinate axes; h is along the unit

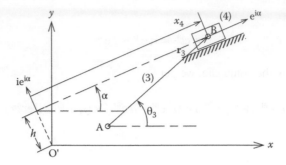

FIGURE 1.63   The engine chain.

normal to vector $\mathbf{u}_n$. The distance of the slider from O' along vector $\mathbf{u}$ is x_4. Notice that vector $\mathbf{u}_n$ is leading vector $\mathbf{u}$; then, h is positive in this direction. In situations where vector $\mathbf{u}$ is leading vector $\mathbf{u}_n$, h is negative.

Analysis:

Consider the vector loop for point B:

$$\mathbf{r}_B = \mathbf{r}_A + \mathbf{r}_3$$
$$= (x_4 + i\, h)\, e^{i\alpha}$$

Or

$$r_3\, e^{i\theta_3} = (x_4 + ih)\, e^{i\alpha} - \mathbf{r}_A \tag{1.22}$$

In order to simplify the analysis, the vector $\mathbf{r}_A$ is resolved into components ξ and η along vectors $\mathbf{u}$ and $\mathbf{u}_n$, respectively. Thus,

$$\mathbf{r}_A = (\xi + i\eta)\, e^{i\alpha}$$

Or

$$x_A + iy_A = (\xi + i\eta)\, e^{i\alpha}$$

In order to find ξ and η, divide both sides by $e^{i\alpha}$. Thus,

$$(x_A + iy_A)\, e^{i\alpha} = (\xi + i\eta)$$
$$(x_A + iy_A)(\cos\alpha - \sin\alpha) = (\xi + i\eta)$$
$$(x_A \cos\alpha + y_A \sin\alpha) + i\,(-x_A \sin\alpha + y_A \cos\alpha) = (\xi + i\eta)$$

Therefore,

$$\xi = x_A \cos\alpha + y_A \sin\alpha$$
$$\eta = -x_A \sin\alpha + y_A \cos\alpha$$

Substituting in Equation 1.21, we get

$$r_3\, e^{i\theta_3} = [(x_4 - x_A \cos\alpha - y_A \sin\alpha) + i\,(h + x_A \sin\alpha - y_A \cos\alpha)]\, e^{i\alpha}$$

Multiplying by conjugates, we get

$$r_3^2 = (x_4 - x_A \cos\alpha - y_A \sin\alpha)^2 + (h + x_A \sin\alpha - y_A \cos\alpha)^2 \qquad (1.23)$$

From Equation 1.23,

$$x_4 = x_A \cos\alpha + y_A \sin\alpha \pm \sqrt{r_3^2 - (h + x_A \sin\alpha - y_A \cos\alpha)^2} \qquad (1.24)$$

The positive and negative signs indicate that there are two possible positions for the slider depending on the configuration of the chain. The positive sign is considered when the projection of vector $\mathbf{r}_3$ on the line of action of the slider is along $e^{i\alpha}$. Otherwise, the negative sign is considered. The value of θ_3 is obtained from Equation 1.22:

$$\cos\theta_3 = \frac{x_4 \cos\alpha - h \sin\alpha - x_A}{r_3} \qquad (1.25a)$$

$$\sin\theta_3 = \frac{x_4 \sin\alpha + h \cos\alpha - y_A}{r_3} \qquad (1.25b)$$

1.9.3.4 Shaper Chain

The chain in its general configuration is shown in Figure 1.64. Link (4) oscillates about point Q, (x_Q, y_Q). The slider (3) slides on link (4). Its position is determined by the distance x_4 from Q. The motion is transmitted to the chain through point A on the slider. Point A is considered to be at a normal distance "h" from link (4) $(h = AB)$. Given:

1. The location of Q, (x_Q, y_Q)
2. The position of A, (x_A, y_A)
3. The normal distance, h, of the slider

Find:

1. The distance, x_4
2. The angle of link (4), θ_4

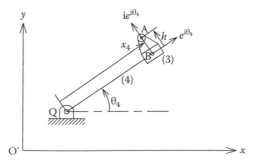

FIGURE 1.64   The shaper chain.

Analysis:

Set the unit vectors $e^{i\theta_4}$ and $i\,e^{i\theta_4}$ along link (4) and normal to it, respectively. It is noted that h is positive when it is along $i\,e^{i\theta_4}$. Consider the vector loop for point A:

$$\mathbf{r}_A = \mathbf{r}_Q + (x_4 + ih)\,e^{i\theta_4}$$

Or

$$(x_4 + ih)\,e^{i\theta_4} = \mathbf{r}_A - \mathbf{r}_Q$$
$$= (x_A - x_Q) + i\,(y_A - y_Q) \tag{1.26}$$

Multiplying by the conjugates, we get

$$x_4^2 + h^2 = (x_A - x_Q)^2 + (y_A - y_Q)^2$$

From which

$$x_4 = \sqrt{(x_A - x_Q)^2 + (y_A - y_Q)^2 - h^2} \tag{1.27}$$

In order to obtain the value of θ_4, divide Equation 1.26 by $(x_4 + ih)$. Thus,

$$e^{i\theta_4} = \frac{(x_A - x_Q) + i\,(y_A - y_Q)}{(x_4 + ih)}$$
$$= \frac{[(x_A - x_Q) + i\,(y_A - y_Q)]\,(x_4 - ih)}{(x_4 + ih)\,(x_4 - ih)}$$

This leads to

$$\cos\theta_4 + i\sin\theta_4 = \frac{[x_4(x_A - x_Q) + h(y_A - y_Q)] + i\,[x_4(y_A - y_Q) - h\,(x_A - x_Q)]}{(x_4^2 + h^2)}$$

Therefore,

$$\cos\theta_4 = \frac{[x_4(x_A - x_Q) + h\,(y_A - y_Q)]}{(x_4^2 + h^2)} \tag{1.28a}$$

$$\sin\theta_4 = \frac{[x_4(y_A - y_Q) - h\,(x_A - x_Q)]}{(x_4^2 + h^2)} \tag{1.28b}$$

1.9.3.5 Tilting Block Chain

A tilting block chain is shown in Figure 1.65. The block is pivoted at point Q, (x_Q, y_Q); link (3) slides inside the block. Point Q is at a distance "h" from link (3). The motion is transmitted to the chain at point A on link (4).

Given:

1. The location of Q, (x_Q, y_Q)
2. The position of A, (x_A, y_A)
3. The normal distance, h

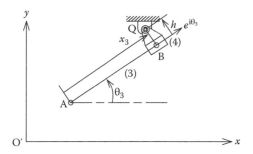

FIGURE 1.65 The tilting block chain.

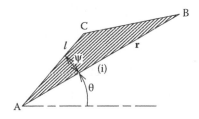

FIGURE 1.66 General link (i).

Find:

1. The distance, x_3
2. The angle of link (3), θ_3

Analysis:

The analysis is very similar to that of the shaper chain; the results can be obtained by replacing point A by point Q and vice versa, and the subscripts 4 by 3 in Equations 1.27 and 1.28. Therefore,

$$x_3 = \sqrt{(x_Q - x_A)^2 + (y_Q - y_A)^2 - h^2} \tag{1.29}$$

$$\cos\theta_3 = \frac{[x_3(x_Q - x_A) + h\,(y_Q - y_A)]}{(x_3^2 + h^2)} \tag{1.30a}$$

$$\sin\theta_3 = \frac{[x_3(y_Q - y_A) - h\,(x_Q - x_A)]}{(x_3^2 + h^2)} \tag{1.30b}$$

1.9.3.6 Position of a Point on a Link

In compound mechanisms, the motion of a chain is transmitted to the next chain through a point on a link. So, it is necessary to obtain the position of this point to use it as an input for the next chain. Consider link (i) in Figure 1.66.

Given:

(x_A, y_A), **l**, and ψ.

Find:

x_c and y_c

Analysis:

$$\mathbf{r}_c = \mathbf{r}_A + \mathbf{l}$$
$$= (x_A + iy_A) + l\,e^{i\,(\theta+\psi)}$$
$$= [x_A + l\cos(\theta+\psi)] + i\,[y_A + l\sin(\theta+\psi)]$$

Therefore,

$$x_c = x_A + l\cos(\theta+\psi) \tag{1.31}$$

$$y_c = y_A + l\sin(\theta+\psi) \tag{1.32}$$

EXAMPLE 1.5

Perform position analysis for the mechanism shown in Figure 1.67.
Analysis:

1. Crank:

$$r_2 = 120,\ \theta_2 = 205°$$

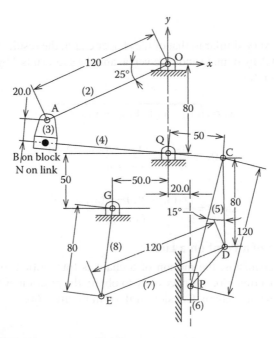

FIGURE 1.67   Mechanism of EXAMPLE 1.5.

From Equation 1.13,

$$x_A = r_2 \cos\theta_2$$
$$= -108.76$$

From Equation 1.14,

$$y_A = r_2 \sin\theta_2$$
$$= -50.71$$

2. Shaper chain:

$x_Q = 0$, $y_Q = -80$, $h = -20$ (**h** is opposite to $e^{i\theta_4}$), $l = 50$, $\psi = 80°$.

From Equation 1.27,

$$x_4 = \sqrt{(x_A - x_Q)^2 + (y_A - y_Q)^2 - h^2}$$
$$= 110.8$$

From Equation 1.28a,

$$\cos\theta_4 = \frac{[x_4(x_A - x_Q) + h(y_A - y_Q)]}{(x_4^2 + h^2)}$$
$$= -0.996$$

From Equation 1.28b,

$$\sin\theta_4 = \frac{[x_4(y_A - y_Q) - h(x_A - x_Q)]}{(x_4^2 + h^2)}$$
$$= 0.08$$
$$\theta_4 = 175.2°$$

From Equation 1.31,

$$x_C = x_Q + l\cos(\theta + \psi)$$
$$= 49.8$$

From Equation 1.32,

$$y_C = y_Q + l\sin(\theta + \psi)$$
$$= -84.2$$

3. Engine chain:

$r_3 = 120$, $\alpha = 270°$, $h = 20$, $l = 80$, $\psi = 15°$, $x_A = x_C = 49.8$,
$y_A = y_C = -84.2$,

From Equation 1.24,

$$x_4 = x_A \cos\alpha + y_A \sin\alpha \pm \sqrt{r_3^2 - (h + x_A \sin\alpha - y_A \cos\alpha)^2}$$
$$= 200.4 \text{ (we choose the + sign from the configuration)}$$

From Equation 1.25a,

$$\cos\theta_3 = \frac{x_4 \cos\alpha - h\sin\alpha - x_A}{r_3}$$

$$= -0.25$$

From Equation 1.25b,

$$\sin\theta_3 = \frac{x_4 \sin\alpha + h\cos\alpha - y_A}{r_3}$$

$$= 0.97$$
$$\theta_3 = 255.6°$$

From Equation 1.31,

$$x_D = x_C + l\cos(\theta + \psi)$$

$$= 50.7$$

From Equation 1.32,

$$y_D = y_C + l\sin(\theta + \psi)$$

$$= -164.2$$

4. Four-bar chain:

$r_3 = 120$, $r_4 = 80$, $x_Q = x_G = -50$, $y_Q = y_G = -130$, $x_A = x_D = 50.7$,
$y_A = y_D = -164.2$

From Equation 1.16,

$$d = \sqrt{(x_Q - x_A)^2 + (y_Q - y_A)^2}$$

$$= 106.3$$

From Equation 1.17a,

$$\sin\theta_d = \frac{y_Q - y_A}{d}$$

$$= 0.322$$

From Equation 1.17b,

$$\cos\theta_d = \frac{x_Q - x_A}{d}$$

$$= 0.947$$
$$\theta_d = 161.2°$$

$$\cos\beta = \frac{r_3^2 + d^2 - r_4^2}{2r_3 d}$$

$$= 0.757$$
$$\beta = 40.84°$$

From Equation 1.15,

$$\theta_3 = \theta_d + \beta \ (+ \text{ from configuration})$$
$$= 202.1°$$

From Equation 1.21a,

$$\sin\theta_4 = \frac{r_3 \sin\theta_3 - d \sin\theta_d}{r_4}$$
$$= -0.991$$

From Equation 1.21b,

$$\cos\theta_4 = \frac{r_3 \cos\theta_3 - d \cos\theta_d}{r_4}$$
$$= -0.132$$
$$\theta_4 = 262.43°$$

The aforementioned analysis is for one crank position. However, it is possible to make position analysis for the links over a complete crank cycle. The relation between the motion of output of link (8) and the crank rotational angle is shown in Figure 1.68.

1.9.3.7 Tracing

Analytical analysis is a powerful tool for tracing a mechanism at different positions. It is possible to trace the path of a point on a moving plane. Consider that point C, (x_C, y_C), is on a link and it is required to trace its path on a plane rotating with another link.

As an example, consider the shaper mechanism shown in Figure 1.69a. At the shown position, the trace of a point on the oscillation link right underneath the tip of the crank on a plane rotating with the crank is shown in Figure 1.69b.

The traced path is given by the following vector:

$$\mathbf{r}_t = (x_C + iy_C)\, e^{i\theta}$$
$$= (x_C \cos\theta + y_C \sin\theta) + i\,(-x_C \sin\theta + y_C \cos\theta)$$

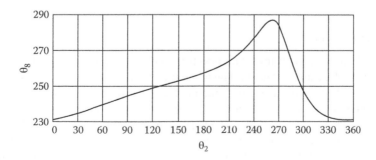

FIGURE 1.68 Plot of the output angle with the input angle.

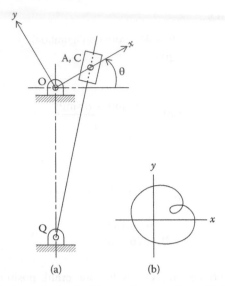

FIGURE 1.69 Tracing the path of point B on a plane rotating with the crank. (a) xOy represents the rotating coordinates (b) trace of a point on a rotating plane.

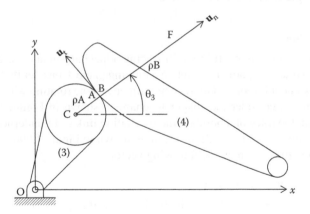

FIGURE 1.70 Sliding links.

1.9.4 SLIDING LINKS

Sliding links are used to transmit motion from a body to another through direct contact, for example, cams and gears. The general configuration of sliding links consists of two bodies; link (3) is the driving link and link (4) is the driven link in Figure 1.70.

The points of contact are A [on link (3)] and B [on link (4)]. Vectors $\mathbf{u}_t$ and $\mathbf{u}_n$ are unit vectors tangent and normal to the surface in contact, respectively. The radii of curvature are ρ_A and ρ_B with the centers of curvatures at points C and F for links (3) and (4), respectively. The radii of curvature are considered instantaneously constant, although they may change when the points of contact change. Thus, AB can be considered as an imaginary link joining the centers of curvature C and F. In order to obtain its inclination θ_3, consider the following vector equation:

$$\mathbf{r}_F = \mathbf{r}_C + (\rho_A + \rho_B)\,\mathbf{u}_n$$

$$= \mathbf{r}_C + (\rho_A + \rho_B)\,e^{i\theta_3}$$

Since the input motion is delivered by link (3), the position of point C and, consequently, vector $\mathbf{r}_C$ are specified. The vector $\mathbf{r}_F$ depends on the type of motion and the configuration of link (4). The cases considered are described in Sections 1.9.4.1 through 1.9.4.4.

1.9.4.1   Cam with a Spherical Oscillating Follower

Figure 1.71 represents a circular cam with radius R actuating an oscillating follower with a spherical tip of radius r. In fact, this system is equivalent to a four-bar mechanism where

–OC represents the crank.

–CF = $R + r$ represents the coupler.

–QF represents the rocker.

When the cam rotates through an angle θ, the analysis of the four-bar chain presented in Section 1.9.3.2 is used to determine the follower angle θ_4.

1.9.4.2   Cam with a Spherical Translating Follower

Figure 1.72 represents a circular cam with radius R actuating a translating follower with a spherical tip of radius r.

$$Y_F = e\,\sin\theta + \sqrt{(R+r)^2 - (e\,\cos\theta)^2}$$

$$y_F = e\,\sin\theta + \sqrt{(R+r)^2 - (e\,\cos\theta)^2}$$

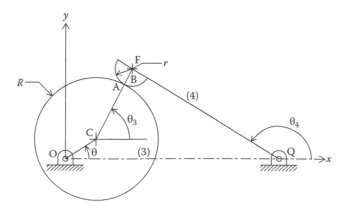

FIGURE 1.71   A cam with oscillating spherical follower.

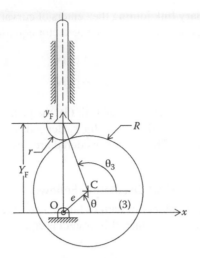

FIGURE 1.72   A cam with translating spherical follower.

The displacement of the follower from its lowest position is given by

$$y = y_F - r_o$$

where $r_o = R - e$

$$\sin\theta_3 = \frac{-e\cos\theta}{R+r}$$

$$\cos\theta_3 = \frac{y_F - e\sin\theta}{R+r}$$

1.9.4.3   Cam with a Flat-Faced Oscillating Follower

Figure 1.73 represents a circular cam with radius R actuating an oscillating flat-faced follower. The face of the follower is at a distance h from its pivot. It is clear that

$$\mathbf{u}_t = e^{i\theta_4}$$

$$\mathbf{u}_n = -ie^{i\theta_4}$$

Consider the vector loop

$$e\,e^{i\theta} + (R+h)\,\mathbf{u}_n = r_1 + x_4\,\mathbf{u}_t$$
$$(x_C - r_1 + i\,y_C) = [x_4 + i\,(R+h)]\,e^{i\theta_4}$$

where

$$x_C = e\cos\theta$$
$$y_C = e\sin\theta$$

Multiplying each side by its conjugate, we get

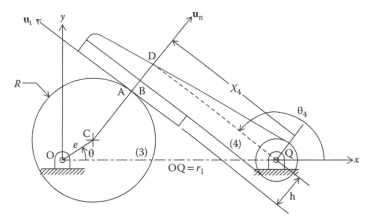

FIGURE 1.73   A cam with oscillating flat-faced follower.

$$x_4 = \sqrt{(x_C - r_1)^2 + y_C^2 - (R+h)^2}$$

Also,

$$e^{i\theta_4} = \frac{(x_C - r_1) + i\,y_C}{x_4 + i\,(R+h)}$$

Multiplying by the conjugate of the denominator, we get

$$e^{i\theta_4} = \frac{[(x_C - r_1) + i\,y_C][x_4 - i\,(R+h)]}{x_4^2 + (R+h)^2}$$

$$\cos\theta_4 = \frac{x_4(x_C - r_1) + y_C(R+h)}{x_4^2 + (R+h)^2}$$

$$\sin\theta_4 = \frac{x_4 y_C - (x_C - r_1)(R+h)}{x_4^2 + (R+h)^2}$$

1.9.4.4   Cam with a Flat-Faced Translating Follower
For the cam with the translating flat-faced follower shown in Figure 1.74,

$$Y_4 = R + e\sin\theta$$

The displacement of the follower from its lowest position is given by

$$y_F = R + e\sin\theta - r_o$$

where $r_o = R - e$. Thus,

$$y_F = e\,(1 + \sin\theta)$$

$$s_4 = e\cos\theta$$

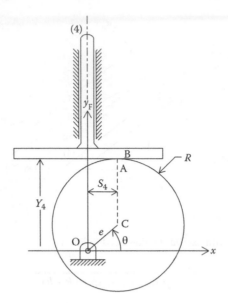

FIGURE 1.74 A cam with flat-faced translating follower.

PROBLEMS

1.1 Determine the DOF for the linkages in Figure P1.1. Which of these linkages represents mechanisms?

1.2 The lengths of the consequent links of a four-bar chain are 40, 120, 100, and 140 mm. Different motions are obtained by fixing one of the links at a time. Plot the relation between the output motion and the input motion in all possible cases.

1.3 If the largest link is fixed in the previous chain, trace the path of a point on the middle of the coupler link. Also trace the path of a point on the middle of the rocker on a plane rotating with the crank.

1.4 The lengths of the crank and the connecting rod in a single-slider crank chain are 60 and 150 mm, respectively. For all possible inversions, plot the output motion against the input motion.

1.5 The distance between the centers of the two blocks of an ellipse trammel is 75 mm. Plot the path of a point on the coupling link 75 mm away from the nearest block. Also, trace the path of a point located at the middle of the distance between the centers of the two blocks.

1.6 For the shaper mechanism shown in Figure P1.6, plot the motion of ram R with the crank rotational angle θ. Also, plot the path of point P on the middle of link BR.

$$OA = 30\,mm, QB = 200\,mm, BR = 150\,mm, BP = 75\,mm.$$

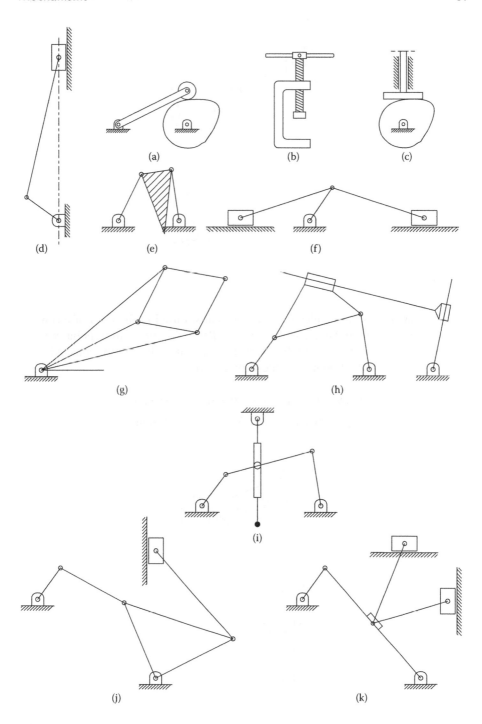

(a) (b) (c)

(d) (e) (f)

(g) (h)

(i)

(j) (k)

FIGURE P1.1

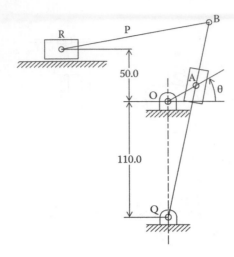

FIGURE P1.6

1.7 Draw the outline skeleton of the mechanism used in the head of a sewing machine to the needle bar (Figure P1.7). Plot the path of point N on
the needle starting from the lowest position and using 16 divisions. Also,
trace the path of point P on the middle of link CN.

$$OA = 40\,mm, AB = 120\,mm, QB = 80\,mm,$$
$$QC = 50\,mm, CN = 150\,mm, CP = 75\,mm$$

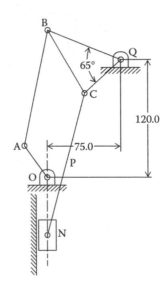

FIGURE P1.7

1.8 Design a four-bar mechanism such that $r_1 = 100$ mm, $r_3 = 80$ mm, the rocker angle is $60°$, and the time ratio is 1.

1.9 Design a four-bar mechanism such that $r_2 = 30$ mm, $r_3 = 70$ mm, the rocker angle is $90°$, and the time ratio is 1.2.

1.10 Design a four-bar mechanism such that rl $= 100$ mm, $r_2 = 30$ mm, $r_3 = 70$ mm, and the time ratio is 1.2.

1.11 For the shaper mechanism of Problem 1.6, obtain the time ratio. When the crank makes $30°$ with the horizontal datum, point C is a point on the oscillatory link under point A. Trace the path of C on a plane rotating with the crank.

1.12 Figure P1.12 shows an outline of the Zoller double-piston engine. Draw the mechanism with the given dimensions. Starting when piston B is at the extreme left position, and using 12 divisions in the cycle, plot the path of C and the displacement diagrams of both pistons C and D on the same diagram.

$OA = 20$ mm, $AC = 30$ mm, $BC = 120$ mm, angle $ACB = 90°$.

$AB = 126$ mm, $CD = 110$ mm.

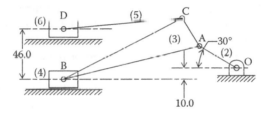

FIGURE P1.12

1.13 For the mechanism of a press machine shown in Figure P1.13, OA is a rotating crank and ABC is a bell crank. The die block is attached to the pin C, which slides along the slot in the link DE. Plot the displacement diagram of point C on the die block with the crank rotational angle θ.

$OA = 30$ mm, $AB = 95$ mm, $BC = 80$ mm, $QB = 90$ mm.

Angle $ABC = 90°$.

1.14 Choose any suitable dimensions to construct Hart, Watt, Peaucellier, and grasshopper straight-line motion mechanisms. Trace the path of points that move on a straight line in each case.

1.15 A point moves on a straight line by means of Watt's mechanism. It is required to magnify its motion three times. Construct such a mechanism.

1.16 Design a Geneva wheel with a center distance of 120 mm. The driven wheel rotates $60°$ for every revolution of the driving shaft.

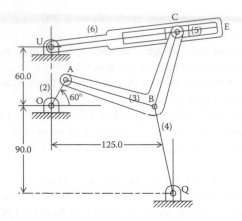

FIGURE P1.13

1.17 A universal joint is used to connect two shafts inclined at 20°, and the driving shaft speed is 1000 rpm. Find the extreme angular velocities of the driven shaft and its maximum acceleration.

1.18 A universal joint connects two shafts whose axes intersect at 15°. The driving shaft rotates uniformly at 1000 rpm. Plot, for one complete revolution of the driving shaft, the angular speed and the angular acceleration of the driven shaft. The driving shaft rotates at a uniform speed of 1000 rpm. Determine the greatest permissible angle between the shaft axes so that the total fluctuation of speed of the driven shaft does not exceed 150 rpm.

1.19 In assembling a double universal joint, the fork on one end of the intermediate shaft makes an angle β with the fork on the other end. Derive the relation between the driven and driving shaft angles of rotation if the angle between the shafts is 20° and the angle between the forks is 30°.

1.20 Sketch Oldham coupling and prove the following:
    a. The center of the intermediate disk describes a circle with diameter equal to the distance between the centerlines of the two shafts.
    b. The absolute angular velocity of the intermediate disk is double the angular velocity of either shaft.

Analytical Method

1.21 Figure P1.21 shows a group of four-bar chains in different configurations. The lengths of the links are $r_3 = 90$ mm and $r_4 = 80$ mm. Find the values of θ_3 and θ_4 for each configuration. The locations of the links are shown in the figure.

1.22 Figure P1.22 shows a group of engine chains in different configurations. The lengths of the links are $r_3 = 150$ mm. Find the values of θ_3 and x_4 for each configuration. The locations of the links are shown in the figure.

1.23 Figure P1.23 shows a group of shaper chains in different configurations. Find the values of θ_4 and the position of the block from the pivot of link (4) for each configuration. The locations of the links are shown in the figure.

1.24 Figure P1.24 shows a group of tilting blocks in different configurations. Find the values of θ_3 and the position of the block from the input point A for each configuration. The locations of the links are shown in the figure.

1.25 Solve Problems 1.2 through 1.6 and 1.11 through 1.13 using the analytical method.

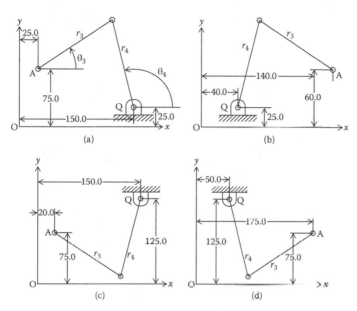

FIGURE P1.21

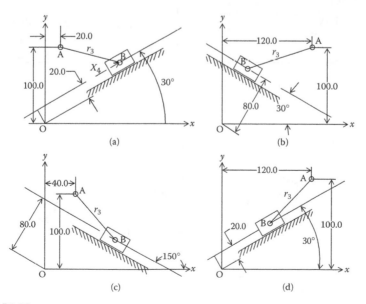

FIGURE P1.22

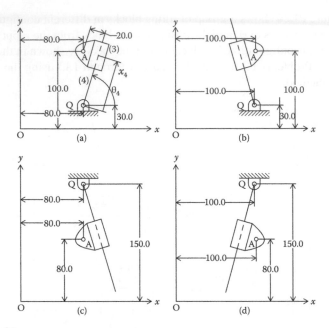

FIGURE P1.23

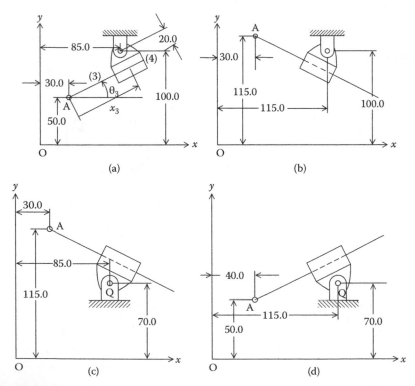

FIGURE P1.24

2 Velocities and Accelerations

The study of kinematics analysis (mainly velocities and accelerations) is rather important for the design engineer. Although the driving member of a mechanism usually rotates at a constant angular speed, other members have acceleration. Accordingly, inertia forces impose additional forces on the members. These forces become serious especially in high-speed machinery and must be considered in the design.

In manipulating the velocities and accelerations in mechanisms, vector algebra, which is explained in Chapter 1, is used. There are two ways to study the kinematics of mechanisms, namely, graphical and analytical methods. Graphical method is simpler and gives a complete picture of the velocities and accelerations of all members in a mechanism. Its disadvantages are that the analysis is performed for one position of the mechanism, needs much labor work, and lacks accuracy. On the other hand, the analytical method is more accurate and the calculations are faster especially if we use computer software. Both methods are presented in this chapter.

2.1 ABSOLUTE PLANE MOTION OF A PARTICLE

A particle moving on a fixed plane has two degrees of freedom. This means that its position can be determined by a vector that has a magnitude and a direction. The representation of vectors is explained in Section 1.9. The use of complex polar vectors is probably the simplest in the analysis.

Suppose that a particle P is moving on a fixed plane on a certain path as shown in Figure 2.1. The velocity of the particle $\mathbf{V}$ is tangent to the path. The position vector of point P is given by

$$\mathbf{r} = r\, e^{i\theta} \tag{2.1}$$

The absolute velocity of P is obtained by differentiating Equation 2.1 with respect to time. Thus,

$$\mathbf{V} = (\dot{r} + i\, r\, \dot{\theta})e^{i\theta} \tag{2.2}$$

The absolute acceleration of P is obtained by differentiating Equation 2.2 with respect to time. Thus,

$$A = \left[(\ddot{r} - r\, \dot{\theta}^2) + i(r\, \ddot{\theta} + 2\, \dot{r}\, \dot{\theta}) \right] e^{i\theta} \tag{2.3}$$

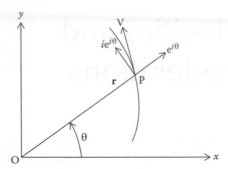

FIGURE 2.1   Path of a point on a fixed plane.

2.2   RELATIVE MOTION

2.2.1   MOTION OF A POINT RELATIVE TO A POINT ON A FIXED PLANE

Suppose that point P is moving on a fixed plane. At some instant, point P is sliding on point Q, which is fixed on the plane. Suppose that we construct the coordinate axes at the center of curvature I of the path as shown in Figure 2.2. The radius of curvature of the path at this instant is ρ. The vectors $\mathbf{u}_n$ and $\mathbf{u}_t$ are unit vectors normal and tangent to the path.

The position of P is given by

$$\mathbf{r} = \rho \mathbf{u}_n$$

The velocity of point P relative to the plane is represented by $\mathbf{V}_{PQ}$ and is given by

$$\mathbf{V}_{PQ} = \dot{\mathbf{r}} = \dot{\rho}\mathbf{u}_n + \rho \dot{\mathbf{u}}_n$$

It can be shown that

$$\dot{\mathbf{u}}_n = \dot{\theta}\mathbf{u}_t \text{ and } \dot{\mathbf{u}}_t = -\dot{\theta}\mathbf{u}_n$$

Thus,

$$\mathbf{V}_{PQ} = \dot{\rho}\mathbf{u}_n + \rho\dot{\theta}\mathbf{u}_t \tag{2.4}$$

But

$$\mathbf{V}_{PQ} = V_{PQ}\mathbf{u}_t \tag{2.5}$$

Comparing Equations 2.4 and 2.5, we conclude that

$$\dot{\rho} = 0 \tag{2.6}$$

$$\dot{\theta} = \frac{V_{PQ}}{\rho} \tag{2.7}$$

The acceleration of P relative to Q ($\mathbf{A}_{PQ}$) is obtained by differentiating Equation 2.5 and using Equation 2.7. Therefore,

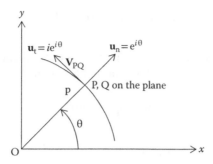

FIGURE 2.2 Tangential and normal components.

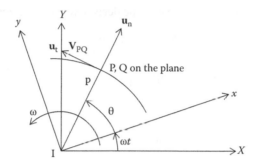

FIGURE 2.3 Motion of a point on a rotating plane.

$$\mathbf{A}_{PQ} = \dot{V}_{PQ}\,\mathbf{u}_{t} - \frac{V_{PQ}^{2}}{\rho}\mathbf{u}_{n} \tag{2.8}$$

It is worth to note that since point Q lies on a fixed plane, Equations 2.5 and 2.8 represent the absolute velocity and acceleration of point P.

2.2.2 MOTION OF A POINT RELATIVE TO A POINT ON A ROTATING PLANE

Suppose that the plane of Figure 2.3 rotates about point I with an angular velocity ω and an angular acceleration α. The motion of P relative to the plane is the same. Axis Ix makes an angle ωt with the fixed axis IX. The position vector of point P becomes

$$\rho_{n} = \rho\,e^{i\,(\theta + \omega t)} \tag{2.9}$$

The absolute velocity and acceleration of point P are obtained by differentiating Equation 2.9 with respect to time. Thus,

$$\mathbf{V}_{P} = i\rho\,(\dot{\theta} + \omega)\,e^{i\,(\theta + \omega t)} \tag{2.10}$$

$$\mathbf{A}_P = \left[-\rho(\dot{\theta} + \omega)^2 + i(\ddot{\theta} + \alpha)\right]e^{i(\theta + \omega t)} \tag{2.11}$$

The absolute velocity and acceleration of point Q are obtained by differentiating Equation 2.9 with respect to time; θ is constant for point Q at this instant.

$$\mathbf{V}_Q = i\rho\omega\, e^{i(\theta + \omega t)} \tag{2.12}$$

$$\mathbf{A}_Q = \left(-\rho\omega^2 + i\alpha\right)e^{i(\theta + \omega t)} \tag{2.13}$$

The velocity of P relative to Q is obtained by subtracting Equation 2.12 from Equation 2.10. Also, the acceleration of P relative to Q is obtained by subtracting Equation 2.13 from Equation 2.11. Since the derivatives of ρ are zero (Equation 2.6),

$$\mathbf{V}_{PQ} = I\rho\dot{\theta}\, e^{I(\theta + \omega t)} \tag{2.14}$$

$$\mathbf{A}_{PQ} = \left(-\rho\dot{\theta}^2 - 2\rho\omega\dot{\theta} + I\ddot{\theta}\right)e^{I(\theta + \omega t)} \tag{2.15}$$

Note that

$$\mathbf{u}_n = e^{I(\theta + \omega t)}$$

$$\mathbf{u}_t = i\, e^{I(\theta + \omega t)}$$

Also, if we use Equation 2.7, then

$$\mathbf{V}_{PQ} = V_{PQ}\mathbf{u}_t \tag{2.16}$$

$$\mathbf{A}_{PQ} = \dot{V}_{PQ}\,\mathbf{u}_t - \left(\frac{V_{PQ}^2}{\rho} + 2V_{PQ}\omega\right)\mathbf{u}_n \tag{2.17}$$

$$= \mathbf{A}_{PQ}^t + \mathbf{A}_{PQ}^c + \mathbf{A}_{PQ}^{cr}$$

As a conclusion,

- The velocity of P relative to Q is a vector with magnitude V_{PQ} and direction tangent to the path.
- The acceleration of P relative to Q has three vector components:
  1. The tangential component $\mathbf{A}_{PQ}^t$ has magnitude $\dot{V}_{PQ}$ and direction tangent to the path.
  2. The centripetal component $\mathbf{A}_{PQ}^c$ has magnitude V_{PQ}^2/ρ and direction toward the center of curvature.
  3. The Coriolis component $\mathbf{A}_{PQ}^{cr}$ has magnitude $2V_{PQ}\omega$ and direction toward the center of curvature. If the directions of V_{PQ} or ω or both

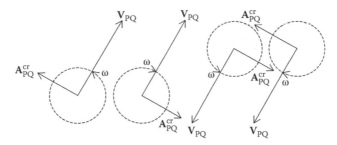

FIGURE 2.4   Direction of the Coriolis components.

change, then the direction of the vector changes. A general rule to obtain the direction of the Coriolis component is to rotate $\mathbf{V}_{PQ}$ 90° in the direction of ω as shown in Figure 2.4.

2.3   APPLICATIONS TO COMMON LINKS

2.3.1   ROTATING LINKS

Rotating links rotate about a fixed center as cranks, the rocker in the four-bar mechanism, the oscillating link in the shaper mechanism, and so forth.

Figure 2.5 shows the rotating link OP with length r and is rotating about O with an angular velocity $\dot{\theta} = \omega$ and an angular acceleration $\ddot{\theta} = \alpha$; both are considered positive in the counterclockwise direction. Point P is at the tip of the link and traces a circular path. Applying Equations 2.2 and 2.3 and put into consideration that

$$r = \text{constant}$$
$$\dot{r} = \ddot{r} = 0$$

$$\mathbf{u}_n = e^{i\theta}$$

$$\mathbf{u}_t = i\,e^{i\theta}$$

Therefore,

$$\mathbf{V}_P = \omega\, r\, \mathbf{u}_t \tag{2.18}$$

$$\mathbf{A}_A = -\omega^2\, r\, \mathbf{u}_n + \alpha\, r\, \mathbf{u}_t \tag{2.19a}$$

or

$$\mathbf{A}_A = -\frac{V_P^2}{r}\, \mathbf{u}_n + \alpha\, r\, \mathbf{u}_t \tag{2.19b}$$

$$\mathbf{A}_A = \mathbf{A}_P^c + \mathbf{A}_P^t \tag{2.20}$$

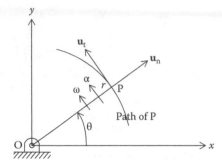

FIGURE 2.5   Kinematics of a rotating link.

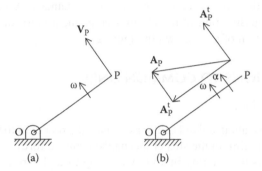

FIGURE 2.6   (a) Velocity of a rotating link (b) acceleration of a rotating link.

Conclusions:

- The velocity of a point on a rotating link is a vector of magnitude equal
  to the angular velocity times the distance from the point to the center of
  rotation, and is normal to the link in the direction of rotation (Figure 2.6a).
- The acceleration of a point on a rotating link is a vector that has two components (Figure 2.6b):
    1. The centripetal component A_P^c with magnitude equal to the square of
       the angular velocity times the distance (or the square of the velocity
       divided by the distance) is directed toward the center of rotation.
    2. The tangential component (also called transverse component) A_P^t with
       magnitude equal to the angular acceleration times the distance from the
       point to the center of rotation is normal to the link in the direction of the
       angular acceleration.

2.3.2   FLOATING LINKS

Floating links are those that are not fixed at any point on the link, such as the
coupler in the four-bar mechanism, the connecting rod in the engine mechanism,
and so forth. Consider link AB with length r (Figure 2.7a). Its initial position is at
A_1B_1 and then moves to AB. This motion can be considered as a combination of

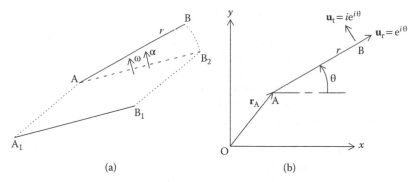

FIGURE 2.7 Kinematics of a floating link. (a) Representation of the equivalent motion (b) velocity and acceleration of a point on a floating link.

translation (from A_1B_1 to AB_2) and a rotation motion about A (in the direction from AB_2 to AB). The link is considered to have an angular velocity ω and an angular acceleration α.

Consider the floating link in Figure 2.7b; the position of point B is given by

$$\mathbf{r}_B = \mathbf{r}_A + AB\, e^{i\theta}$$

The velocity and acceleration of point B are (AB is constant)

$$\mathbf{V}_B = \mathbf{V}_A + i\,\omega\, AB\, e^{i\theta}$$
$$= \mathbf{V}_A + \omega\, AB\, \mathbf{u}_t$$

$$\mathbf{V}_B = \mathbf{V}_A + \mathbf{V}_{BA} \tag{2.21}$$

$$\mathbf{A}_B = \mathbf{A}_A + \left(-\omega^2\, AB + i\,\omega\, r\right) e^{i\theta}$$
$$= \mathbf{A}_A - \omega^2\, AB\, \mathbf{u}_n + \alpha\, AB\, \mathbf{u}_t$$

$$= \mathbf{A}_A - \frac{V_{BA}^2}{AB}\,\mathbf{u}_n + \alpha\, AB\, \mathbf{u}_t$$

$$\mathbf{A}_B = \mathbf{A}_A + \mathbf{A}_{BA} \tag{2.22}$$

where

$$\mathbf{A}_{BA} = \mathbf{A}_{BA}^c + \mathbf{A}_{BA}^t \tag{2.23a}$$

$$\mathbf{A}_{BA}^c = -\frac{V_{BA}^2}{AB}\,\mathbf{u}_n \tag{2.23b}$$

From the above analysis, we arrive at two important conclusions:

1. The velocity (or acceleration) of a point on a link is equal to the velocity (or acceleration) of another point on the same link plus the velocity (or acceleration) of the first point relative to the other.
2. The relative motion of two points on the same link is considered as if one is rotating with respect to the other. The relative velocity and acceleration are the same as those of a rotating link.

Note that the first conclusion is general and can be applied to any two points not necessary on the same link. This is called the law of relative motion. The second conclusion applies only for points on the same link.

2.3.3 LINKS MOVING ON A FIXED PLANE

The slider in Figure 2.8 slides inside a circular surface with an absolute velocity V_P. The path traced by point P is a circular arc with a radius R. The velocity and acceleration of point P are given by Equations 2.5 and 2.8 and considering point Q to be fixed.

$$\mathbf{V}_P = V_P \, \mathbf{u}_t \tag{2.24}$$

$$\mathbf{A}_P = \dot{V}_P \, \mathbf{u}_t - \frac{V_P^2}{R} \, \mathbf{u}_n \tag{2.25}$$

If the surface is flat (Figure 2.9), like in the case of the piston in engine mechanism, the radius of curvature of the path, $R = \infty$, then

$$\mathbf{A}_P = \dot{V}_P \, \mathbf{u}_t \tag{2.26}$$

That is, the velocity and acceleration of P are in the direction of motion. If point B is another point on the link, its path is also a straight line parallel to that of P. Therefore,

$$\mathbf{V}_B = \mathbf{V}_P \tag{2.27}$$

$$\mathbf{A}_B = \mathbf{A}_P \tag{2.28}$$

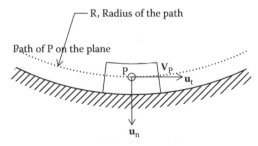

FIGURE 2.8 Motion of a slider on a fixed curved surface.

2.3.4 Links Sliding on a Rotating Link

In sliding links, the contact between the two links is usually a point or line contact as described in the higher pairs. Mechanisms incorporating sliding action are cams, gears, shapers, and similar mechanisms. To perform kinematics analysis for such mechanisms, we use the equations derived in Section 2.2.2. First of all, we have to determine the path of the point of contact of one link on a plane rotating with the other link, determine the radius of curvature of the path, and determine the directions of the common normal and the common tangent of the path. In some cases (Figure 2.10a), the path is predetermined such as for cams with knife-edge followers, cams with roller followers (we deal with the center of the roller), and blocks sliding on rotating links as in the shaper mechanisms. The path has the same configuration of the links on which sliding occurs.

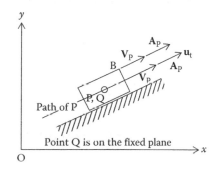

FIGURE 2.9 Motion of a slider on a flat surface.

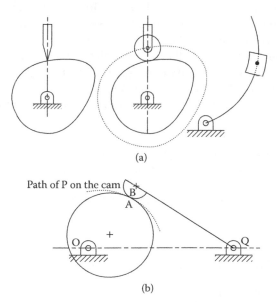

FIGURE 2.10 Sliding motion relative to rotating links. (a) Predetermined path (b) path is to be determined.

In other cases, like cams with flat-faced, spherical, roller (if the motion of the roller is to be considered), and gears, the path of the point of contact is to be determined (Figure 2.10b).

The process of obtaining the path is tedious and lacks accuracy. However, there are two approaches to overcome this difficulty. The first is to treat the mechanism as an equivalent system, as explained in Section 1.9.4. The drawback of this method is that it does not give the relative motion at the point of contact directly. The second is to use the approach developed by the author of this chapter, which is explained in the following section.

2.3.4.1   Mostafa's Theory

The theory states that, in sliding links, the acceleration of a point B on a link relative to another point A on the other link has two components: a normal component, A_{BA}^n, along the common normal and a sliding component, A_{BA}^{SL}, along the common tangent such that

$$\mathbf{A}_{BA} = A_{BA}^n + A_{BA}^{SL} \tag{2.29a}$$

$$\mathbf{A}_{BA}^{SL} = A_{BA}^{SL}\, \mathbf{u}_t \tag{2.29b}$$

$$\mathbf{A}_{BA}^n = \frac{1}{\rho_A + \rho_B}\left[\rho_A\rho_B(\omega_A - \omega_B)^2 - V_{BA}^2 - 2V_{BA}(\rho_A\, \omega_A + \rho_B\, \omega_B)\right]\mathbf{u}_n \tag{2.30}$$

where

- ρ_A is the radius of curvature of the link on which point A lies.
- ρ_B is the radius of curvature of the link on which point B lies.
- ω_A is the angular velocity of the link on which point A lies.
- ω_B is the angular velocity of the link on which point B lies.
- V_{BA} is the velocity of B relative to A. $\mathbf{V}_{BA}$ is along $\mathbf{u}_t$.
- $\mathbf{u}_t$ is a unit vector along the common tangent.
- $\mathbf{u}_n$ is a unit vector along the common normal directed from A to B. $\mathbf{u}_t$ is leading $\mathbf{u}_n$ by 90°.

Equation 2.30 is based on that V_{BA} is positive in the direction of $\mathbf{u}_t$, and ω_A and ω_B are positive in the counterclockwise direction. It can be put as

$$\mathbf{A}_{BA}^n = \mathbf{A}_{BA}^{RL} + \mathbf{A}_{BA}^{RT} + \mathbf{A}_{BA}^{MA} + \mathbf{A}_{BA}^{MB} \tag{2.31}$$

where

$-\mathbf{A}_{BA}^{RL}$ is termed the rolling component, with a magnitude of $\dfrac{\rho_A\rho_B(\omega_A - \omega_B)^2}{\rho_A + \rho_B}$ and a direction from point A (or the center of curvature of A) to point B (or the center of curvature of B). It is not affected by the direction of ω_A and ω_B.

$-A_{BA}^{RT}$ is termed the rotational component, with a magnitude of $\dfrac{V_{BA}^2}{\rho_A + \rho_B}$ and
a direction from B (or the center of curvature of B) to A (or the center of curvature of A). It is not affected by the direction of V_{BA}.

$-A_{BA}^{MA}$ is termed the "Mostafa A" component, with a magnitude of $\dfrac{2 V_{BA}\, \rho_A \omega_A}{\rho_A + \rho_B}$
and a direction from B to A. It can be determined by rotating V_{BA} 90° with ω_A as in the case of the Coriolis component.

$-A_{BA}^{MB}$ is termed the "Mostafa B" component, with a magnitude of $\dfrac{2 V_{BA}\, \rho_B\, \omega_B}{\rho_A + \rho_B}$
and a direction from B to A. It can be determined by rotating V_{BA} 90° with ω_B as in the case of the Coriolis component.

The proof of the theory is outlined as follows:

- Let ω_A and ω_B, and α_A and α_B be the angular velocities and the angular accelerations of links (3) and (4) (Figure 2.11).
- Let ρ_A and ρ_B be the radii of curvature of the surfaces at the point of contact.
- Let points C and F be the centers of curvature of links (3) and (4) at the point of contact.
- Set the unit vectors u_n and u_t along the common normal and the common tangent to the surfaces; u_t is leading u_n by 90° in the counterclockwise direction. u_n is directed from A to B. The velocity of B relative to A is assumed positive in the direction of u_t. Thus,

$$V_{BA} = V_{BA}\, u_t \tag{2.32}$$

Use the law of relative motion; then,

$$V_A = V_C + \omega_A\, \rho_A\, u_t \tag{2.33}$$

$$V_B = V_F - \omega_B\, \rho_B\, u_t \tag{2.34}$$

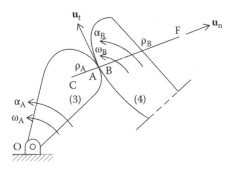

FIGURE 2.11   Representation of Mostafa's theory.

Subtract Equation 2.33 from Equation 2.34; then,

$$\mathbf{V}_{BA} = \mathbf{V}_B - \mathbf{V}_A$$
$$= \mathbf{V}_F - \mathbf{V}_C - (\omega_A \, \rho_A + \omega_B \, \rho_B) \, \mathbf{u}_t$$

or

$$\mathbf{V}_{FC} = \mathbf{V}_{BA} + (\omega_A \, \rho_A + \omega_B \, \rho_B) \, \mathbf{u}_t$$
$$= (V_{BA} + \omega_A \, \rho_A + \omega_B \, \rho_B) \mathbf{u}_t \tag{2.35}$$

The acceleration of B relative to A has normal and tangential components; that is,

$$A_{BA} = A_{BA}^n + A_{BA}^t \tag{2.36}$$

Now

$$\mathbf{A}_A = \mathbf{A}_C - \rho_A \omega_A^2 \mathbf{u}_n + \rho_A \alpha_A \mathbf{u}_t \tag{2.37}$$

$$\mathbf{A}_B = \mathbf{A}_F + \rho_B \, \omega_B^2 \, \mathbf{u}_n - \rho_B \, \alpha_B \, \mathbf{u}_t \tag{2.38}$$

Subtracting the above two equations, we get

$$\mathbf{A}_{BA} = \mathbf{A}_{FC} + (\rho_A \, \omega_A^2 + \rho_B \, \omega_B^2) \, \mathbf{u}_n - (\rho_A \, \alpha_3 + \rho_B \, \alpha_4) \, \mathbf{u}_t \tag{2.39}$$

To obtain $\mathbf{A}_{FC}$, consider the imaginary link CF. According to Equation 2.23

$$\mathbf{A}_{FC} = -\frac{V_{FC}^2}{\rho_A + \rho_B} \mathbf{u}_n + A_{FC}^t \, \mathbf{u}_t \tag{2.40}$$

Substituting Equation 2.40 into Equation 2.39, we get

$$\mathbf{A}_{BA} = \left(-\frac{V_{FC}^2}{\rho_A + \rho_B} + \rho_A \omega_A^2 + \rho_B \omega_B^2 \right) \mathbf{u}_n + \left(A_{FC}^t - \rho_A \alpha_3 - \rho_B \alpha_4 \right) \mathbf{u}_t \tag{2.41}$$

Comparing Equations 2.36 and 2.41, we can determine $\mathbf{A}_{BA}^n$. Obtaining $\mathbf{A}_{BA}^t$ in this stage is meaningless since it is usually determined from the analysis of the system as a whole. Thus,

$$\mathbf{A}_{BA}^n = \left(-\frac{V_{FC}^2}{\rho_A + \rho_B} + \rho_A \, \omega_A^2 + \rho_B \, \omega_B^2 \right) \mathbf{u}_n$$

Using Equation 2.35 and rearranging terms, we arrive at Equation 2.30

$$A_{BA}^n = \frac{1}{\rho_A + \rho_B} \left[\rho_A \rho_B (\omega_A - \omega_B)^2 - V_{BA}^2 - 2 V_{BA} (\rho_A \, \omega_A + \rho_B \, \omega_B) \right] \mathbf{u}_n \tag{2.30}$$

Special attention should be paid in setting the direction of the unit vectors $\mathbf{u}_n$ and $\mathbf{u}_t$ according to which point is related to which. The direction of $\mathbf{u}_n$ is from the point which motion is related to, to the other point; $\mathbf{u}_t$ is leading $\mathbf{u}_n$ in the counterclockwise direction with 90°. When one of the surfaces is concaved, its radius of curvature is negative. If the surface of one of the links is a point, the direction of $\mathbf{u}_n$ is decided according to the centers of curvature with the same sequence described.

2.4 ANALYSIS OF MECHANISMS: GRAPHICAL METHOD

In the preceding sections, we derived the equations necessary to make kinematics analysis for the planer mechanisms. In this section, we present applications to a group of mechanisms as examples.

2.4.1 ENGINE MECHANISM

Figure 2.12a represents an engine mechanism with crank OA rotating clockwise with a uniform angular velocity ω_2. It is required to determine the velocity and the acceleration of piston B and the angular velocity and the angular acceleration of the connecting rod.

2.4.1.1 Velocity Analysis

Point A is at the end of a rotating link. Its velocity is obtained from Equation 2.18. Since the crank rotates clockwise, ω_2 is negative. Thus,

$$V_A = \omega_2\, OA, \text{ normal to line OA with } \omega_2$$

Point B lies on floating link AB. To determine the velocity of point B, we use Equation 2.21.

$$\mathbf{V}_B = \mathbf{V}_A + \mathbf{V}_{BA}$$

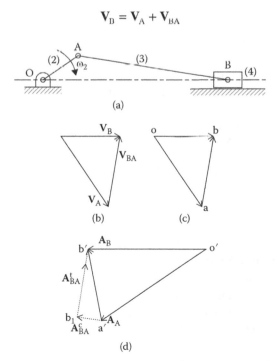

FIGURE 2.12 (a) Engine mechanism (b) vector velocity polygon (c) velocity polygon using notations (d) acceleration polygon.

This equation is a vector equation that contains three vectors; each has a magnitude and a direction. It can be solved only for two unknowns. In our case, V_A is known in magnitude and direction. Vector V_{BA} is unknown in magnitude. Its direction, according to conclusion 2 (Section 2.3.2), is normal to AB. Point B lies on link AB as well as on the piston. The piston has translation motion. Thus, according to Section 2.3.3, its direction is along the direction of motion of the piston. The direction of V_B is known, while its magnitude is unknown. Thus, the unknown quantities in Equation 2.21 are the magnitudes of V_B and V_{BA}.

To determine the unknown quantities, a vector polygon is constructed. Using a suitable scale, the vector V_A is drawn as shown in Figure 2.12b. From the starting point of this vector, a line parallel to the direction of motion of the piston is drawn. From the end of the vector V_A, a line perpendicular to AB is drawn to intersect the previous line at a point as shown in Figure 2.10b. The triangle obtained forms the velocity polygon. The magnitudes of V_B and V_{BA} are thus determined. Their proper directions are determined by placing the arrows according to the vector equation. In such an equation, the vectors on the right-hand side are added together in a sense opposite to that of the vectors on the other side.

The velocity polygon obtained so far can be clarified by the velocity polygon notation shown in Figure 2.12c. Pole o is the origin of the polygon from where all absolute velocities branch. Lower-case letters at the end of each line passing through o represent the absolute velocities of the corresponding upper-case letters in the mechanism. For example,

- Line oa represents V_A.
- Line ob represents V_B.

Lines passing through points other than o represent relative velocities of the corresponding points in the mechanism. For example,

- Line ab represents V_{BA}.
- Line ba represents V_{AB}.

The angular velocity of the connecting rod (3) may be obtained from V_{BA} or V_{AB}. Both yield the same result and is given by:

$$\omega_3 = \frac{V_{BA}}{AB} \quad \text{Counterclockwise}$$

2.4.1.2 Acceleration Analysis

Since crank OA rotates with a uniform angular velocity, the transverse component of the acceleration of point A is zero. According to Equation 2.19, A_A has only a centripetal component equal to ω_2^2 OA and is directed toward O. The acceleration of B is determined by applying Equations 2.22 and 2.23; that is,

$$A_B = A_A + A_{BA}$$
$$= A_A + A_{BA}^c + A_{BA}^t$$

The vector quantities in the above equation can be listed as follows:

Vector	Magnitude	Direction
$\mathbf{A}_A$	$\omega_2^2\,OA$	Parallel to AO
$\mathbf{A}_{BA}^c$	$\omega_3^2 \times AB = \dfrac{V_{BA}^2}{AB}$	Parallel to BA
$\mathbf{A}_{BA}^t$	Unknown	Perpendicular to BA
$\mathbf{A}_B$	Unknown	In the direction of motion of point B

The two unknown quantities can be determined by drawing the acceleration polygon as shown in Figure 2.12d. Line $o'a'$ is drawn parallel to AO to represent $\mathbf{A}_A$ to a suitable scale. From point a', line $a'b'_1$ is drawn parallel to BA to represent $\mathbf{A}_{BA}^c$ with the same scale. A line normal to BA is drawn from b_1 to meet a line through o' parallel to the direction of motion of B at b'. Line $o'b'$ represents the acceleration of B, while line b_1b' represents $\mathbf{A}_{BA}^t$. The angular acceleration of link (3) is given by

$$\alpha_3 = \frac{A_{BA}^t}{AB}$$

where A_{BA}^t is the absolute value of the vector $\mathbf{A}_{BA}^t$. The direction of the angular acceleration α_3 of link (3) is counterclockwise as indicated by the direction of the vector $\mathbf{A}_{BA}^t$ in Figure 2.12d.

2.4.2 FOUR-BAR MECHANISM

For the four-bar mechanism shown in Figure 2.13a, crank OA rotates with an angular velocity of 30 rad/s clockwise and with an angular acceleration of 200 rad/s² counterclockwise. It is required to find the angular velocities and the angular accelerations of links (3) and (4). The lengths of the links are

$$OA = 5 \text{ cm}, AB = 7.5 \text{ cm}, QB = 9 \text{ cm}, \text{ and } OQ = 10 \text{ cm}.$$

Velocity Analysis
The velocity of A is given by

$$V_A = \omega_2 \times OA = 30 \times 5 = 150 \text{ cm/s}$$

The velocity of B is obtained by the vector equation

$$\mathbf{V}_B = \mathbf{V}_A + \mathbf{V}_{BA}$$

where

Vector	Magnitude (cm/s)	Direction
$\mathbf{V}_A$	150	Perpendicular to OA with ω_2
$\mathbf{V}_{BA}$	Unknown	Perpendicular to AB
$\mathbf{V}_B$	Unknown	Perpendicular to QB

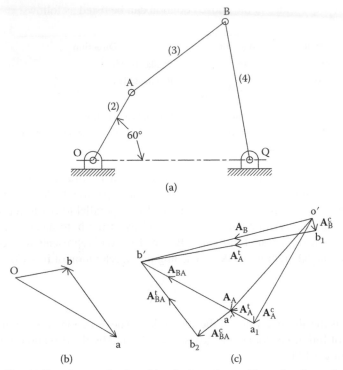

FIGURE 2.13   (a) Four-bar mechanism (b) velocity polygon (c) acceleration polygon.

The velocity polygon is shown in Figure 2.13b. The magnitude of $\mathbf{V}_{BA}$ is 108.3 cm/s, while the magnitude of $\mathbf{V}_B$ is 65.4 cm/s. Thus,

$$\omega_3 = \frac{108.3}{7.5} = 14.44 \text{ rad/s} \quad \text{Counterclockwise}$$

$$\omega_4 = \frac{65.4}{9} = 7.27 \text{ rad/s} \quad \text{Clockwise}$$

Acceleration Analysis

The acceleration of A has centripetal and transverse components. Their magnitudes are

$$A_A^c = \omega_2^2 \times OA = 30^2 \times 5 = 4500 \text{ cm/s}^2$$

$$A_A^t = \alpha_2 \times OA = 200 \times 5 = 1000 \text{ cm/s}^2$$

The acceleration of point B is obtained by applying the vector equation

$$\mathbf{A}_B = \mathbf{A}_A + \mathbf{A}_{BA}$$

$$A_B^c + A_B^t = A_A^c + A_A^t + A_{BA}^c + A_{BA}^t$$

where

Vector	Magnitude (cm/s²)	Direction
A_A^c	4500	Parallel to AO
A_A^t	1000	Perpendicular to AO
A_{BA}^c	$\dfrac{V_{BA}^2}{AB} = 1564$	Parallel to BA
A_{BA}^t	Unknown	Perpendicular to BA
A_B^c	$\dfrac{V_B^2}{QB} = 475.24$	In the direction of BQ
A_A^t	Unknown	Perpendicular to BQ

The acceleration polygon is drawn as shown in Figure 2.13c.

- Line $o'a_1$ represents the centripetal component of point A.
- Line a_1a' represents the transverse component of point A.
- Line $a'b_2$ represents the centripetal component of B relative to A.
- Line $o'b_1$ represents the centripetal component of point B.
- From point b_1 and point b_2, lines perpendicular to QB and AB are drawn, respectively, to intersect at b'.
- Line b_2b' represents the transverse component of B relative to A and is equal to 3456 cm/s². This yields an angular acceleration of 460.8 rad/s² counterclockwise for link (3).
- Line b_1b' represents the transverse component of B and is equal to 3456 cm/s². This yields an angular acceleration of 736 rad/s² counterclockwise for link (4).

2.4.2.1 Velocity and Acceleration Images

Referring to Figure 2.13b and c, it is seen that link AB is represented by line ab in the velocity polygon and by line $a'b'$ in the acceleration polygon. Line ab is denoted as the velocity image of link AB, while $a'b'$ is its acceleration image. Thus, each link in any mechanism has an image in the velocity and acceleration polygons. According to the notation previously used, in these polygons, the images are denoted by small letters, while the corresponding links are denoted by the same letters in capital. The velocity and acceleration of any point on a link can be simply obtained by using the concept of images. Suppose that link AB (Figure 2.14a) has an angular velocity ω and an angular acceleration α. The velocity of B relative to A is a vector normal to AB and is represented by line ab as indicated in Figure 2.14b, such that

$$ab \times scale = \omega \times AB$$

or

$$\frac{ab \times scale}{AB} = \omega$$

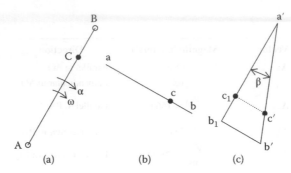

FIGURE 2.14   (a) A general link (b) velocity image (c) acceleration image.

Similarly, if C is a point on the same link, the velocity of C relative to A is represented by line ac so that

$$\frac{ac \times \text{scale}}{AC} = \omega$$

Hence,

$$\frac{ac}{AC} = \frac{ab}{AB} = \omega \text{ (to scale)} \tag{2.42}$$

The acceleration of B relative to A has centripetal and transverse components as shown in Figure 2.14c.

The acceleration image of link AB is represented by line a'b', where

$$a'b' \times \text{scale} = \sqrt{\omega^4 + \alpha^2}\,AB$$

or

$$\frac{a'b' \times \text{scale}}{AB} = \sqrt{\omega^4 + \alpha^2}$$

Line a'b' makes an angle β with BA such that

$$\tan\beta = \frac{\alpha'}{\omega^2}$$

The acceleration of C relative to A is represented by a'c', which is given by

$$\frac{a'c' \times \text{scale}}{AB} = \sqrt{\omega^4 + \alpha^2} = \frac{a'b' \times \text{scale}}{AB} \tag{2.43}$$

From Equations 2.42 and 2.43, we conclude that the locations of points c and c' on the velocity and acceleration images have the same proportions as the location of point C on the link.

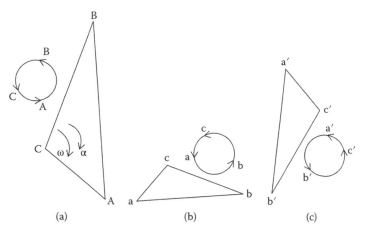

FIGURE 2.15 Orientation of the velocity and acceleration images. (a) The link (b) velocity image (c) acceleration image.

Let us consider link ABC with a triangular shape as shown in Figure 2.15a and has an angular velocity ω and an angular acceleration α. The velocity image is triangle abc (Figure 2.15b). Lines ab, bc, and ca are the velocity images of AB, BC, and CA, respectively, such that

$$\frac{ab}{AB} = \frac{bc}{BC} = \frac{ac}{AC} = \omega \text{ (to scale)} \tag{2.44}$$

Also, triangle a'b'c' (Figure 2.15c) is the acceleration image of link ABC, where

$$\frac{a'b'}{AB} = \frac{b'c'}{BC} = \frac{a'c'}{AC} = \sqrt{\omega^4 + \alpha^2} \text{ (to scale)} \tag{2.45}$$

According to Equations 2.44 and 2.45, triangles ABC, abc, and a'b'c' are similar. It is important to note that the orientation of the letters on the three triangles must be in the same sense as indicated by the circles in Figure 2.15.

In conclusion, when the velocity and acceleration of any two points on a link are known, the velocity and acceleration of any other point on the same link can be obtained by constructing the images that have similar shapes as the link. This is demonstrated by the following example.

2.4.3 COMPOUND MECHANISM

Figure 2.16a represents a toggle mechanism. Crank OA rotates at a constant speed of 300 rpm clockwise. It is required to determine the velocity and acceleration of ram F at the shown position. The dimensions of the links are as follows:

OA = 25 cm, AB = 80 cm, QB = 30 cm, QC = 35 cm,

CD = 100 cm, PD = PE = 22.5 cm, DE = 25 cm, EF = 45 cm.

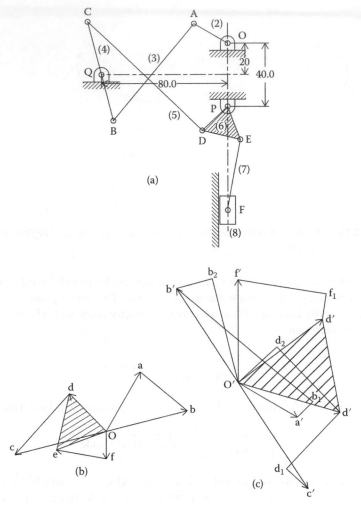

FIGURE 2.16 Velocity and acceleration polygons of a compound mechanism. (a) Mechanism (b) velocity polygon (c) acceleration polygon.

Analysis:

1. *Velocity*

$$\omega_2 = \frac{2 \times \pi \times 300}{60} = 31.4 \text{ rad/s}$$

The velocity of point A is

$$V_A = 31.4 \times 25 = 785 \text{ cm/s}, \quad \text{normal to OA with } \omega_2.$$

The linkage OABQ is a four-bar mechanism. The velocity image of link OB is represented by ob as shown in Figure 2.16b. Since BQC is one link and point C is on the extension of BQ, its velocity is obtained in the velocity polygon by extending line bo, and then locating at point c with the same proportions as the link. The linkage QCDP is also a four-bar mechanism in which the velocity of point C is known. Hence, the velocity of point D can be determined. The velocity of point E is obtained by drawing triangle ode in the velocity polygon similar to PDE. The arrangement of the letters on both triangles must be in the same orientation. The linkage PEF is similar to an engine mechanism and the velocity of point E is known. The velocity of F can then be determined. From the velocity polygon shown in Figure 2.16b, the velocity of F is 307.7 cm downward.

2. Acceleration

Point A has only a centripetal component.

$$A_A = 31.4^2 \times 25 = 2.456 \times 10^4 \, \text{cm/s}^2$$

Following the same steps as in the velocity, we can construct the acceleration polygon as shown in Figure 2.16c. The acceleration of point F is equal to 3.766×10^4 cm/s².

2.4.4 MECHANISM WITH SLIDING LINKS

The analysis of sliding links is presented in Section 2.3.5. Some examples are presented in the following sections.

2.4.4.1 Shaper Mechanism

Consider the shaper mechanism shown in Figure 2.17a. Block (3) is sliding on a circular arc link (4) with a radius R and center at point C. The crank rotates with an angular velocity ω_2 and an angular acceleration α_2. It is required to determine the sliding velocity, the sliding acceleration, the angular velocity, and the angular acceleration of link (4).

Analysis:

1. *Velocity*

$$V_A = \omega_2 \times OA$$
$$\mathbf{V}_B = \mathbf{V}_A + \mathbf{V}_{BA}$$

Vector	Magnitude	Direction
$\mathbf{V}_A$	$\omega_2 \times OA$	Perpendicular to O_2 A with ω
$\mathbf{V}_{BA}$	Unknown	Tangent to the arc or perpendicular to line OC
$\mathbf{V}_B$	Unknown	Perpendicular to Q B

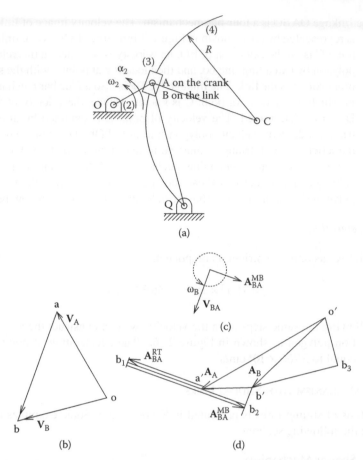

FIGURE 2.17   Velocity and acceleration polygons of the shaper mechanism. (a) Mechanism (b) velocity polygon (c) determination of the M-components (d) acceleration polygon.

The velocity polygon is shown in Figure 2.17b. The velocity of point B relative to point A is represented by ab. The angular velocity of link (4) is given by

$$\omega_4 = \frac{ob \times scale}{QB} \quad Counterclockwise$$

2. *Acceleration*

$$A_A = \omega_2^2 \times OA$$
$$\mathbf{A}_B = \mathbf{A}_A + \mathbf{A}_{BA}$$
$$\mathbf{A}_B = \mathbf{A}_B^c + \mathbf{A}_{BA}^t$$

The acceleration of B relative to A, $\mathbf{A}_{BA}$, is obtained by applying Equations 2.29 through 2.31.

$$\mathbf{A}_{BA} = \mathbf{A}_{BA}^n + \mathbf{A}_{BA}^{SL} = \mathbf{A}_{BA}^{RL} + \mathbf{A}_{BA}^{RT} + \mathbf{A}_{BA}^{MA} + \mathbf{A}_{BA}^{MB} + \mathbf{A}_{BA}^{SL}$$

Thus,

$$\mathbf{A}_B^c + \mathbf{A}_{BA}^t = \mathbf{A}_{BA}^{RL} + \mathbf{A}_{BA}^{RT} + \mathbf{A}_{BA}^{MA} + \mathbf{A}_{BA}^{MB} + \mathbf{A}_{BA}^{SL}$$

It should be noted that

$$\rho_A = 0$$
$$\rho_B = R$$
$$\omega_A = \omega_2$$
$$\omega_B = \omega_4$$

The components of the vector equations are presented in the following table. The acceleration polygon is drawn as shown in Figure 2.17d.

Vector	Magnitude	Direction
$\mathbf{A}_A$	$\omega_2^2\, OA$	Parallel to AO
$\mathbf{A}_{BA}^{RL}$	$\dfrac{\rho_A\rho_B\left(\omega_A - \omega_B\right)^2}{\rho_A + \rho_B} = 0$	
$\mathbf{A}_{BA}^{RT}$	$\dfrac{V_{BA}^2}{R}$	Parallel to CA
$\mathbf{A}_{BA}^{MA}$	$\dfrac{2\,V_{BA}\rho_A\omega_A}{\rho_A + \rho_B} = 0$	
$\mathbf{A}_{BA}^{MB}$	$2\,V_{BA}\omega_B$	Parallel to line AC (obtained by rotating $\mathbf{V}_{BA}$ $90°$ with ω_4; Figure 2.15c).
$\mathbf{A}_{BA}^{SL}$	Unknown	Perpendicular to CA
$\mathbf{A}_B^c$	$\dfrac{V_B^2}{QB}$	In the direction of BQ
$\mathbf{A}_A^t$	Unknown	Perpendicular to BQ

- Line o'a' is parallel to AO; it represents $\mathbf{A}_A$.
- Line a'b$_1$ is parallel to CA; it represents $\mathbf{A}_{BA}^{RT}$.
- Line b$_1$b$_2$ is parallel to AC; it represents $\mathbf{A}_{BA}^{MB}$.
- Line o'b$_3$ is parallel to BQ; it represents $\mathbf{A}_B^c$.

From point b$_2$ a line perpendicular to AC and from point b$_3$ a line perpendicular to QB are drawn to intersect at point b'.

2.4.4.2 Cam Mechanism

A circular cam with a flat-faced translating follower is shown in Figure 2.18a. The cam rotates with a uniform angular speed ω counterclockwise. It is required to determine the velocity and acceleration of the follower.

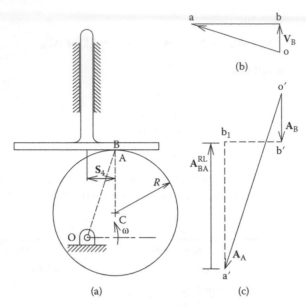

FIGURE 2.18 Velocity and acceleration polygons of a cam with translating flat-faced follower. (a) Cam mechanism (b) velocity polygon (c) acceleration polygon.

Analysis:

1. *Velocity*

$$V_A = \omega_2 \times OA$$
$$\mathbf{V}_B = \mathbf{V}_A + \mathbf{V}_{BA}$$

The velocity polygon is shown in Figure 2.18b.

Vector	Magnitude	Direction
$\mathbf{V}_A$	$\omega \times OA$	Perpendicular to OA with ω_2
$\mathbf{V}_{BA}$	Unknown	Tangent to the cam (normal CA)
$\mathbf{V}_B$	Unknown	Vertical

2. *Acceleration*

$$A_A = \omega^2 \times OA$$

$$\mathbf{A}_B = \mathbf{A}_A + \mathbf{A}_{BA}^{RL} + \mathbf{A}_{BA}^{RT} + \mathbf{A}_{BA}^{MA} + \mathbf{A}_{BA}^{MB} + \mathbf{A}_{BA}^{s}$$

It should be noted that

$$\rho_A = R$$
$$\rho_B = \infty$$
$$\omega_A = \omega$$
$$\omega_B = 0$$

The components of the vector equations are presented in the following table.

Vector	Magnitude	Direction
A_A	$\omega^2 \times R$	Parallel to line AO
A_{BA}^{RL}	$R\omega_A^2$	Parallel to line CA
A_{BA}^{RT}	$\dfrac{V_{BA}^2}{R+\infty}=0$	
A_{BA}^{MA}	$\dfrac{2V_{BA}\,R\omega_A}{R+\infty}=0$	
A_{BA}^{MB}	$2V_{BA}\omega_B=0$	
A_{BA}^{SL}	Unknown	Perpendicular to CA
A_B	Unknown	Vertical

The acceleration polygon is drawn as shown in Figure 2.18c.
- Line o′a′ is parallel to OA; it represents A_A.
- Line a′b₁ is parallel to CA; it represents A_{BA}^{RT}.
- Line b₁b₂ is parallel to AC; it represents A_{BA}^{MA}.
- Line o′b₃ is parallel to QB; it represents A_B^c.

From point b₁, a line perpendicular to line AC is drawn. From point o′, a vertical line is also drawn to intersect the previous line at point b′.

2.5 METHOD OF INSTANTANEOUS CENTERS FOR DETERMINING THE VELOCITIES

2.5.1 INSTANTANEOUS CENTER OF A BODY

The concept of the instantaneous centers states the following: Any displacement of a body having plane motion may be considered, at a given instant, as a pure rotation around some point called the instantaneous center (point I, Figure 2.19). The velocity

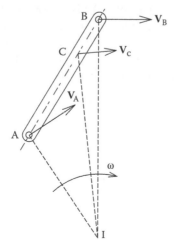

FIGURE 2.19 Instantaneous center of a link.

of any point on this body is similar to that obtained for rotating links previously discussed in Section 2.3.1. It is then necessary to locate the instantaneous center and to determine the angular velocity of the body.

Consider the floating link AB (Figure 2.19). Suppose that the velocity V_A of point A is known in magnitude (V_A) and direction. Also, suppose that the direction of motion of point B is also known. The velocity of point A is normal to IA, while the direction of motion of point B is normal to IB. Point I is then can be located by drawing lines from points A and B; each is perpendicular to the direction of motion of the corresponding point.

The angular velocity ω is given by

$$\omega = \frac{V_A}{IA}$$

The direction of ω is determined according to the direction of V_A. The velocity of point B is given by

$$V_B = \omega \times IB$$
$$= V_A \frac{IB}{IA}$$

Also, if point C is on the link, then

$$V_C = V_A \frac{IC}{IA}$$

In another situation, when a link has a translation motion (Figure 2.20), its instantaneous center is at infinity on a line normal to the direction of motion. This is because the link is considered to be rotating about a center at infinity. Its angular velocity is zero.

2.5.2 Instantaneous Center of a Pair of Links

The instantaneous center of a pair of links is defined as the point about which one link is considered to be rotating with respect to the other at a given instant. Thus, the turning joint connecting links (1) and (2) (Figure 2.21) represents their instantaneous center and is denoted as I_{12} or I_{21}. It is important to note that I_{12} is located at the center of the turning joint. This point, in fact, is a pair of coincident points, one on each

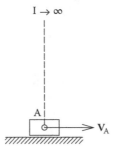

FIGURE 2.20 Instantaneous center of a sliding link.

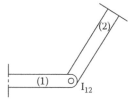

FIGURE 2.21   Instantaneous center of a pair of links.

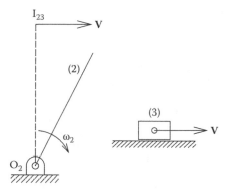

FIGURE 2.22   Instantaneous center of two general links.

link, such that the relative velocity between them is zero; otherwise, the two links are separate, or, in other words, the two points have the same absolute velocity.

A more general definition for the instantaneous center of a pair of links is as follows:

> The instantaneous center of a pair of links is at the common position of that pair of coincident points, one on each link, that have the same absolute velocity.
>
> The instantaneous center of the floating link of the preceding section is actually the center of this link relative to the fixed frame (not shown) whose velocity is zero. Accordingly, the velocity of I is zero, which is obvious. As an example, consider the system shown in Figure 2.22.

Link (3) has a translation motion and is moving with a velocity V. Link (2) is rotating about O_2 with an angular velocity ω_2, say clockwise. In order to find the instantaneous center of (2) relative to (3), I_{32}, it is necessary to locate a point, which when considered to move with link (3) or rotate with link (2), which in turn has the same velocity V. The directions of the velocity of the points rotating with link (2) are normal to the line joining them with point O_2. The points that have velocities in the direction of V lie on a line passing through point O_2 and are normal to V. On this line, there is a point, I_{23}, whose velocity is exactly equal to V such that

$$V = \omega_2 \times (O_2 - I_{23})$$

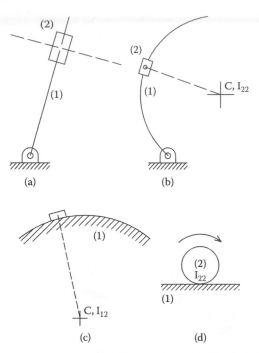

FIGURE 2.23 Instantaneous center of sliding and rolling links. (a) Straight link (b) curved link (c) curved link (d) rolling motion.

or

$$O_2 - I_{23} = \frac{V}{\omega_2}$$

If a block slides on a link, their instantaneous center is at the center of curvature of the link, which lies on the common normal to the surfaces of contact (Figure 2.23a, b, and c). If the link is straight as in (a), the center is at infinity on the normal line. When two bodies have pure rolling, the point of contact is their instantaneous center as in (d).

2.5.3 LAW OF THREE CENTERS

The law of three centers states the following:

For any three links having plane motion, their three centers lie on a straight line.

Consider now the three links connected as shown in Figure 2.24. Link (2) is hinged to link (1) at point O_2 and has a relative angular velocity ω_{21}. Link (3) is hinged to link (1) at point O_3 and has a relative angular velocity ω_{31}. It is clear that points O_2 and O_3 represent the instantaneous centers I_{21} and I_{31}, respectively. The relative motion between the three links is the same whether link (1) is fixed or not fixed. For the sake of simplicity, we assume that link (1) is fixed. The instantaneous center of links (2) and (3) is obtained by locating a point that, when rotating with link (2) or link (3), will have the same absolute velocity. If we assume a point such as

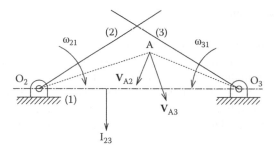

FIGURE 2.24 Representation of the law of three centers.

point A, the directions of its velocity about each of O_2 or O_3 are not the same. Points rotating about O_2 or O_3 and having the same direction are on the line joining O_2 or O_3. That is, they lie on the line joining I_{21} and I_{31}. The law of the three centers is therefore verified. The exact location of I_{23} is such that

$$\omega_{21}\left(I_{21}\cdot I_{23}\right)=\omega_{31}\left(I_{31}\cdot I_{23}\right)$$

or

$$\frac{I_{21}\,I_{23}}{I_{31}\,I_{23}}=\frac{\omega_{31}}{\omega_{21}}$$

Generally,

$$\frac{\omega_{ij}}{\omega_{jk}}=\frac{jk-ik}{ij-ik} \tag{2.46}$$

where ij denotes the instantaneous center of links (i) and (j) after dropping I. This notation will be used from now on. The line jk – ik represents the distance between the centers jk and ik.

2.5.4 Locating the Instantaneous Centers of a Mechanism

When the number of links in a mechanism is n, then the total number of centers N in the mechanism is given by

$$N=n(n-1)$$

The procedure is outlined as follows:

1. Make a list of all the centers in the form

12	13	14	15	1n
	23	24	25	2n
		34	35	3n
			....	
				$(n-1)\,n$

2. There are some centers that can be located by inspection such as the joint of the links.

3. The rest of the centers are obtained by applying the law of three centers. The process is simplified by using a bookkeeping diagram. All centers are represented by points. The known centers are joined by solid lines. The centers to be determined are connected by dashed lines. To determine any center, the dashed line representing them must lie inside a four-sided diagonal with solid lines. This procedure is illustrated by the following examples.

EXAMPLE 2.1

Find the instantaneous centers of the four-bar mechanism shown in Figure 2.25, and find the angular velocities of links (3) and (4) if ω_2 is known.

SOLUTION

The number of centers N is given by

$$N = 4 \times (3.1)6$$

The links are represented on the bookkeeping diagram by points 1, 2, 3, and 4 as shown in the figure.

The list of the centers is

12	13	14
	23	24
		34

The known centers are 12 (point O_2), 23 (point A), 34 (point B), and 41 (point O_4). They are connected by solid lines. The centers to be determined are

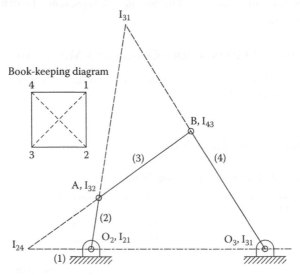

FIGURE 2.25   Instantaneous center of a four-bar mechanism.

31 and 24. Center 31 is located on the line joining centers 12 and 23 [considering links (1), (2), and (3)]. It is also located on the line joining centers 14 and 43 [considering links (1), (3), and (4)]. The intersection of the two lines locates the position of 31. Similarly, the location of center 42 is at the intersection of lines 21 – 14 and 23 – 34.

Once the instantaneous centers are located, the angular velocity of any link can be determined. Applying Equation 2.46, thus

$$\frac{\omega_{41}}{\omega_{21}} = \frac{21-42}{41-42}$$

Also,

$$\frac{\omega_{31}}{\omega_{21}} = \frac{21-32}{31-32}$$

To obtain the linear velocity of a point on a link, consider the following example.

EXAMPLE 2.2

Find the instantaneous centers of the engine mechanism shown in Figure 2.26, and then find the angular velocity of link (3) and the velocity of point B; ω_2 is known.

SOLUTION

The centers are obtained as outlined in Example 2.1.

The angular velocity of link (3) is given by

$$\frac{\omega_{31}}{\omega_{21}} = \frac{21-32}{31-32}$$

The velocity of point B may be obtained by two alternative methods.

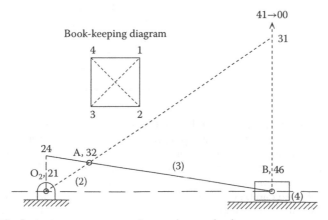

FIGURE 2.26 Instantaneous centers of an engine mechanism.

1. Point B is considered as a point on link (4). If we locate center 42, its absolute velocity, when considered to rotate with link (2), is the same as the absolute velocity of any point moving with link (4) [link (4) has a translation motion]. Therefore,

$$V_B = V_{42} \times (O_2 - 42)$$

2. Point B is considered as a point on link (3). ω_{31} was already obtained. Therefore,

$$V_B = \omega_{31} \times (B - 31)$$

For a compound mechanism, it is not necessary to obtain all the centers. We obtain only the centers that are needed to determine the required velocities as demonstrated by the following example.

EXAMPLE 2.3

In the toggle mechanism analyzed in Section 2.4.3 and shown in Figure 2.27, the crank rotates at an angular velocity of 31.4 rad/s. Find the velocity of the slider (8).

SOLUTION

The mechanism consists of eight links. Thus, the number of centers is

$$N = 8 \times 7 = 56$$

The list of the centers is

12	13	**14**	15	**16**	17	**18**
	23	24	25	26	27	28
		34	35	36	37	38
			35	46	47	48
				56	57	58
					67	68
						78

The known centers are marked by bold numbers as shown in the list and are joined by solid lines in the bookkeeping diagram. It is clear that it is a waste of time to obtain all the remaining centers. We locate the minimum number of the centers required to determine the velocity of point B. If we consider point B to be on link (8), we need to locate center 82. To do this, we need to locate 24, 46, and 68. Center 48 then can be obtained, and, finally, center 82 is determined.

Center 24 is at the intersection of line 12 – 14 and line 23 – 34. The procedure is organized in the following form:

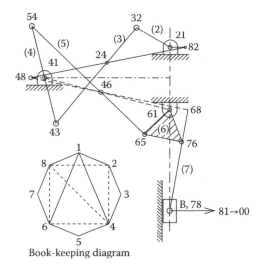

Book-keeping diagram

FIGURE 2.27   Instantaneous centers of a compound mechanism.

$$24\begin{cases}12-14 \\ 23-34\end{cases} \quad 46\begin{cases}14-16 \\ 45-56\end{cases} \quad 68\begin{cases}67-78 \\ 16-18\end{cases}$$

$$48\begin{cases}46-68 \\ 14-18\end{cases} \quad 28\begin{cases}12-18 \\ 24-48\end{cases}$$

The distance between points O_2 and the center 28 is 9.8 cm. The velocity of slider B is given by

$$V_B = 31.4 \times 9.8 = 307.7 \, \text{cm/s}$$

Point B may be considered as a point on link (7). In this case, we determine ω_{71}.

$$\frac{\omega_{71}}{\omega_{21}} = \frac{21-72}{71-72}$$

Thus, it is necessary to determine center 72. To do so, centers 24, 46, and 71 are determined first, then 47, and then 72.

2.6   ANALYTICAL ANALYSIS

In Section 1.9, the basis of the analytical analysis was laid down by using polar complex vectors. In addition, we pointed out that analytical analysis requires dividing compound mechanisms into basic chains. These chains are the crank, the four-bar, the engine, the shaper, and the tilting block chains. The position analysis for each chain was derived. The next step is to deduce equations to determine the velocities and accelerations of the links of each chain. This is carried out in the forth coming sections.

2.6.1   CRANK

Usually, the crank (Figure 2.28) is the driving link. Its position, angular velocity, and angular acceleration are specified.

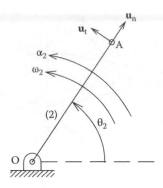

FIGURE 2.28 Analysis of the crank.

Given:

$$r_2, \theta_2, \omega_2, \alpha_2$$

Find:

$$V_A^x, V_A^y, A_A^x, A_A^y$$

Analysis:
According to Equations 2.18 and 2.19,

$$\mathbf{V}_A = \omega\, r\, \mathbf{u}_t$$
$$\mathbf{A}_A = \omega^2\, r\, \mathbf{u}_n + \alpha\, r\, \mathbf{u}_t$$

where

$$\mathbf{u}_n = e^{i\theta_2}$$
$$\mathbf{u}_t = i\, e^{i\theta_2}$$

Thus,

$$\mathbf{V}_A = i\omega_2\, r_2\, e^{i\theta_2}$$
$$\mathbf{A}_A = -i\omega_2^2\, r_2\, e^{i\theta_2} + \alpha_2\, r_2\, e^{i\theta_2}$$

Expanding and separating the real and imaginary parts, we get

$$V_A^x = -\omega_2 r_2 \sin\theta_2 \qquad\qquad (2.47a)$$

$$V_A^y = \omega_2 r_2 \cos\theta_2 \tag{2.47b}$$

$$A_A^x = r_2(-\omega_2^2 \cos\theta_2 - \alpha_2 \sin\theta_2) \tag{2.48a}$$

$$A_A^y = r_2(-\omega_2^2 \sin\theta_2 + \alpha_2 \cos\theta_2) \tag{2.48b}$$

2.6.2 Four-Bar Chain

The four-bar chain is shown in Figure 2.29. The position analysis is presented in Section 1.9.3.2.

Given:

$$V_A^x, \ V_A^y, \ A_A^x, \ A_A^y$$

Find:

$$\omega_3, \omega_4, \alpha_3, \alpha_4$$

Analysis:
For the velocity analysis, consider Equation 2.21:

$$\mathbf{V}_B = \mathbf{V}_A + \mathbf{V}_{BA}$$

Link (4) is a rotating link and link (3) is a floating link. Thus,

$$\mathbf{V}_B = \omega_4 \, r_4 i \, e^{i\theta_4}$$
$$\mathbf{V}_{BA} = \omega_3 \, r_3 i \, e^{i\theta_3}$$

Substituting into Equation 2.24, we get

$$\omega_4 \, r_4 \, i \, e^{i\theta_4} = \left(V_A^x + i V_A^y\right) + \omega_3 \, r_3 \, i \, e^{i\theta_3}$$

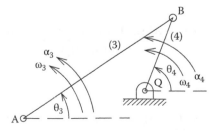

FIGURE 2.29 Analysis of the four-bar chain.

Multiplying both sides by $e^{-i\theta_4}$ and considering the real parts leads to

$$0 = V_A^x \cos\theta_4 + V_A^y \sin\theta_4 - \omega_3 \, r_3 \sin(\theta_3 - \theta_4)$$

Therefore,

$$\omega_3 = \frac{V_A^x \cos\theta_4 + V_A^y \sin\theta_4}{r_3 \sin(\theta_3 - \theta_4)} \tag{2.49}$$

The same procedure can be applied to obtain ω_4:

$$\omega_4 = \frac{-V_A^x \cos\theta_3 - V_A^y \sin\theta_3}{r_4 \sin(\theta_4 - \theta_3)} \tag{2.50}$$

For the acceleration analysis, consider Equation 2.22:

$$\mathbf{A}_B = \mathbf{A}_A + \mathbf{A}_{BA}$$

Using the values of the acceleration of rotating links, we arrive at

$$\mathbf{A}_B = r_4(-\omega_4^2 + i\alpha_4)e^{i\theta_4}$$
$$\mathbf{A}_{BA} = r_3(-\omega_3^2 + i\alpha_3)e^{i\theta_3}$$

Substituting in the vector equation gives

$$r_4(-\omega_4^2 + i\alpha_4)e^{i\theta_4} = (A_A^x + i A_A^y) + r_3(-\omega_3^2 + i\alpha_3)e^{i\theta_3} \tag{2.51}$$

Multiplying Equation 2.51 by $e^{-i\theta_4}$ and considering the real parts, we can obtain α_3:

$$-r_4\,\omega_4^2 = A_A^x \cos\theta_4 + A_A^y \sin\theta_4 - r_3\omega_3^2 \cos(\theta_3 - \theta_4) - r_3\alpha_3 \sin(\theta_3 - \theta_4)$$

Therefore,

$$\alpha_3 = \frac{A_A^x \cos\theta_4 + A_A^y \sin\theta_4 + r_4\,\omega_4^2 - r_3\,\omega_3^2 \cos(\theta_3 - \theta_4)}{r_3 \sin(\theta_3 - \theta_4)} \tag{2.52}$$

Similarly,

$$\alpha_4 = \frac{-A_A^x \cos\theta_3 - A_A^y \sin\theta_3 + r_3\,\omega_3^2 - r_4\,\omega_4^2 \cos(\theta_4 - \theta_3)}{r_4 \sin(\theta_4 - \theta_3)} \tag{2.53}$$

2.6.3 ENGINE CHAIN

The engine chain is shown in Figure 2.30. The position analysis is presented in Section 1.9.3.3. It should be noted that the angle of the line of action in the position analysis was denoted by α.

Given:

$$V_A^x, V_A^y, A_A^x, A_A^y$$

Find:

$$\omega_3, V_4, \alpha_3, A_4$$

Analysis:

$$\mathbf{V}_B = \mathbf{V}_A + \mathbf{V}_{BA}$$

But

$$\mathbf{V}_4 = \mathbf{V}_B = V_4\, e^{i\beta}$$
$$\mathbf{V}_{BA} = \omega_3\, r_3\, i\, e^{i\theta_3}$$

Thus,

$$V_4\, e^{i\beta} = (V_A^x + iV_A^y) + \omega_3\, r_3\, i\, e^{i\theta_3} \tag{2.54}$$

Multiply both sides of Equation 2.54 by $e^{-i\beta}$. From the imaginary parts,

$$0 = -V_A^x \sin\beta + V_A^y \cos\beta + \omega_3\, r_3 \cos(\theta_3 - \beta)$$

Thus,

$$\omega_3 = \frac{V_A^x \sin\beta - V_A^y \cos\beta}{r_3 \cos(\theta_3 - \beta)} \tag{2.55}$$

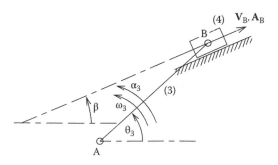

FIGURE 2.30 Analysis of the engine chain.

From the real parts,

$$V_4 = V_A^x \cos\beta + V_A^y \sin\beta + \omega_3\, r_3 \sin(\theta_3 - \beta) \tag{2.56}$$

For the acceleration,

$$\mathbf{A}_4 = \mathbf{A}_B = A_4\, e^{i\beta}$$
$$\mathbf{A}_{BA} = r_3\left(-\omega_3^2 + i\alpha_3\right)e^{i\theta_3}$$

Thus,

$$A_4\, e^{i\beta} = \left(A_A^x + iA_A^y\right) + r_3\left(-\omega_3^2 + i\alpha_3\right)e^{i\theta_3} \tag{2.57}$$

Multiplying Equation 2.57 by $e^{-i\beta}$ and considering the imaginary parts, we get

$$0 = -A_A^x \sin\beta + A_A^y \cos\beta + \omega_3^2 r_3 \sin(\theta_3 - \beta) + \alpha_3 \cos(\theta_3 - \beta) \tag{2.58}$$

Hence,

$$\alpha_3 = \frac{A_A^x \sin\beta - A_A^y \cos\beta + \omega_3^2\, r_3 \sin(\theta_3 - \beta)}{r_3 \cos(\theta_3 - \beta)} \tag{2.59}$$

From the real parts,

$$A_4 = A_A^x \cos\beta + A_A^y \sin\beta - \omega_3^2 r_3 \cos(\theta_3 - \beta) - \alpha_3 r_3 \sin(\theta_3 - \beta) \tag{2.60}$$

2.6.4 Shaper Chain

For the shaper chain shown in Figure 2.31, links (3) and (4) are sliding links. Let point B is on block (3), while point C is on link 4. $\mathbf{u}_t$ is along link (4), while $\mathbf{u}_n$ is lagging $\mathbf{u}_t$ by 90°.

Given:

$$V_A^x, V_A^y, A_A^x, A_A^y$$

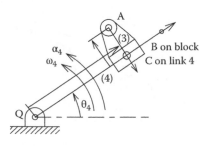

FIGURE 2.31 Analysis of the shaper chain.

Find:

$$\omega_4, \alpha_4, V_{CB}, A_{CB}$$

Analysis:

It is clear that the block and the link have the same angular velocity and angular acceleration. That is,

$$\omega_3 = \omega_4$$
$$\alpha_3 = \alpha_4$$

For the velocity analysis, consider the vector loop

$$\mathbf{V}_C = \mathbf{V}_A + \mathbf{V}_{BA} + \mathbf{V}_{CB}$$

It is clear that

$$\mathbf{V}_C = \omega_4 x_4\, i\, e^{i\theta_4}$$
$$\mathbf{V}_{BA} = \omega_4\, h\, e^{i\theta_4}$$
$$\mathbf{V}_{CB} = V_{CB}\, \mathbf{u}_t$$
$$= V_{CB}\, e^{i\theta_4}$$

Substituting into the vector equation of the velocities leads to

$$\omega_4\, x_4\, i\, e^{i\theta_4} = (V_A^x + iV_A^y) + \omega_4\, h\, e^{i\theta_4} + V_{CB}\, e^{i\theta_4}$$

or

$$\omega_4(-h + i\, x_4)e^{i\theta_4} = (V_A^x + iV_A^y) + V_{CB}\, e^{i\theta_4}$$

Multiplying both sides by $e^{-i\theta_4}$ and considering the imaginary parts, we arrive at

$$\omega_4 = \frac{-V_A^x \sin\theta_4 + V_A^y \cos\theta_4}{x_4} \tag{2.61}$$

From the real parts,

$$V_{CB} = -\omega_4\, h - V_A^x \cos\theta_4 - V_A^y \sin\theta_4 \tag{2.62}$$

For the acceleration, consider the loop

$$\mathbf{A}_C = \mathbf{A}_A + \mathbf{A}_{BA} + \mathbf{A}_{CB}$$
$$\mathbf{A}_C = x_4(-\omega_4^2 + i\alpha)e^{i\theta_4} \tag{2.63}$$
$$\mathbf{A}_{BA} = h(\alpha_4 + i\omega_4^2)e^{i\theta_4}$$

According to Equation 2.29,

$$\mathbf{A}_{CB} = \mathbf{A}_{CB}^n + \mathbf{A}_{CB}^{SL}$$

The tangential, called sliding, component $\mathbf{A}_{BA}^{SL}$ is given by

$$\mathbf{A}_{CB}^{SL} = A_{CB}^{SL}\,\mathbf{u}_t$$
$$= A_{CB}^{SL}\,e^{i\theta_4}$$

The normal component is determined according to Equation 2.30 ($\rho_A = 0$ and $\rho_B = \infty$):

$$\mathbf{A}_{CB}^n = -2V_{CB}\,\omega_C\,\mathbf{u}_n$$

where ω_C is the angular velocity of the link (4). Thus,

$$\omega_C = \omega_4$$

and

$$\mathbf{A}_{CB}^n = 2V_{CB}\,\omega_4\,i\,e^{i\theta_4}$$

Substituting in Equation 2.62 gives

$$x_4\left(-\omega_4^2 + i\alpha_4\right)e^{i\theta_4} = \left(A_A^x + i\,A_A^y\right) + h\left(\alpha_4 + i\,\omega_4^2\right)e^{i\theta_4} + 2V_{CB}\,\omega_4\,i\,e^{i\theta_4} + A_{CB}^{SL}\,e^{i\theta_4}$$

$$(2.64)$$

Multiplying by $e^{-i\theta_4}$ and considering the imaginary part leads to

$$\alpha_4 = \frac{-A_A^x \sin\theta_4 + A_A^y \cos\theta_4 + h\omega_4^2 + 2V_{CB}\omega_4}{x_4} \qquad (2.65)$$

Considering the real parts, we get

$$A_{CB}^{SL} = -A_A^x \cos\theta_4 - A_A^y \sin\theta_4 - x_4\omega_4^2 - h\alpha_4 \qquad (2.66)$$

2.6.5 TILTING BLOCK CHAIN

The tilting block chain is shown in Figure 2.32.

Given:

$$V_A^x, V_A^y, A_A^x, A_A^y$$

Find:

$$\omega_3, \alpha_3, V_{BC}, A_{BC}$$

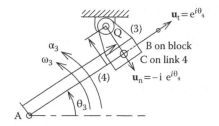

FIGURE 2.32   Analysis of the tilting block chain.

Analysis:

For the velocity analysis, consider the vector loop

$$\mathbf{V}_B = \mathbf{V}_A + \mathbf{V}_{CA} + \mathbf{V}_{BC}$$

As in the shaper chain,

$$\omega_3 = \omega_4 = \omega_B$$
$$\alpha_3 = \alpha_4 = \alpha_B$$

Substituting with the value of each vector gives

$$\omega_3 \, h \, e^{i\theta_3} = \left(V_A^x + i V_A^y \right) + \omega_3 \, x_3 \, i \, e^{i\theta_3} + V_{BC} \, e^{i\theta_3}$$

Multiplying by $e^{-i\theta_3}$ and considering the imaginary parts, we get

$$\omega_3 = \frac{V_A^x \sin\theta_3 - V_A^y \cos\theta_3}{x_3} \tag{2.67}$$

From the real parts,

$$V_{BC} = \omega_3 \, h - V_A^x \cos\theta_3 - V_A^y \sin\theta_3 \tag{2.68}$$

For the acceleration, consider the loop

$$\mathbf{A}_B = \mathbf{A}_A + \mathbf{A}_{CA} + \mathbf{A}_{BC}$$

Then,

$$h\left(\alpha_4 + i\omega_4^2\right)e^{i\theta_3} = \left(A_A^x + i A_A^y\right) + x_3\left(-\omega_3^2 + i\alpha_3\right)e^{i\theta_3} + 2\,V_{BC}\,\omega_3\,i\,e^{i\theta_3} + A_{BC}^{SL}\,e^{i\theta_3}$$

Multiplying by $e^{-i\theta_3}$ and considering the imaginary parts leads to

$$\alpha_3 = \frac{A_A^x \sin\theta_3 - A_A^y \cos\theta_3 + h\omega_3^2 + 2V_{BC}\,\omega_3}{x_3} \tag{2.69}$$

Considering the real parts, we get

$$A_{BC}^{SL} = -A_A^x \cos\theta_3 - A_A^y \sin\theta_3 + x_3\,\omega_3^2 + h\alpha_3 \tag{2.70}$$

2.6.6 KINEMATICS OF A POINT ON A LINK

Consider link ABC (Figure 2.33) for which the angular velocity ω and the angular acceleration α are known. AC is of length l and makes an angle ψ with AB. Suppose that the velocity and acceleration of point A are also known. The velocity and acceleration of point C can be determined as follows:

$$\begin{aligned}\mathbf{V}_C &= \mathbf{V}_A + \mathbf{V}_{CA}\\ &= V_A^x + iV_A^y + \omega l i\, e^{i(\theta+\psi)}\\ &= V_A^x + i + \omega l\left[-\sin(\theta+\psi) + i\cos(\theta+\psi)\right]\end{aligned}$$

Therefore,

$$V_C^x = V_A^x - \omega l \sin(\theta+\psi) \tag{2.71a}$$

$$V_C^y = V_A^y + \omega l \cos(\theta+\psi) \tag{2.71b}$$

For the acceleration,

$$\begin{aligned}\mathbf{A}_C &= \mathbf{A}_A + \mathbf{A}_{CA}\\ &= A_A^x + i A_A^y + l(-\omega^2 + i\alpha)e^{i(\theta+\psi)}\end{aligned}$$

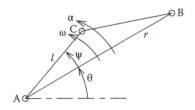

FIGURE 2.33 Analysis of a general link.

Therefore,

$$A_C^x = A_A^x - l\left[\omega^2 \cos(\theta + \psi) + \alpha \sin(\theta + \psi)\right] \tag{2.72a}$$

$$A_C^y = A_A^y - l\left[\omega^2 \sin(\theta + \psi) + \alpha \cos(\theta + \psi)\right] \tag{2.72b}$$

2.6.7 Application to a Compound Mechanism

For the mechanism in Example 1.5 (Figure 2.34), if crank OA rotates at 10 rad/s, determine the velocities and accelerations of all links.

The position of all links was determined in Chapter 1. For the kinematics analysis, we consider the chains included in the mechanism.
Analysis:

1. *Crank*
   From Equations 2.47 and 2.48

$$V_A^x = 507.1 \text{ cm/s} \qquad V_A^y = -1088 \text{ cm/s}$$
$$A_A^x = 10880 \text{ cm/s}^2 \qquad A_A^y = 5071 \text{ cm/s}^2$$

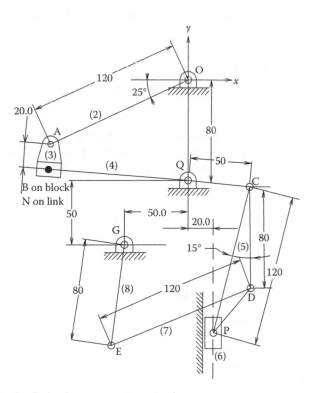

FIGURE 2.34 Analysis of a compound mechanism.

2. *Shaper chain*

Using Equations 2.61, 2.62, 2.65, and 2.66

$$\omega_4 = 9.4 \text{ rad/s} \qquad V_{NB}^{SL} = 785 \text{ cm/s}$$
$$\alpha_4 = 63.22 \text{ rad/s}^2 \qquad A_{NB}^{SL} = 1900 \text{ cm/s}^2$$

From Equations 2.71 and 2.72

$$V_C^x = 39.6 \text{ cm/s} \qquad V_C^y = 467.9 \text{ cm/s}$$
$$A_C^x = -4135 \text{ cm/s}^2 \qquad A_C^y = 3528 \text{ cm/s}^2$$

3. *Engine chain*

For the engine chain, we use Equations 2.55, 2.56, 2.59, and 2.60. In these equations, we replace V_A^x by V_C^x, V_A^y by V_C^y, A_A^x by A_C^x, and A_A^y by A_C^y. Also, we change the subscript 3 by 5, 4 by 6, and B by E. We obtain

$$\omega_5 = -0.34 \text{ rad/s} \quad V_6 = -478.2 \text{ cm/s}$$
$$\alpha_5 = 35.5 \text{ rad/s}^2 \quad A_6 = -2478.0 \text{ cm/s}^2$$

From Equations 2.70 and 2.71, and replacing the subscript A by C and C by D, we get

$$V_D^x = 12.36 \text{ cm/s} \qquad V_D^y = 467.57 \text{ cm/s}$$
$$A_D^x = -1289.0 \text{ cm/s}^2 \quad A_D^y = 35662.0 \text{ cm/s}^2$$

3. *Four-bar chain*

For the four-bar chain, we use Equations 2.49, 2.50, 2.52, and 2.53. In these equations, we replace V_A^x by V_D^x, V_A^y by V_D^y, A_A^x by A_D^x, and A_A^y by A_D^y. Also, we change the subscript 3 by 7 and 4 by 8. We obtain

$$\omega_7 = 4.46 \text{ rad/s} \qquad \omega_8 = 2.69 \text{ rad/s}$$
$$\alpha_7 = 37.98 \text{ rad/s}^2 \quad \alpha_8 = 78.53 \text{ rad/s}^2$$

2.6.8   CAM WITH A SPHERICAL OSCILLATING FOLLOWER

The circular cam with a radius R that actuates an oscillating follower with a spherical tip with a radius r is shown in Figure 2.35. The position analysis was presented in Section 1.9.4.2. This means that θ_3 and θ_4 are determined. Suppose that the cam rotates with a uniform angular speed ω. It is required to obtain the angular velocity and the angular acceleration of the follower. Also, it is required to determine the sliding velocity and the sliding acceleration between the follower and the cam.

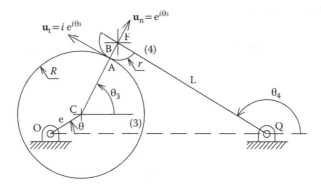

FIGURE 2.35 A cam with a spherical oscillating follower.

Analysis:

1. *Velocity*
   We set unit vectors $\mathbf{u}_n$ and $\mathbf{u}_t$; $\mathbf{u}_t$ is leading $\mathbf{u}_n$ by 90°. OC is equivalent to a crank. Thus,

$$V_C^x = -\omega e \sin\theta$$
$$V_C^y = \omega e \cos\theta$$

For point A,

$$\mathbf{V}_A = \mathbf{V}_C + \mathbf{V}_{AC}$$
$$V_A^x = -\omega e \sin\theta - \omega R \sin\theta_3$$
$$V_A^y = \omega e \cos\theta + \omega R \cos\theta_3$$

For point B,

$$\mathbf{V}_B = \mathbf{V}_A + \mathbf{V}_{BA}$$
$$= \mathbf{V}_F + \mathbf{V}_{BF}$$

The distance QF is equal to L. Thus,

$$V_A^x + i V_A^y + V_{BA}\mathbf{u}_t = L\omega_4\, i\, e^{i\theta_4} - r\omega_4\, i\, e^{i\theta_3}$$
$$V_A^x + i V_A^y + V_{BA}\, i\, e^{i\theta_3} = L\omega_4\, i\, e^{i\theta_4} - r\omega_4\, i\, e^{i\theta_3}$$

Multiplying both sides by $e^{-i\theta_3}$ and considering the real parts, we arrive at

$$\omega_4 = -\frac{V_A^x \cos\theta_3 + V_A^y \sin\theta_3}{L\sin(\theta_4 - \theta_3)}$$

From the imaginary parts, we get

$$V_{BA} = V_A^x \sin\theta_3 - V_A^y \cos\theta_3 + \omega_4\left[-r + L\cos(\theta_4 - \theta_3)\right]$$

2. *Acceleration*

$$A_C^x = -\omega^2 e\cos\theta$$
$$A_C^y = -\omega^2 e\sin\theta$$
$$A_{AC}^x = -\omega^2 R\cos\theta_3$$
$$A_{AC}^y = -\omega^2 R\sin\theta_3$$

$$A_A^x = -\omega^2 e\cos\theta - \omega^2 R\cos\theta_3$$
$$A_A^y = -\omega^2 e\sin\theta - \omega^2 R\sin\theta_3$$

$$\mathbf{A}_B = \mathbf{A}_A + \mathbf{A}_{BA}$$
$$= \mathbf{A}_F + \mathbf{A}_{BF}$$

According to Equations 2.28 through 2.30,

$$\mathbf{A}_{BA} = \mathbf{A}_{BA}^n + \mathbf{A}_{BA}^{SL}$$
$$\mathbf{A}_{BA} = A_{BA}\,\mathbf{u}_t$$
$$\mathbf{A}_{BA}^n = A_{BA}^n\,\mathbf{u}_n$$
$$A_{BA}^n = \frac{1}{R+r}\left[Rr(\omega_3 - \omega_4)^2 - V_{BA}^2 - 2V_{BA}(R\omega_3 + r\omega_4)\right]$$

Thus,

$$A_A^x + i A_A^y + \left(A_{BA}^n + i A_{BA}^n\right)e^{i\theta_3} = L\left(-\omega_4^2 + i\alpha_4\right)e^{i\theta_4} + r\left(-\omega_4^2 + i\alpha_4\right)\left(-e^{i\theta_3}\right)$$

Multiplying both sides by $e^{-i\theta_3}$ and considering the real parts, we get

$$\alpha_4 = -\frac{A_A^x \cos\theta_3 + A_A^y \sin\theta_3 + A_{BA}^n - r\omega_4^2 + L\omega_4^2 \cos(\theta_4 - \theta_3)}{L\sin(\theta_4 - \theta_3)}$$

From the imaginary parts,

$$A_{BA}^{SL} = A_A^x \sin\theta_3 - A_A^y \cos\theta_3 + \alpha_4\left[L\cos(\theta_4 - \theta_3) - r\right] - L\omega_4^2 \sin(\theta_4 - \theta_3)$$

2.6.9 CAM WITH A SPHERICAL TRANSLATING FOLLOWER

The circular cam of the previous example actuates a translating follower with a spherical tip with a radius r as shown in Figure 2.36. The position analysis was presented in Section 1.9.4.3. This means that θ_3 is determined. Suppose that the cam rotates with a uniform angular speed ω. It is required to obtain the velocity and the acceleration of the follower. Also, it is required to determine the sliding velocity and the sliding acceleration between the follower and the cam.

Analysis:

The analysis is mostly similar to that of the previous example. The differences is that

$$V_A^x = -\omega\,e\sin\theta - \omega\,R\sin\theta_3$$
$$V_A^y = \omega\,e\cos\theta + \omega\,R\cos\theta_3$$
$$A_A^x = -\omega^2\,e\cos\theta - \omega^2 R\cos\theta_3$$
$$A_A^y = -\omega^2\,e\sin\theta - \omega^2 R\sin\theta_3$$

$$\mathbf{V}_B = i\,V_B$$
$$\mathbf{A}_B = i\,A_B$$

Thus,

$$V_A^x + i\,V_A^y + V_{BA}\,i\,e^{i\theta_3} = i\,V_B$$

From the real parts,

$$V_{BA} = \frac{V_A^x}{\sin\theta_3}$$

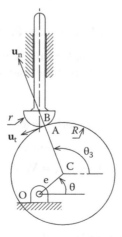

FIGURE 2.36 A cam with translating spherical follower.

From the imaginary parts,

$$V_B = V_A^y + V_{BA} \cos\theta_3$$

For the acceleration,

$$A_{BA}^n = \frac{1}{R+r}\left[Rr\omega^2 - V_{BA}^2 - 2V_{BA}R\omega\right]$$

$$A_A^x + i\,A_A^y + (A_{BA}^n + i\,A_{BA}^{SL})e^{i\theta_3} = i\,A_B$$

From the real parts,

$$A_{BA}^{SL} = \frac{A_A^x + A_{BA}^n \cos\theta_3}{\sin\theta_3}$$

From the imaginary parts,

$$A_B = A_A^y + A_{BA}^{SL} \cos\theta_3 + A_{BA}^n \sin\theta_3$$

When the spherical tip is replaced by a roller with the same radius, the angular velocity and acceleration of the roller are obtained by considering $V_{FB} = V_{BA}$ and $A_{FB}^t = A_{BA}^{SL}$. Therefore,

$$\omega_r = \frac{V_{BA}}{r}$$

$$\alpha_r = \frac{A_{FB}^t}{r}$$

2.6.10 Cam with a Flat-Faced Oscillating Follower

The circular cam of the previous example actuates an oscillating flat-faced follower (Figure 2.37). Point F is the intersection of the common normal with a line parallel to $\mathbf{u}_t$ from point Q. For this cam,

$$V_A^x = -\omega e \sin\theta - \omega R \sin\theta_4$$
$$V_A^y = \omega e \cos\theta + \omega R \cos\theta_4$$
$$\mathbf{V}_B = \mathbf{V}_A + \mathbf{V}_{BA} = \mathbf{V}_F + \mathbf{V}_{BF}$$

Then,

$$V_A^x + i V_A^y + V_{BA}\, e^{i\theta_4} = -h\omega_4\, e^{i\theta_4} + x_4\omega_4 i\, e^{i\theta_4}$$

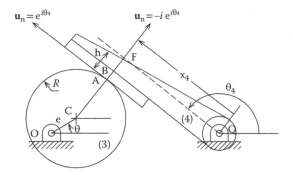

FIGURE 2.37 A cam with oscillating flat-faced follower.

Multiplying both sides by $e^{-i\theta_4}$ and considering the imaginary parts, we arrive at

$$\omega_4 = \frac{V_A^y \cos\theta_4 - V_A^x \sin\theta_4}{x_4}$$

From the imaginary parts, we get

$$V_{BA} = -V_A^x \cos\theta_4 - V_A^y \sin\theta_4 - \omega_4 h$$

For the acceleration,

$$A_A^x = -\omega^2 e\cos\theta - \omega^2 R\cos\theta_4$$
$$A_A^y = -\omega^2 e\sin\theta - \omega^2 R\sin\theta_4$$
$$\rho_A = R, \quad \rho_B = \infty$$

$$A_{BA}^n = R(\omega_3 - \omega_4)^2 - 2V_{BA}\omega_4$$

$$A_A^x + i A_A^y + \left(-i A_{BA}^n + A_{BA}^{SL}\right)e^{i\theta_4} = x_4\left(-\omega_4^2 + i\alpha_4\right)e^{i\theta_4} + h\left(-\omega_4^2 + i\alpha_4\right)\left(-ie^{i\theta_4}\right)$$

Multiplying both sides by $e^{-i\theta_4}$ and considering the imaginary parts leads to

$$\alpha_4 = \frac{-A_A^x \sin\theta_4 + A_A^y \cos\theta_4 - A_{BA}^n - h\omega_4^2}{x_4}$$

From the real parts,

$$A_{BA}^{SL} = -A_A^x \cos\theta_4 - A_A^y \sin\theta_4 + \alpha_4 h - x_4\omega_4^2$$

2.6.11   CAM WITH A FLAT-FACED TRANSLATING FOLLOWER

The circular cam of the previous example actuates a translating flat-faced follower (Figure 2.38). Point F is the intersection of the common normal with a line parallel to $\mathbf{u}_t$ from point Q.

For the velocity,

$$V_A^x = -\omega e \sin\theta - \omega R$$

$$V_A^y = \omega e \cos\theta$$

$$iV_B = V_A^x + iV_A^y + V_{BA}$$

Thus,

$$V_B = V_A^y = \omega e \cos\theta$$

$$V_{BA} = -V_A^x = \omega e \sin\theta + \omega R$$

For the acceleration,

$$A_A^x = -\omega^2 e \cos\theta$$

$$A_A^y = -\omega^2 e \sin\theta - \omega^2 R$$

$$A_{BA}^n = R\omega^2$$

$$iA_B = A_A^x + i A_A^y + A_{BA}^{SL} + i A_{BA}^n$$

Therefore,

$$A_B = A_A^y + A_{BA}^n = -\omega^2 e \sin\theta$$

$$A_{BA}^{SL} = -A_A^x$$

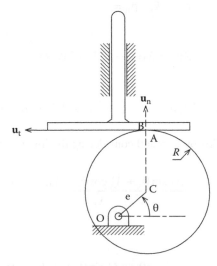

FIGURE 2.38   A cam with translating flat-faced follower.

PROBLEMS

Graphical Method

2.1 A particle moves such that

$$r = b(1 - \cos\theta)$$
$$\theta = 4t$$

Find its velocity and acceleration.

2.2 Write down the components of the velocity and the acceleration for the following cases:

a. A particle moves around a circle with a radius 25 cm and center at the origin with a constant angular velocity of 20 rad/s and an angular acceleration of 100 rad/s².

b. A particle moves around a circle with a radius 20 cm and center at the origin with a constant angular velocity of 5 rad/s.

c. A particle moves with a speed of 10 m/s and an acceleration of 15 m/s² along a straight line that makes 60° with the x-axis and at a distance 20 cm from it.

d. A particle is moving on an Archimedean spiral

$$r = 10\theta$$
$$\theta = 20t$$

2.3 A particle is moving relative to the xOy plane such that

$$x = 3\cos t$$
$$y = 2t$$

Find the velocity, the sliding acceleration, and the normal acceleration. Also, find the value of the radius of curvature for $t = 0, 1, 2$, and 3 seconds.

2.4 In Problem 2.3, the xOy plane rotates about O with a constant angular velocity of 5 rad/s clockwise. Find the Coriolis component when $t = 4$ seconds and then obtain the relative acceleration between the particle and the plane.

2.5 A particle moves with a constant radial speed of 2 cm/s away from the center of a disk rotating with a uniform angular velocity of 10 rad/s clockwise. Find the acceleration of the particle relative to the disk when $t = 5$ seconds and then obtain the value of its absolute acceleration.

2.6 The crank of a single-cylinder diesel engine is 12 cm long. The length of the connecting rod is 36 cm, and the line of stroke of the piston passes through the crank bearing. The engine runs at 2000 rpm clockwise.

a. Draw the velocity and the acceleration polygons when the crank makes 60° with the line of stroke. Also, determine the velocity and the acceleration of the piston.

  b. Determine the velocity and the acceleration of point C on the connecting rod 15 cm from the piston pin. Also, find the radius of curvature of the curve traced by this point on a fixed plane at the same position.

2.7 Repeat Problem 2.6 when the crank has an acceleration of 300 rad/s².

2.8 The lengths of the consequent links in a four-bar mechanism are 8, 4, 7, and 6 cm; the 8-cm link is fixed. If the crank rotates with a uniform speed of 3000 rpm counterclockwise, find the angular acceleration of the rocker when the crank makes 45° with the horizontal datum and at the extreme right position of the rocker.

2.9 For the Watt's mechanism shown in Figure P2.9, locate point P that moves on an approximate straight line. Demonstrate this by obtaining the direction of the velocity and the acceleration of P at several positions. Assume a unit angular velocity for link OA. At the shown position, AB is normal to OA and QB. AP/PB = QB/OA.

$$OA = 80\,mm, AB = 60\,mm, QB = 120\,mm$$

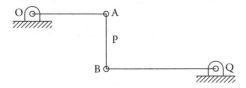

FIGURE P2.9

2.10 For the Peaucellier mechanism shown in Figure P2.10, find the velocity and the acceleration of point P if link (2) rotates with an angular velocity of 5 rad/s and an angular acceleration of 1 rad/s² (both are clockwise). Choose any position for the mechanism.

$$OQ = QA = 50\,mm, OB = OC = 80\,mm, AB = BD = AC = CD = 120\,mm$$

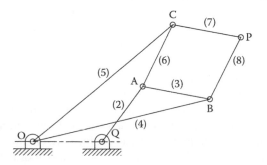

FIGURE P2.10

2.11 For the mechanism shown in Figure P2.11, the piston has a vertical velocity downward of 200 cm/s and an upward acceleration of 2800 cm/s².

Find the angular velocity and the angular acceleration of crank OA. Also, find the velocity and acceleration of point D and the angular velocities and the angular accelerations of the consequent links of the mechanism in magnitude and direction.

OA = 50 mm, AB = 200 mm, QB = QC = 80 mm, angle BQC = 90°,
CE = 150 mm, CD = ED = 80 mm

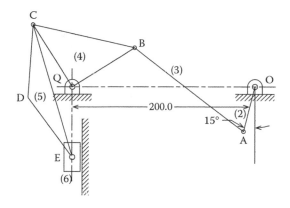

FIGURE P2.11

2.12 For the mechanism shown in Figure P2.12, crank OA rotates with a uniform speed of 42 rad/s. Find the velocity and the acceleration for both sliders B and D when the crank makes 60° as shown in the figure.

OA = 50 mm, AB = 250 mm, AC = 100 mm,
CB = 175 mm, CD = 200 mm

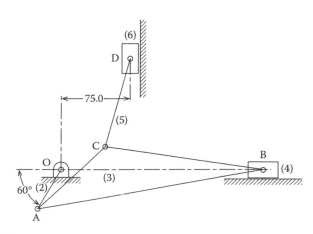

FIGURE P2.12

2.13 The mechanism shown in Figure P2.13 is used in a two-cylinder 60°
      V-engine. Crank OA rotates with a uniform speed of 2000 rpm clock-
      wise. When the crank is horizontal, find the velocity and the acceleration
      of both sliders B and D. Also, find the magnitude and the direction of the
      angular velocity and the angular acceleration of link CD.

$$OA = 50\,mm, AB = 150\,mm, QB = QC = 150\,mm,$$
$$CE = 150\,mm, CD = ED = 80\,mm$$

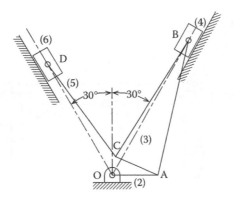

FIGURE P2.13

2.14 For the mechanism shown in Figure P2.14, find the velocity and the
      acceleration of slider N when the crank makes 120° with the horizontal
      datum. The crank rotates at a uniform speed of 300 rpm counterclock-
      wise and makes an angle 60°.

$$OA = 40\,mm, AB = 120\,mm, QB = 80\,mm, QC = 50\,mm, CN = 150\,mm$$

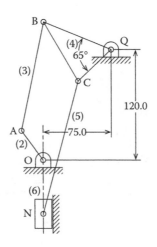

FIGURE P2.14

2.15 Crank OA of the crossed link mechanism shown in Figure P2.15 rotates counterclockwise at a uniform speed of 1000 rpm. Determine the velocity and the acceleration of block C. Also, determine the angular velocity and the angular acceleration of links (3), (4), and (5).

OA = 60 mm, AB = 280 mm, QB = 120 mm, OQ = 300 mm,

BC = 300 mm, OQ = 270 mm

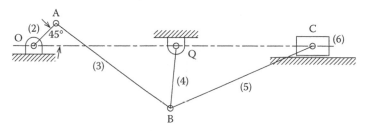

FIGURE P2.15

2.16 For the mechanism shown in Figure P2.16, link (2) rotates clockwise at a constant speed of 600 rpm. Find the angular velocity and the angular acceleration of links (3), (5), and (6).

OA = 80 mm, AC = CB = 120, OQ = 400 mm, QD = 120 mm,

DC = 260 mm

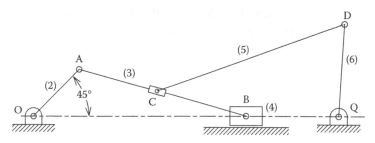

FIGURE P2.16

2.17 Figure P2.17 shows the skeleton outline of an air pump to produce a stroke four times the crank. The crank [link (2)] rotates counterclockwise with a constant speed of 300 rpm. Find the velocity and the acceleration of piston D.

OA = 40 mm, AB = AC = 40, QB = 130 mm, CD = 150 mm

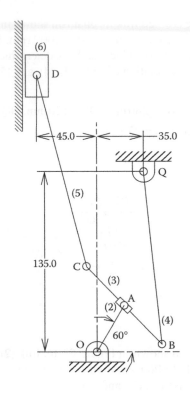

FIGURE P2.17

2.18 For the crossed-link mechanism shown in Figure P2.18, crank OA rotates
at 900 rpm clockwise and with an angular acceleration of 50 rad/s² counterclockwise. Determine the velocity and the acceleration of block D. Also,
find the angular velocity and angular acceleration of links (4) and (5).

OA = 60 mm, AB = 200 mm, QB = QC = 100 mm,

angle BQC = 60°, CD = 200 mm

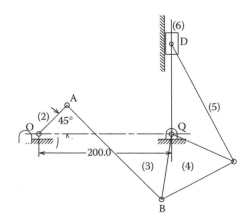

FIGURE P2.18

2.19 Figure P2.19 shows the skeleton outline of the Atkinson gas engine. Crank OA rotates uniformly at 500 rpm counterclockwise. Find the velocity and the acceleration of block D when the crank makes 30°.

$OA = 60\,mm$, $AB = 180\,mm$, $BC = 20\,mm$, $AC = 180\,mm$,

$QC = 80\,mm$, $CD = 180\,mm$

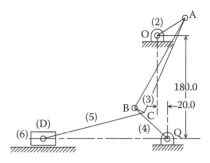

FIGURE P2.19

2.20 Find the angular velocity and angular acceleration of link (6) (Figure P2.20) if link (2) rotates clockwise with a constant speed of 300 rpm.

$OA = 60\,mm$, $AB = 230\,mm$, $QB = QC = 1350\,mm$, $BC = 100\,mm$,

$CD = 270\,mm$, $QD = 180\,mm$

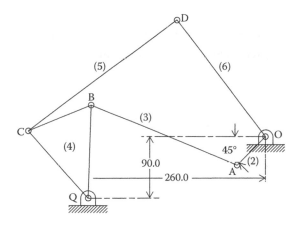

FIGURE P2.20

2.21 Find the angular velocity and the angular acceleration of link (7) of the eight-bar linkage shown in Figure P2.21. Also, find the velocity and the acceleration of point F on the slider (8). The angular velocity of the crank

(2) is 1 rad/s counterclockwise. The angular acceleration of the crank is 2 rad/s² clockwise.

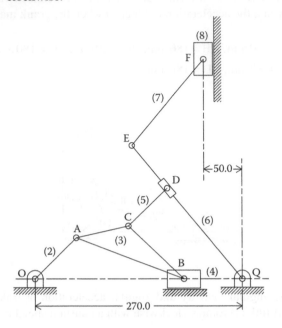

FIGURE P2.21

2.22 Figure P2.22 shows a double-slider mechanism. Crank OA rotates clockwise with a uniform speed of 300 rpm. Find the velocity and the accelerations of the sliders.

$$OA = 40\,mm, AB = 120\,mm, AC = 30\,mm,$$
$$angle\ ACB = 90°, CD = 120\,mm$$

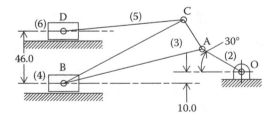

FIGURE P2.22

2.23 Figure P2.23 shows a toggle mechanism with eight links. Crank OA rotates clockwise with a uniform speed of 200 rpm clockwise. Find the velocity and the accelerations of the sliders. Also, find the angular velocities and the angular accelerations of all links.

OA = 25 mm, AB = 80 mm, QB = 30 mm, QC = 35 mm,

CD = 100 mm, UD = UE = DE = 20 mm, EF = 45 mm

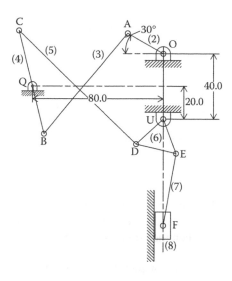

FIGURE P2.23

2.24 For the crank shaper mechanism shown in Figure P2.24, crank OA rotates at 300 rpm clockwise. Determine the velocity and the acceleration of ram C.

OA = 30 mm, AC = 190 mm, CD = 200 mm

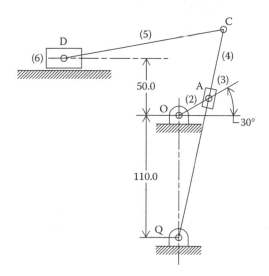

FIGURE P2.24

2.25   The crank of the tilting block mechanism shown in Figure P2.25 rotates counterclockwise at 600 rpm. Find the angular velocity and the angular acceleration of link AC.

$$OA = 25\,\text{mm}$$

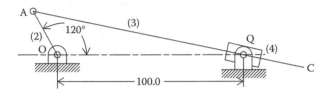

FIGURE P2.25

2.26   The yoke mechanism shown in Figure P2.26 actuates slider C. The crank rotates counterclockwise at 500 rpm. Find the velocity and the acceleration of C when the crank makes 45°.

$$OA = 50\,\text{mm, } BC = 100\,\text{mm}$$

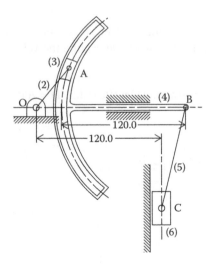

FIGURE P2.26

2.27   Make a complete velocity and acceleration analysis for the mechanism shown in Figure P2.27. The angular velocity of crank OA is 24 rad/s clockwise. What is the absolute velocity and acceleration of point B?

$$OA = 25\,\text{mm, } BC = 125\,\text{mm, } ABC \text{ is one link, angle } ABC = 90°$$

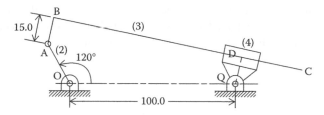

FIGURE P2.27

2.28 For the mechanism shown in Figure P2.28, the velocity of point A is 30 cm/s to the right. Find the velocity and the acceleration of C. Also, find the angular velocity and the angular acceleration of link ABC.

OA = 25 mm, BC = 125 mm, ABC is one link, QN = 15 mm, angle ABC = 90°

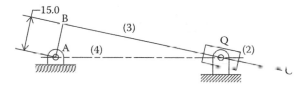

FIGURE P2.28

2.29 In Figure P2.29, link (4) is guided to move horizontally at a constant speed of 50 cm/s to the left. Determine the angular velocity and the angular acceleration of link (2). Also, find the velocity and the acceleration of block C.

OB = 170 mm, BC = 120 mm

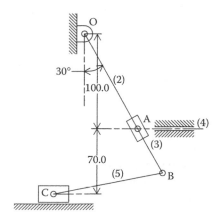

FIGURE P2.29

2.30 Determine the velocity and the acceleration of ram B of the mechanism shown in Figure P2.30. The crank is 30 mm long and rotates at 300 rpm clockwise. At the shown position, the crank makes 30°.

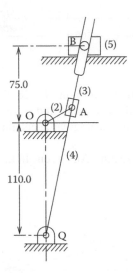

FIGURE P2.30

2.31 Crank (2) of the single-slider crank inversion shown in Figure P2.31 rotates counterclockwise at 600 rpm. Find the sliding velocity and acceleration of the piston, and the angular velocity and the angular acceleration of the cylinder. The crank is 20 mm long and makes an angle 45°. The piston rod AB is 60 mm long.

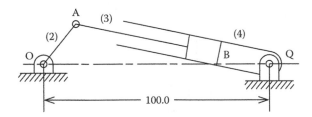

FIGURE P2.31

2.32 Link (4) in Figure P2.32 is rotating at 30 rad/s counterclockwise. Determine the velocity and the acceleration of point C, and the angular velocity and the angular acceleration of link (2).

$$OA = 120\,mm, QB = 180\,mm, BC = 160\,mm$$

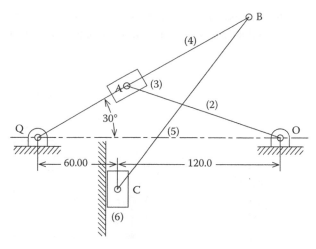

FIGURE P2.32

2.33 Block (5) is hinged at the end of link (6) and slides on link (3) of the four-bar mechanism, as shown in Figure P2.33. If link (2) rotates clockwise at 200 rpm, determine the angular velocity and the angular acceleration of link (6).

$$OA = 80\,mm, AB = 240\,mm, QB = 160\,mm, UC = 200\,mm$$

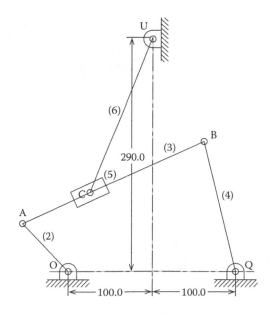

FIGURE P2.33

2.34 In the mechanism shown in Figure P2.34, link (2) rotates clockwise at 100 rpm. Find the velocity and acceleration of block 5. Crank OA is 120 mm.

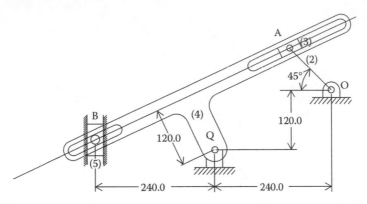

FIGURE P2.34

2.35 In the mechanism shown in Figure P2.35, link (2) rotates clockwise at 100 rpm. Find the velocity and the acceleration of block (5).

OA = 60 mm, AB = 240 mm, AC = 80 mm,
CD = 240 mm. DE = 140 mm

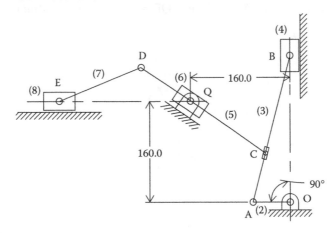

FIGURE P2.35

2.36 For the mechanism shown in Figure P2.36, the angular velocity of the crank is 72 rad/s. Calculate the angular velocity and the angular acceleration of the link UE when the crank is horizontal.

OA = 40 mm, AB = 250 mm, AF = 170 mm, QB = 120 mm,
FC = 50 mm, CD = 30 mm, UE = 180 mm

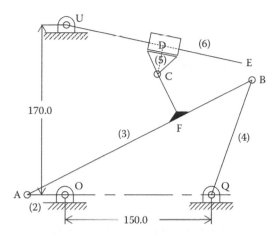

FIGURE P2.36

2.37 Figure P2.37 shows two slotted links, OA and QB, each of which are driven independently. The angular speed of OA is 30 rad/s clockwise and the angular speed of QB is 20 rad/s clockwise. Find the absolute velocity and the absolute acceleration of pin P.

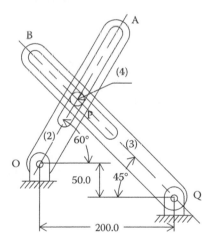

FIGURE P2.37

2.38 In the mechanism shown in Figure P2.38, crank OA rotates uniformly at 120 rpm clockwise. Find the angular velocity and the angular acceleration of link DE and the sliding velocity and the sliding acceleration of block C.

OA = 30 mm, AB = 95 mm, QB = 90 mm, BC = 80 mm

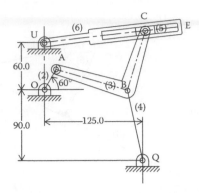

FIGURE P2.38

2.39 In Figure P2.23, link OA is 100 mm long and rotates at a uniform speed
      of 15 rad/s counterclockwise. Find the angular velocity and the angular
      acceleration of link (4).

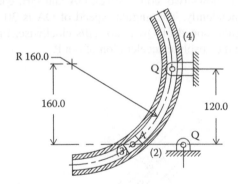

FIGURE P2.39

2.40 The cam shown in Figure P2.40 moves to the left with a constant speed
      of 10 m/s. Find the angular velocity and the angular acceleration of the
      follower.

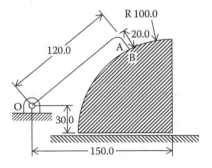

FIGURE P2.40

2.41 Repeat Problem 2.40 when the tip of the follower is spherical (Figure P2.41).

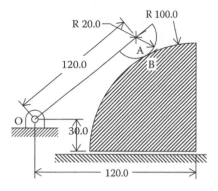

FIGURE P2.41

2.42 The straight-sided cam shown in Figure P2.42 rotates clockwise at 600 rpm and actuates an oscillating follower. Find the angular velocity and angular acceleration of the follower and the roller.

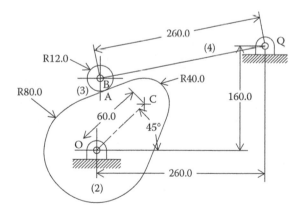

FIGURE P2.42

2.43 The large roller of the mechanism shown in Figure P2.43 rotates about center O with a speed of 150 rpm clockwise. The arm OQ rotates at 200 rpm clockwise and carries a small roller, which rotates freely about center Q. The contact between the two rollers is pure rolling (no sliding). There is a slip between the small roller and the follower. Determine the magnitude and the direction of the angular velocities and the angular accelerations of the follower and the small roller.

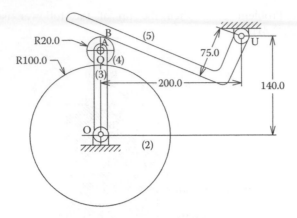

FIGURE P2.43

2.44 The mechanism shown in Figure P2.44 is driven through crank OA at a speed of 10 rad/s counterclockwise. There is a pure rolling contact between the roller and the follower. Find the magnitude and the direction of angular velocity and the angular acceleration of the follower and the roller.

OA = 100 mm, AB = 400 mm, QB = 200 mm, AC = BC = 225 mm

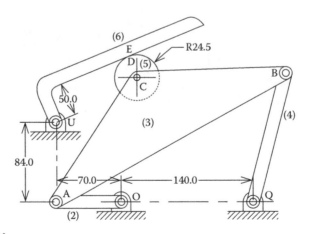

FIGURE P2.44

2.45 Repeat Problem 2.44 if pure sliding occurs between the roller and the follower [the roller is fixed with link (3)].

2.46 Crank OA of the mechanism shown in Figure P2.46 rotates with a uniform speed of 10 rad/s clockwise. Find the angular velocity and the angular acceleration of link (6) analytically.

$$r_2 = 60\,\text{mm},\ r_3 = 200\,\text{mm},\ r_5 = 150\,\text{mm},\ r_6 = 150\,\text{mm}$$

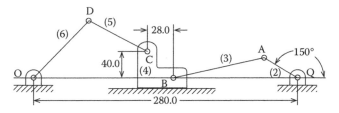

FIGURE P2.46

2.47 Solve Problems 2.6 to 2.45 analytically.

3 Cams

3.1 INTRODUCTION

A cam is a machine element that gives a specified periodic motion to another machine element, called the follower, by direct contact. Usually, the cam has rotational motion, although in some special cases, its motion may be reciprocation or oscillation. A typical cam and follower system is shown in Figure 3.1.

3.2 TYPES OF CAMS

There are several types of cams; some of them are described in the following sections.

3.2.1 DISK CAMS

Disk cams are sometimes called plate cams (Figure 3.1). They are widely used. In this type, contact between the cam and the follower is maintained by external means such as a spring, an external load, or both.

3.2.2 WEDGE CAM

The wedge cam shown in Figure 3.2 is actually a disk cam except that it is in the form of a wedge and has a reciprocating motion.

3.2.3 CYLINDRICAL FACE CAM

The cam is shown in Figure 3.3. The follower is placed at a distance from the axis of the cam and its motion is normal to the plane of rotation.

3.2.4 CYLINDRICAL CAM

The cam profile is engraved on the surface of a cylinder (Figure 3.4). The follower is guided to have a transverse motion. The follower has a small roller, which is inserted in the groove of the cam.

3.2.5 DISK FACE CAM

They are actually disk cams with the follower riding in a groove in the face of the cam (Figure 3.5).

3.2.6 YOKE CAM

The yoke cam (Figure 3.6) is simply an eccentric disk bounded by a frame, called a yoke, which has only a reciprocating motion.

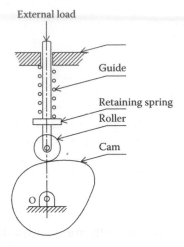

FIGURE 3.1   Typical disk cam.

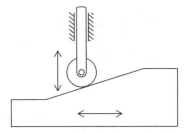

FIGURE 3.2   Wedge cam.

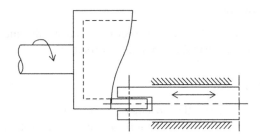

FIGURE 3.3   Cylindrical face cam.

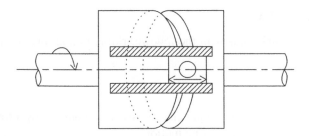

FIGURE 3.4   Cylindrical cam.

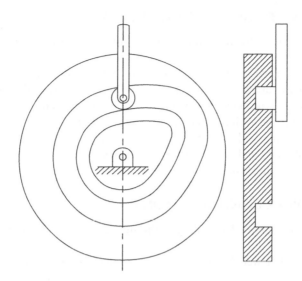

FIGURE 3.5   Face cam.

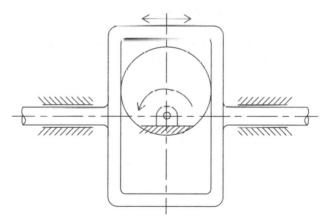

FIGURE 3.6   Yoke cam.

The last three types of cams are positive cams. They are sometimes used to secure the contact between the cam and the followers.

3.2.7 OTHER KINDS OF CAMS

There are several types of cams that have different configurations and different performance, for example, axially moving cams with translating followers [56]. They are actually solid surfaces with different forms and actuate any type of the follower. When a cam of this kind is moved axially, the action of the follower changes. Another kind of cam is the space cam. This kind allows the follower to operate in any direction.

3.3   MODES OF INPUT/OUTPUT MOTION

1. Rotating cam with translating follower such as the cams in Figures 3.1 through 3.6.
2. Rotating cam with rotating follower. The followers of the cams of Figure 3.1 may have an oscillatory motion as in Figure 3.7. Also, the cams shown in Figures 3.3 through 3.5 may have oscillating followers.
3. Translating cam with translating follower, as in Figure 3.2.
4. Stationary cam with rotating follower. The cams of the first type can be made stationary and the follower system revolves with respect to the center-line of the vertical shaft.

3.4   FOLLOWER CONFIGURATIONS

The followers are classified according to the surface in contact with the cam.

1. Knife edge follower (Figure 3.8a). In fact, it does not have any practical value since it has point contact. The stress is very high and hence the wear is excessive.
2. Roller follower (Figure 3.8b).
3. Flat-faced (Figure 3.8c).
4. Spherical-faced follower (Figure 3.8d).

The flat-faced follower and the spherical follower shown are used for cam profiles that have large curvatures and where the space is limited.

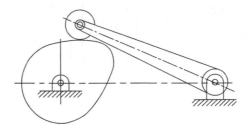

FIGURE 3.7   A cam with oscillating follower.

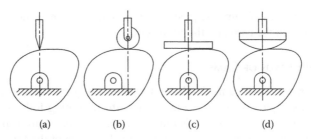

FIGURE 3.8   Types of followers. (a) Knife edge (b) roller (c) flat (d) spherical.

3.4.1 FOLLOWER ARRANGEMENT

For roller translating followers, the position of the centerline of the follower has two arrangements.

1. Inline (or radial) follower. The centerline of the follower passes through the centerline of the cam shaft.
2. Offset follower. The centerline of the follower does not pass through the centerline of the cam shaft. The amount of offset is the distance between these two centerlines. The offset causes a reduction of the side thrust present in the roller follower during transmission of the load.

3.5 CLASSES OF CAMS

Generally, there are two classes of cams. The first is the specified motion cams where the motion of the follower is given by a certain displacement diagram, and then, the cam contour is laid off and manufactured accordingly. The second is known as the specified contour cam where the cam is made of simple geometrical curves such as circular arcs and straight lines that are easy to produce.

3.5.1 SPECIFIED MOTION DISK CAMS

3.5.1.1 Cam Nomenclature

Cam profile: The working surface of a cam, which is in contact with the follower. It is also called cam contour. It is formed in a way to give the follower the specified motion required (Figure 3.9).

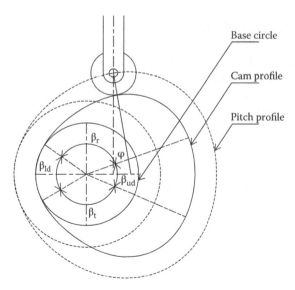

FIGURE 3.9 Cam contour.

Pitch contour: It is a contour parallel to cam profile. The distance between the cam and the pitch contours is equal to the radius of the roller. It is the path generated by the center of the roller as the follower is rotated about the cam when it is stationary.

Base circle: It is the smallest circle in the cam. It is the circle upon which the cam contour is constructed. The size of the cam depends on the radius of the base circle.

Rise: It is the part of the cam profile that causes the follower to rise.

Rise angle, β_r: It is the cam rotational angle during which the follower rises.

Lift, s: It is the maximum distance that the follower raises. It is a displacement for translating followers and an angle for oscillating followers.

Upper dwell: It is the part of the cam profile at which the follower stays at its upper most position. It has the form of an arc of a circle whose center is the center of the base circle.

Upper dwell angle, β_{ud}: It is the cam rotational angle during the upper dwell.

Return: It is the part of the cam profile that causes the follower to return back to its lower position.

Return angle, β_t: It is the cam rotational angle during which the follower returns.

Lower dwell: It is the part of the cam profile at which the follower stays at its lower most position. It is a part of the base circle.

Lower dwell angle, β_{ld}: It is the cam rotational angle during the lower dwell.

Pressure angle, φ: It is the angle at any point between the common normal between the cam and the surface of the follower with the instantaneous direction of the follower motion. This angle is important in cam design because it represents the steepness of the cam profile. It also affects the transmitted force.

3.5.1.2 Displacement Diagram

The displacement diagram is a specification for the motion desired for the follower as a function of the time. When cams rotate at a constant speed, time is proportional to the cam rotational angle θ. Since the motion of the follower is repeated every cam revolution, the diagram is, then, drawn against θ with length equivalent to one complete cam revolution, that is, 360°. Its height represents the total follower displacement (lift) from the lowest position. The diagram consists basically of four portions (Figure 3.10).

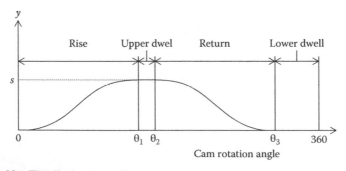

FIGURE 3.10   The displacement diagram.

During the rise, the follower moves away from the cam center. This takes place over an angle equal the rise angle β_r degrees of the cam rotation. The upper dwell occurs after the rise. The follower is at a constant distance from the cam center during the upper dwell angle β_{ud}. During the return angle β_t, the follower returns to its original position and stays there for the lower dwell angle β_{ld}. The variable θ_1 is the end of the rise stroke; it is equal to β_r. The variable θ_2 is the end of the upper dwell; $\beta_{ud} = \theta_2 - \theta_1 - \theta_3$ is the end of the return stroke; $\beta_{ld} = \theta_3 - \theta_2$.

During the rise and the return strokes, the follower's motion may follow one of the standard basic motions described in Section 3.5.1.3. However, the motion may be of any nature that fulfills certain requirements regarding the velocity and acceleration of the follower.

3.5.1.3   Basic Motions

3.5.1.3.1   Uniform Motion (Constant Velocity)

The displacement of the follower is proportional to the cam rotational angle θ. If β_r denotes the rise angle, the displacement of the follower y is represented by the equation of a straight line, which is in the form

$$y = C_1 + C_2 \theta \tag{3.1}$$

where C_1 and C_2 are arbitrary constants and are determined from the conditions at the beginning and end of the rise, which are

$$\text{At } t = 0, y = 0$$
$$t = \beta_r, y = s$$

where s is the lift of the follower. By applying the first condition, C_1 is zero. According to the second condition,

$$C_2 = \frac{s}{\beta_r}$$

Substituting the values of the constants in Equation 3.1,

$$y = \frac{s}{\beta_r} \theta \tag{3.2}$$

The velocity of the follower is obtained by differentiating Equation 3.2 with respect to the time.

$$V = \frac{dy}{dt} = \frac{dy}{d\theta} \omega = \frac{s}{\beta_r} \omega$$

where ω is the angular velocity of the cam and is usually constant. The velocity is constant during the rise (Figure 3.11b), and consequently, the acceleration is zero. However, since the follower is at rest at the lower and the upper dwells, the velocity rises suddenly and the acceleration theoretically rises to infinity causing a shock in the system. The follower, displacement, velocity, and acceleration are illustrated in Figure 3.11c.

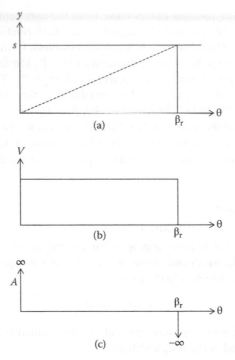

FIGURE 3.11 The uniform motion. (a) Displacement (b) velocity (c) acceleration of the follower.

3.5.1.3.2 Modified Uniform Motion (Constant Velocity)

The sudden rise of the velocity for the uniform motion is rectified by using a modified uniform motion. It is a modification for the uniform motion to eliminate the sudden change in the velocity. A gradual change is accomplished by using smooth curves, usually arcs of a circle at the beginning and end of the motion. The straight line is tangent to both arcs (Figure 3.12a).

The velocity and the acceleration of the follower are shown in Figure 3.12b and c.

3.5.1.3.3 Parabolic Motion (Constant Acceleration)

The displacement diagram, which yields a parabolic motion for the follower, consists of two inverted parabolas with an inflection point at an angle, say α, which is called the inflection angle (Figure 3.12). Each parabola is represented by,

For $0 \leq \theta \leq \alpha$,

$$y_1 = a_1 + a_2\,\theta + a_3\,\theta^2 \tag{3.3}$$

For $\alpha \leq \theta \leq \beta_r$,

$$y_2 = b_1 + b_2\,\theta + b_3\,\theta^2 \tag{3.4}$$

The variables a_1, a_2, a_3, b_1, b_2, and b_3 are arbitrary constants that are determined from the boundary conditions. We should notice that at the inflection point, the

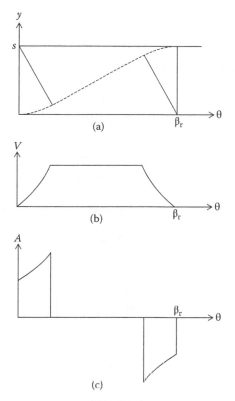

FIGURE 3.12   Modified uniform motion. (a) Displacement (b) velocity (c) acceleration.

values of the follower displacement and velocity are the same for the two regions. Thus, the boundary conditions are,

$$\text{At } \theta = 0, \quad y_1 = 0 \text{ and } V_1 = 0$$
$$\text{At } \theta = \alpha, \quad y_1 = y_2 \text{ and } V_1 = V_2$$
$$\text{At } \theta = \beta_r, \quad y_2 = s \text{ and } V_2 = 0$$

Applying the boundary conditions we get,

$$a_1 = a_2 = 0$$
$$a_3 = \frac{s}{\alpha \beta_r}$$

$$b_1 = -\frac{s\,\alpha}{\beta_r - \alpha}$$
$$b_2 = \frac{2\,s}{\beta_r - \alpha}$$

$$b_3 = -\frac{s}{\beta_r(\beta_r - \alpha)}$$

Substituting in Equations 3.3 and 3.4,

$$y_1 = \frac{s\theta^2}{\beta_r \alpha} \tag{3.5}$$

$$y_2 = s\left[1 - \frac{(\beta_r - \theta)^2}{\beta_r(\beta_r - \alpha)}\right] \tag{3.6}$$

The velocity and acceleration of the follower for the two regions are given by,

$$V_1 = \frac{2s\theta\omega}{\beta_r \alpha} \tag{3.7}$$

$$V_2 = 2s\omega\left(\frac{\beta_r - \theta}{\beta_r(\beta_r - \alpha)}\right) \tag{3.8}$$

$$A_1 = \frac{2s\omega^2}{\beta_r \alpha} \tag{3.9}$$

$$A_2 = -\frac{2s\omega^2}{\beta_r(\beta_r - \alpha)} \tag{3.10}$$

When the inflection point is at the middle of the rise stroke, the displacement, velocity, and acceleration of follower in the two regions are

For $0 \le \theta \le \dfrac{\beta_r}{2}$,

$$y_1 = 2s\left(\frac{\theta}{\beta_r}\right)^2$$

$$V_1 = 4s\omega\frac{\theta}{\beta_r^2}$$

$$A_1 = \frac{4s\omega^2}{\beta_r^2}$$

For $\dfrac{\beta_r}{2} \le \theta \le \beta_r$,

$$y_2 = s\left[1 - 2\left(1 - \frac{\theta}{\beta_r}\right)^2\right]$$

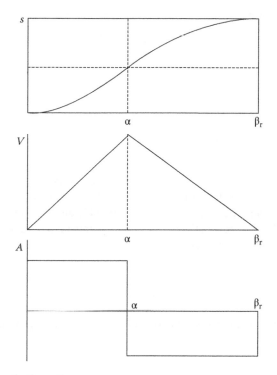

FIGURE 3.13   Parabolic motion.

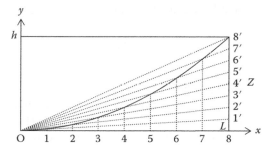

FIGURE 3.14   Graphical construction of the parabola.

$$V_2 = 4 s \omega \left(\frac{\beta_r - \theta}{\beta_r^2} \right)$$

$$A_2 = -\frac{2 s \omega^2}{\beta_r^2}$$

The displacement, velocity, and acceleration of the follower for the parabolic motion are shown in Figure 3.13.

A graphical method for constructing a parabola inside a rectangle of length L and height h is shown in Figure 3.14.

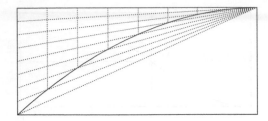

FIGURE 3.15   Inverted parabola.

The length L is divided to equal number of intervals by points denoted by 1, 2, 3, …. Also, the right side is divided by the same number of equal intervals and denoted by the points 1′, 2′, 3′, …. The rays O1′, O2′, and O3′ intersect the vertical lines through 1, 2, 3, … respectively, which are points on a parabola. To prove this, the points are obtained such that

$$\frac{x}{L} = \frac{z}{h} \tag{3.11}$$

where z is the height of a ray. But the y-coordinate of any point on the ray is given by

$$y = \frac{x\,z}{L} \tag{3.12}$$

Substituting Equation 3.12 in Equation 3.13,

$$y = h\frac{x^2}{L^2}$$

This is the equation of a parabola.

To construct the other part of the parabola, we use the same procedure except using the right upper corner and the left side of the rectangle (Figure 3.15).

Note: Unless otherwise stated, the inflection angle is equal to one half the rise angle.

3.5.1.3.4   Simple Harmonic Motion

The motion of the follower is a sinusoidal function of the cam rotational angle θ. Its velocity is zero at the beginning and at the end of the rise stroke and is gradually changing in between. The harmonic function that provides this condition is in the form,

$$V = C\sin\frac{\pi\theta}{\beta_r} \tag{3.13}$$

where C is a constant. But,

$$V = \frac{dy}{dt} = \omega \frac{dy}{d\theta}$$

Thus,

$$\frac{dy}{d\theta} = \frac{C}{\omega} \sin \frac{\pi \theta}{\beta_r} \tag{3.14}$$

The follower displacement is obtained by integrating Equation 3.14, hence

$$y = -\frac{C \beta_r}{\pi \omega} \sin \frac{\pi \theta}{\beta_r} + C_1 \tag{3.15}$$

The constants C and C_1 are determined from the boundary conditions of the motion which are,

$$\text{At } \theta = 0, \quad y = 0$$
$$\text{At } \theta = \beta_r, \quad y = s$$

Apply these conditions to Equation 3.15, then

$$C_1 = \frac{s}{2}$$

$$C = \frac{s \pi \omega}{2 \beta_r}$$

Substituting the values of the constant in Equation 3.15,

$$y = \frac{s}{2}\left(1 - \cos \frac{\pi \theta}{\beta_r}\right) \tag{3.16}$$

The harmonic motion diagram is constructed graphically by drawing a semi-circle with diameter equal to s on the y-axis. The circumference of the semicircle is divided into a number of equal sectors by points that are denoted by $1'$, $2'$, $3'$ ..... The rise angle is divided into the same number or equal intervals by the lines 1, 2, 3 .... Each point on the semicircle is projected on the corresponding line as shown in Figure 3.16.

The velocity of the follower is obtained by differentiating Equation 3.16 with respect to time.

$$V = \frac{s \pi \omega}{2 \beta_r} \sin \frac{\pi \theta}{\beta_r} \tag{3.17}$$

The acceleration of the follower is obtained by differentiating Equation 3.17 with respect to time.

$$A = \frac{s}{2}\left(\frac{\pi\omega}{\beta_r}\right)^2 \cos\frac{\pi\theta}{\beta_r} \qquad (3.18)$$

The displacement, velocity, and acceleration of the follower for the harmonic motion are shown in Figure 3.17.

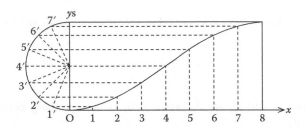

FIGURE 3.16   Graphical construction of the harmonic motion.

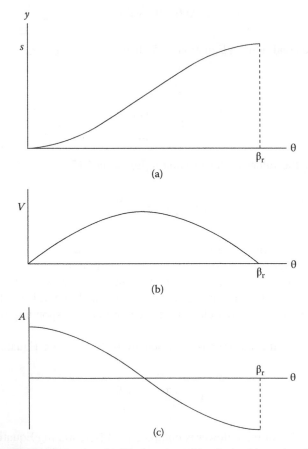

FIGURE 3.17   Harmonic motion. (a) Displacement (b) velocity (c) acceleration.

3.5.1.3.5 Cycloid Motion

The parabolic and the simple harmonic motions have certain acceleration at the beginning and at the end of the motion. The inertia force of the follower is suddenly applied, which may cause serious shocks especially at high speeds. For this reason, they are suitable only for low or moderate speeds. For high-speed cams, the optimum motion is when the acceleration is zero at the beginning and at the end of the rise stroke. A formula that satisfies this requirement is in the form,

$$A = C_1 \sin \frac{2\pi\theta}{\beta_r} \tag{3.19}$$

The velocity is obtained by integrating Equation 3.19

$$V = -\frac{C_1 \beta_r}{2\omega\pi} \cos \frac{2\pi\theta}{\beta_r} + C_2 \tag{3.20}$$

Also, the displacement is given by,

$$y = -C_1 \left(\frac{\beta_r}{2\omega\pi}\right)^2 \sin \frac{2\pi\theta}{\beta_r} + C_2 \frac{\theta}{\omega} + C_3 \tag{3.21}$$

where C_1, C_2, and C_3 are constants. The boundary conditions are

$$\text{At } \theta = 0, \quad y = 0 \text{ and } V = 0$$
$$\text{At } \theta = \beta_r, \quad y_2 = s \text{ and } V = 0$$

Appling these conditions to Equations 3.20 and 3.21, we arrive at

$$C_1 = 0$$

$$C_1 = 2s\left(\frac{\beta_r}{\omega}\right)^2$$

$$C_2 = -s\frac{\beta_r}{\omega}$$

Therefore,

$$y = s\left(\frac{\theta}{\beta_r} - \frac{1}{2\pi} \sin \frac{2\pi\theta}{\beta_r}\right) \tag{3.22}$$

$$V = \frac{s\omega}{\beta_r}\left(1 - \cos \frac{2\pi\theta}{\beta_r}\right) \tag{3.23}$$

$$A = 2\pi s\left(\frac{\omega}{\beta_r}\right)^2 \sin \frac{2\pi\theta}{\beta_r} \tag{3.24}$$

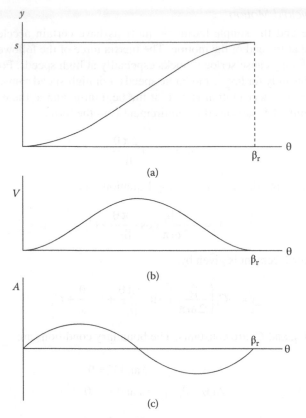

FIGURE 3.18   The cycloid motion. (a) Displacement (b) velocity (c) acceleration.

The displacement, velocity, and acceleration of the follower for the cycloid motion are shown in Figure 3.18.

The displacement diagram for the cycloid can be constructed graphically by using the following analysis.

For Equation 3.22, if we replace $\dfrac{2\pi\theta}{\beta s}$ by α, it takes the form,

$$y = \frac{s}{2\pi}\left(\alpha - \frac{1}{2\pi}\sin\alpha\right)$$

This is the equation of a cycloid resulting from rolling a circle of radius equal to $\dfrac{s}{2\pi}$ on the y-axis. The displacement diagram can thus be constructed by using this idea. A convenient method is illustrated in Figure 3.19. A circle of radius r equal to $\dfrac{s}{2\pi}$ is drawn at the upper left corner (at point Q). The diameters parallel and normal to the displacement axis are drawn. Starting from 0', on the normal diameter, the circumference of the circle is divided to an equal number of intervals by points 1', 2', ..., which are projected on the diameter parallel to the y-axis. The rise angle β_r is also

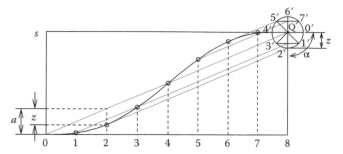

FIGURE 3.19 Construction of cycloid motion.

divided to the same number of equal intervals by the lines 1, 2, ... From the projections, lines parallel to the diagonal 0Q are drawn to intersect the corresponding lines at points that are on the cycloid curve. This is proved as follows:

Since the circle is divided to the same number of intervals as those of β_r,

$$\frac{\alpha}{2\pi} = \frac{\theta}{\beta_r}$$

where α is the angle of a division from 0'. Refer to Figure 3.18.

$$\alpha = \frac{2\pi\theta}{\beta_r}$$

$$a = \frac{s\theta}{\beta_r}$$

$$r = \frac{s}{2\pi}$$

$$z = r\sin\alpha$$

But,

$$y = a - z$$

Therefore,

$$y = s\left(\frac{\theta}{\beta_r} - \frac{1}{2\pi}\sin\frac{2\pi\theta}{\beta_r}\right)$$

3.5.1.3.6 High-Speed Polydyne Cams

The motion of the cam is represented mathematically as a polynomial for the selection of the proper criterion to minimize the acceleration and, consequently, the inertia forces. The polynomial is in the form,

$$y = s(C_0 + C_1\theta^2 + C_1\theta^2 + ...) \tag{3.25}$$

The constants C_0, C_1, ... are determined from the boundary conditions. Thus, the number of these constants is as many as the number of the boundary conditions. Suppose that the conditions to be fulfilled are

$$\text{At } \theta = 0, \quad y = 0, V = 0, A = 0$$
$$\text{At } \theta = \beta_r, \quad y = s, V = 0, A = 0$$

The number of the condition is six and hence the order of the polynomial in Equation 3.25. Thus, the motion is represented by

$$y = s(C_0 + C_1\,\theta + C_2\,\theta^2 + C_3\,\theta^3 + C_4\,\theta^4 + C_5\,\theta^5)$$
$$\dot{y} = s\omega(C_1 + 2\,C_2\,\theta + 3\,C_3\,\theta^2 + 4\,C_4\,\theta^3 + 5\,C\,\theta^4)$$
$$\ddot{y} = s\omega^2(C_2 + 6\,C_3\,\theta + 12\,C_4\,\theta^2 + 20\,C_5\,\theta^3)$$

By applying the boundary conditions, we arrive at

$$C_0 = C_1 = C_2 = 0$$

$$C_3 = \frac{10\,s}{\beta^3}$$

$$C_4 = -\frac{15\,s}{\beta^4}$$

$$C_5 = \frac{6\,s}{\beta^5}$$

The displacement is given by

$$y = s\left[10\left(\frac{\theta}{\beta}\right)^3 - 15\left(\frac{\theta}{\beta}\right)^4 + 6\left(\frac{\theta}{\beta}\right)^5\right]$$

3.5.1.3.7   Follower Motions during the Return

The basic motions were presented for the rise stroke. The displacement diagram for the return stroke can be constructed as outlined earlier in Section 3.5.1.3 by shifting the displacement axis to the end of the rise (at θ_3, Figure 3.10) and measuring the rotational angle, call it γ, backwards as shown in Figure 3.20.

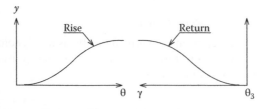

FIGURE 3.20   Motions during the return stroke.

In this case, the equations representing the displacement obtained for the rise can be used by replacing θ by γ. For example, if the displacement during the rise is in the form $y = f(\theta)$, then, for the return, it is represented by,

$$y = f(\gamma)$$

But,

$$\gamma = \theta_3 - \theta$$

Therefore,

$$y = f(\theta_3 - \theta) \tag{3.26}$$

The velocity and acceleration of the follower during the return cycle are given by,

$$V = \frac{df(\theta_3 - \theta)}{dt} \tag{3.27}$$

$$A = \frac{df^2(\theta_3 - \theta)}{dt^2} \tag{3.28}$$

As an example, let the motion during the return be cycloid. Therefore, the motion of the follower during return is given by

$$y = s\left(\frac{(\theta_3 - \theta)}{\beta_t} - \frac{1}{2\pi}\sin\frac{2\pi(\theta_3 - \theta)}{\beta_t}\right)$$

$$V = -\frac{s\omega}{\beta_t}\left(1 - \cos\frac{2\pi(\theta_3 - \theta)}{\beta_t}\right)$$

$$A = 2\pi s\left(\frac{\omega}{\beta_t}\right)^2 \sin\frac{2\pi(\theta_3 - \theta)}{\beta_t}$$

3.5.1.4 Layout of the Cam Profile

The cam profile must be such that the motion described by the displacement diagram is imparted to the follower. A conventional method for the cam layout is to hold the cam stationary and rotate the follower in the direction opposite to the actual direction of the cam rotation. When the cam moves a certain angle, the follower is translated a certain distance according to the displacement diagram. The size of the cam is controlled by the base circle. Its center is the center of cam and it is the smallest circle touching the cam surface.

To demonstrate how to layout a cam profile, consider a cam rotating clockwise. Let the rise angle be 120°, the upper dwell angle be 30°, and the return angle be 120°. The rise is 5 cm, the diameter of the base circle is 8 cm, and the follower motion during the rise and the return is simple harmonic. The displacement diagram is shown in Figure 3.21.

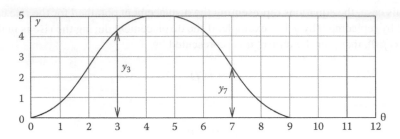

FIGURE 3.21   The displacement diagram for cam layout.

3.5.1.4.1   Cams with Translating Radial Knife Edge Follower

The procedure is illustrated in Figure 3.22 and is outlined by the following steps:

1. The displacement diagram is constructed. The abscissa, which represents the cam rotational angle θ, can be any length, while the follower displacement axis must be drawn to the same scale as the cam. The rise angle β_r is divided to an equal number of divisions (Figure 3.22).
2. The base circle is divided into four regions representing the rise, the upper dwell, the return, and the lower dwell by the lines OA, OB, OC, and OD. Point A is the start of the rise.
3. The rise angle is divided to the same number of equal intervals in the displacement diagram by the lines 1, 2, 3, … and in the base circle by the rays O1′, O2′, O3′, ….
4. The distances y_1, y_2, y_3, … in the displacement diagram are copied on the rays starting from the circumference of the base circle and out. The smooth curve passing through the points on the rays is the cam profile during the rise.
5. The cam profile during the upper dwell is an arc of a circle between the rays OB and OC. Its radius is equal to the base circle radius plus the lift.
6. The cam profile during the return is obtained by following the same steps as in the rise.
7. The cam profile during the lower dwell is the part of the base circle between the rays OD and OA.

3.5.1.4.2   Cams with Translating Radial Roller Follower

When the follower has a roller tip as shown in Figure 3.23, its center, at the beginning of the rise, is on a circle called the pitch circle. The roller center can be regarded as a knife edge. The pitch circle is divided as mentioned earlier in Section 3.5.1.4. The follower displacements are placed on the rays from the pitch circle. These represent the centers of the roller at different positions. The curve passing through all the centers is the pitch contour. Using these positions as centers, circles representing the roller are drawn. The smooth curve that is tangent to all the circles is the cam contour.

3.5.1.4.3 Face Cams with Translating Roller Follower

The profile of a face disk cam is obtained by adding a second curve tangent to the circles from outside as shown in Figure 3.24.

3.5.1.4.4 Cams with Translating Offset Roller Follower

In this type of cams, the centerline of the follower is shifted a distance h from the center of the cam (Figure 3.25). When the cam is held stationary and the follower is rotated, the centerline of the follower is tangent to a circle, called offset circle,

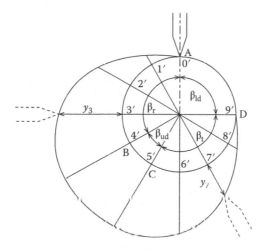

FIGURE 3.22 A cam with radial knife edge translating follower.

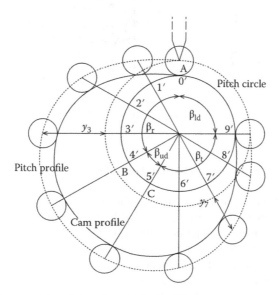

FIGURE 3.23 A cam with radial roller translating follower.

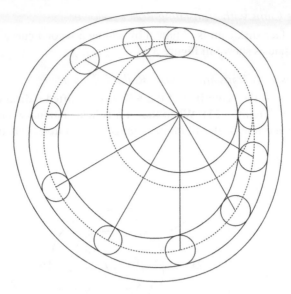

FIGURE 3.24   Face cam.

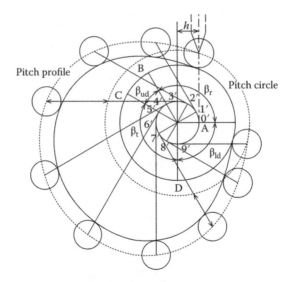

FIGURE 3.25   A cam with offset roller translating follower.

with radius h. At the beginning of the rise, the centerline of the follower is tangent
to this circle at point $0'$. This circle is divided by to the same number of equal divisions as in the displacement diagram by the rays $O1'$, $O2'$, …. From points $1'$, $2'$, …,
tangents to the offset circle are drawn. The distances y_1, y_2, y_3, … in the displacement diagram are copied on these tangents staring from the pitch circle to obtain the
centers of the roller circle. The curve joining the centers is the pitch contour and the
curve that is tangent to the circles is the cam profile. The cam profile for the upper
dwell is an arc of a circle. The rest of the profile is constructed with the same steps.

3.5.1.4.5 Cams with Translating Flat-Faced Follower

At the beginning of the rise, the follower is in contact with the base circle at point A as shown in Figure 3.26. The procedure for obtaining the contour is similar to that of the knife edge follower until determining the points on the rays. In this case, they represent points on the face of the follower, which is normal to the rays. Thus, from each point, a normal to the corresponding ray is drawn until finally we have a set of intersecting lines. The cam profile is a smooth curve that is tangent to all these lines.

3.5.1.4.6 Cams with Oscillating Roller Follower

The displacement diagram for oscillating followers (Figure 3.27) represents the relation between the follower rotational angle φ and the cam rotational angle θ. The angle φ may be drawn to any scale and the values are estimated according to this scale.

As an example, consider a cam rotating clockwise and operating a roller oscillating follower. The length of the follower is 12 cm, the distance between the follower pivot and the cam center is 12 cm, and the radius of the roller is 1 cm. For the displacement diagram, let the rise angle be 120°, the upper dwell angle be 30°, and the return angle be 120°. The follower motion during the rise and the return is simple harmonic.

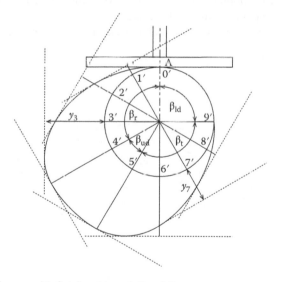

FIGURE 3.26   A cam with flat-faced translating follower.

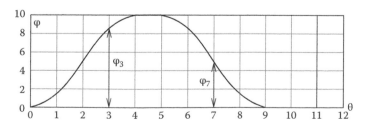

FIGURE 3.27   Displacement diagram for oscillating follower.

To lay out the cam contour, we assume that the cam is fixed and follower pivot rotates about the cam center in the opposite direction to the cam. The following steps are followed (Figure 3.28):

1. The displacement diagram is divided into equal divisions as mentioned earlier in Section 3.5.1.4
2. The center of the cam is at point O and the center of the cam pivot is at point Q.
3. A circle with radius OQ and center at O is drawn.
4. This circle is divided to the same number of divisions as the displacement diagram by lines O1, O2, O3, ... respectively.
5. At the start of the rise, the follower is located at Q; an arc with radius QA is drawn. The angle of this line is φ_0. The consequent follower angles are drawn from line OA and are denoted by lines O1′, O2′, ... respectively.
6. From point O, arcs are draw with radii O1′, O2″, O3′, ..., to intersect the arcs from points 1, 2, 3, ... with radius equal to the length of the follower respectively.
7. The intersection points represent the centers of the roller. The curve passing through these points is the pitch contour.
8. The curve that is tangent to the roller circles is the cam profile.

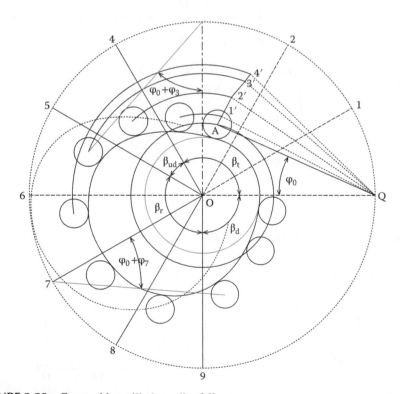

FIGURE 3.28 Cams with oscillating roller follower.

3.5.1.4.7 Cams with Oscillating Flat-Faced Follower

Any point, say point A, on the face of the follower is located. It replaces the center of the roller in the previous case. Steps 1 through 6 are the same. Lines from the intersection points to points 1, 2, 3, ... are drawn (Figure 3.29). These lines represent the face of the follower at different positions. The smooth curve that is tangent to the lines is the cam profile.

When the cam face is a distance h from the follower pivot, a line from Q is drawn parallel to the face of the follower. We use this line as described. The lines obtained are parallel to the face of the follower at different positions. So, lines are drawn parallel to them at a distance h, which represent the face of the follower.

3.5.1.5 Exact Cam Contour

The graphical method for determining the cam contour for the roller follower requires drawing a curve tangent to the circles. For the flat-faced follower, the curve is tangent to the lines representing the follower face at different positions. This process is quite approximate since we do not know the exact location of the tangent point. However, it is possible to obtain a better profile if we locate the points on the contour. The analysis is performed for each type of follower at a time.

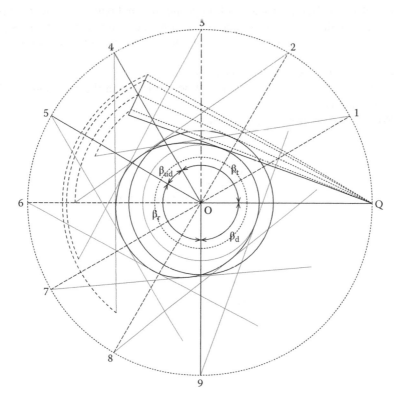

FIGURE 3.29 Cams with oscillating flat-faced follower.

3.5.1.5.1 Translating Roller Follower

Figure 3.30a illustrates a disk cam actuating a translating roller follower. Suppose the cam rotates with an angular velocity ω clockwise. The follower can be simulated as a knife edge follower actuated by the pitch contour.

The velocity polygon to scale ω of the system is shown in Figure 3.30. The velocity of point A is represented by line oa.

$$V_A = OA \times \omega = oa \times \omega$$

Thus, oa = OA. Also, the sliding velocity is represented by line ab and is along the common tangent. The velocity of point B is represented by ob such that

$$ob = \frac{V_B}{\omega} = vs$$

The quantity v is termed as the reduced velocity and is given by,

$$v = \frac{dy}{d\theta}$$

If the velocity polygon is rotated 90° opposite to ω and point o is placed on point O, it coincides with triangle OAb. Line Ab is along the common normal. This property is used to locate the exact tangent point. One follows the following steps:

- Calculate v for each follower position.
- From the center of the cam, draw a line equal to v in the direction of follower motion.
- Rotate v 90° opposite to the direction of rotation.

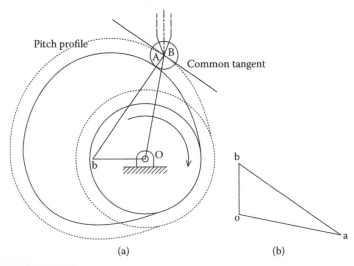

(a)   (b)

FIGURE 3.30   (a) Common normal of the contour (b) velocity polygon.

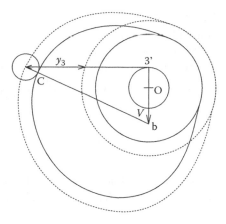

FIGURE 3.31 Locating exact cam contour for roller followers.

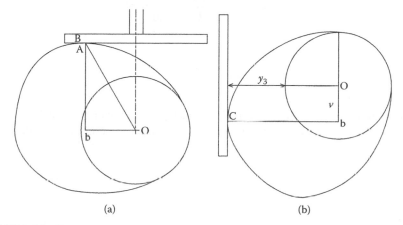

        (a)   (b)

FIGURE 3.32 Exact contour for flat-faced followers. (a) During the rise (b) during the return.

- Draw a line joining the center of the roller with the end of v. This line is along the common normal.
- The intersection of this line with the circle of the roller is a point on the contour.

These steps are performed for position 3′ for the cam in Section 3.5.1.4.4. Point C is an exact point on the cam profile (Figure 3.31).

3.5.1.5.2 Translating Flat-Faced Follower

Figure 3.32a shows a cam with a translating flat-faced follower. The velocity polygon drawn to scale ω and rotated 90° opposite to the direction of rotation of the cam is represented by triangle OAb. Line Ab is along common normal. To locate the exact point of contact, one has to follow the steps of the previous example except the last step. From point b, a line is drawn normal to the face of the follower to intersect it at point C. These steps are performed for position 3′ for the cam in Section 3.5.1.4.5. Point C is the exact point on the cam profile (Figure 3.32b).

3.5.1.5.3 Oscillating Roller Follower

Figure 3.33a shows a cam with an oscillating roller follower of length L and the distance between pivots is S. The velocity polygon drawn to scale ω and rotated $90°$ opposite to the direction of rotation of the cam is represented by triangle OAb. The velocity of the center of the roller is equal to $V_B = L\dfrac{d\varphi}{dt} = L\dfrac{d\varphi}{d\theta}\omega$ and is represented by line Ob.

$$Ob = \frac{V_B}{\omega} = L\frac{d\varphi}{d\theta}$$

Line Ob is parallel to follower while line Ab is along the common normal. Line Ab is extended to intersect line OQ extended at point D. Triangles ObD and QBD are similar. Thus,

$$\frac{Ob}{L} = \frac{OD}{QD} = \frac{OD}{S+OD} = \frac{d\varphi}{d\theta} = \varphi'$$

Thus,

$$OD = \frac{S\varphi'}{1-\varphi'}$$

Since line DA is the common normal, the exact point of contact can be determined. The steps to determine the exact point on the contour are outlined as follows:

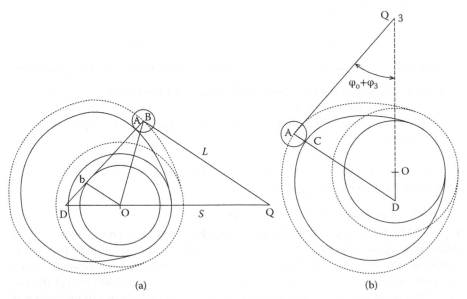

FIGURE 3.33 Exact cam contour for oscillating roller followers. (a) Locating the common normal (b) locating the exact cam contour.

- For one position, draw line OD $= \dfrac{S\varphi'}{1-\varphi'}$ along the lines of centers.
- From the center of the follower pivot, draw line QA with length L and make an angle $\varphi_o + \varphi$ with the lines of centers.
- Join points D and A. From point A, draw the roller circle to intersect line DA at point C; point C is an exact point on the contour.

These steps are performed for position 3 for the cam in Section 3.5.1.4.6.

3.5.1.5.4 Oscillating Flat-Faced Follower

Figure 3.34a shows a cam with an oscillating flat-faced follower. The velocity polygon drawn to scale ω and rotated 90° opposite to the direction of rotation of the cam is represented by triangle OAb. Line Ab is along the common normal. To locate the exact point of contact, one should follow the steps of the previous example except the last step. From point D, a line is drawn normal to the face of the follower to intersect it at point C, which is a point on the cam contour (Figure 3.34b).

3.5.1.6 Analytical Method for Contours

In Section 3.5.1.4, the cam contour is obtained graphically. The process is not accurate because the tangent points for the follower are approximately determined. In this section, exact points on the contour were determined, which improves the accuracy to some extent. This method, although better than the previous, still lacks precision because of the limited number of points that can be located, besides the extra labor and inaccuracy of the graphical method. In the following sections, equations representing the cam contour for all types of followers are presented in Cartesian coordinates, which were developed by the author.

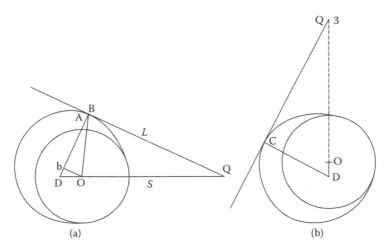

FIGURE 3.34 Exact cam contour for oscillating flat-faced followers. (a) Locating the common normal (b) locating the exact cam contour.

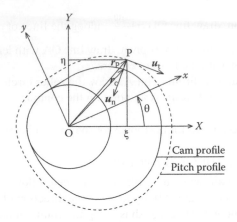

FIGURE 3.35   Coordinate system for the analytical method.

Consider two coordinate systems: the first is XOY, which is fixed and describes the follower motion, and the second is xOy, which rotates with the cam (Figure 3.35).

Consider point P on the pitch contour, which is also a point on the follower with coordinates (ξ, η). Its position vector with respect to the fixed coordinate system (in complex numbers) is given by,

$$r_P = \xi + i\eta \tag{3.28a}$$

The position vector with respect to the rotating coordinate system is given by (see Section 1.9.1.3),

$$r_P = (\xi + i\eta)\, e^{i\theta} \tag{3.28b}$$

The tangent to the pitch contour is obtained by differentiating Equation 3.28b with respect to θ.

$$\frac{dr_P}{d\theta} = (\xi' + i\eta')\, e^{-i\theta} - i(\xi + i\eta)\, e^{-i\theta}$$

$$= \left[(\xi' + \eta) + i(\eta' - \xi)\right] e^{-i\theta} \tag{3.29}$$

The dash represents differentiation with respect to θ. The unit tangent vector u_t is obtained by dividing Equation 3.29 by the absolute value of the vector. Thus,

$$u_t = \frac{\left[(\xi' + \eta) + i(\eta' - \xi)\right] e^{-i\theta}}{N} \tag{3.30}$$

where N is the absolute value of the vector and is given by,

$$N = \sqrt{(\xi' + \eta)^2 + (\eta' - \xi)^2} \tag{3.31}$$

The unit normal vector u_n is obtained by rotating u_t 90° in the clockwise direction. That is, multiplying Equation 3.30 by i. Therefore,

$$u_n = \frac{(\eta' - \xi) - i(\xi' + \eta)}{N} e^{-i\theta} \tag{3.32}$$

Suppose that the normal distance between the cam contour and the pitch contour is R. Thus, the position vector of point C on the cam contour is given by,

$$r_C = \left[\left(\xi + \frac{R(\eta' - \xi)}{N} \right) + i \left(\eta - \frac{R(\xi' + \eta)}{N} \right) \right] e^{-i\theta} \tag{3.33}$$

Therefore, the Cartesian coordinates of a point on the cam contour is obtained by expanding Equation 3.33.

$$x_C = \left(\xi + \frac{R(\eta' - \xi)}{N} \right) \cos\theta + \left(\eta - \frac{R(\xi' + \eta)}{N} \right) \sin\theta \tag{3.34}$$

$$y_C = -\left(\xi + \frac{R(\eta' - \xi)}{N} \right) \sin\theta + \left(\eta - \frac{R(\xi' + \eta)}{N} \right) \cos\theta \tag{3.35}$$

Equations 3.34 and 3.35 can be applied to the different types of followers.

3.5.1.6.1 Translating Roller Follower

Figure 3.36 shows a cam with translating roller follower. The cam has a base circle radius r_0 and the follower has a roller with radius R. The centerline of the follower is offset from the center of the cam at distance h. The center of the roller at its lowest position is at a distance y_0 from the x-axis.

$$y_0 = \sqrt{(r_0 + R)^2 - h^2}$$

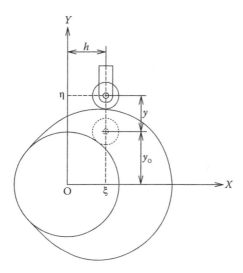

FIGURE 3.36   Translating roller follower.

The follower moves a distance y from its lowest position with one of the motions described in Section 3.5.1.2.

To obtain the coordinates of the cam contour, one uses

$$\xi = h$$
$$\eta = y_o + y$$
$$\xi' = 0$$
$$\eta' = y'$$

Applying to Equations 3.31, 3.34, and 3.35,

$$x_C = \left(h + \frac{R(y' - h)}{N}\right)\cos\theta + (y_o + y)\left(1 - \frac{R}{N}\right)\sin\theta \tag{3.36}$$

$$y_C = -\left(h + \frac{R(y' - h)}{N}\right)\sin\theta + (y_o + y)\left(1 - \frac{R}{N}\right)\cos\theta \tag{3.37}$$

where,

$$N = \sqrt{(y_o + y)^2 + (y' - h)^2} \tag{3.38}$$

EXAMPLE 3.1

Draw the profile of a disk cam actuating a translating roller follower with the following data:

- Base circle radius $r_o = 40$ mm.
- The rise angle $\beta_r = 120°$.
- The upper dwell angle $\beta_{ud} = 30°$.
- The return angle $\beta_t = 120°$.
- The lift is 30 mm.
- The motion of the follower is harmonic during the rise and the return.
- The roller radius $R = 10$ mm.
- The amount of offset $h = 15$ mm opposite to the direction of rotation.
- The cam rotates counterclockwise.

ANALYSIS

$$h = 15\text{mm}$$

$$y_o = \sqrt{(40 + 10)^2 - 15^2} = 47.7\text{mm}$$

For the rise, $0 \leq \theta \leq 120°$.

$$y = 15\left[1 - \cos\frac{3\theta}{2}\right]$$

$$y' = \frac{15 \times 3}{2}\sin\frac{3\theta}{2}$$

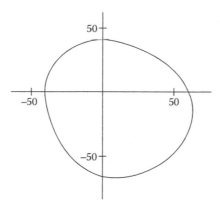

FIGURE 3.37 Cam contour for EXAMPLE 3.1.

For the upper dwell, $120° \leq \theta \leq 150°$.

$$y = 30\,\text{mm}$$
$$y' = 0$$

For the return, one replaces θ by $(\beta_r + \beta_{ud} + \beta_t - \theta)$ in Equations 3.36 through 3.38. For the return, $150° \leq \theta \leq 270°$.

$$y = 15\left[1 - \cos\frac{3(270 - \theta)}{2}\right]$$
$$y' = -\frac{15 \times 3}{2 \times 2}\sin\frac{3(270 - \theta)}{2}$$

For the lower dwell, $270° \leq \theta \leq 360°$.

$$y = 0$$
$$y' = 0$$

Equations 3.36 through 3.38 are used to obtain the profile. Math software is used to plot the profile as shown in Figure 3.37.

3.5.1.6.2 Oscillating Roller Follower

Figure 3.38 shows a cam with an oscillating roller follower. The cam has a base circle radius r_o. The follower has a length L and the radius of the roller is R. The center distance between the pivots is S on the x-axis. The center of the roller at its lowest position makes an angle φ_o with the x-axis such that

$$\varphi_o = \cos^{-1}\left(\frac{L^2 + S^2 - (r_o + R)^2}{2\,S\,L}\right)$$

The follower rotates an angle φ from its lowest position with one of the motions described in Section 3.5.1.2.

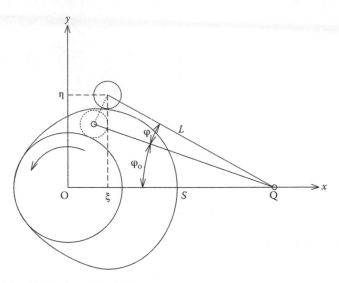

FIGURE 3.38   Oscillating roller follower.

To obtain the coordinates of the cam contour, one uses

$$\xi = S - L\cos(\varphi_o + \varphi)$$
$$\eta = L\sin(\varphi_o + \varphi)$$
$$\xi' = L\sin(\varphi_o + \varphi)\varphi'$$
$$\eta' = L\cos(\varphi_o + \varphi)\varphi'$$

Applying to Equations 3.31, 3.34, and 3.35,

$$x_C = \left(S - L\cos(\varphi_o + \varphi) + \frac{R\left(L\cos(\varphi_o + \varphi)(1 + \varphi') - S \right)}{N} \right)\cos\theta$$
$$+ \left(L\sin(\varphi_o + \varphi) - \frac{LR\sin(\varphi_o + \varphi)(1 + \varphi')}{N} \right)\sin\theta \qquad (3.39)$$

$$y_C = -\left(S - L\cos(\varphi_o + \varphi) + \frac{R\left(L\cos(\varphi_o + \varphi)(1 + \varphi') - S \right)}{N} \right)\sin\theta$$
$$+ \left(L\sin(\varphi_o + \varphi) - \frac{LR\sin(\varphi_o + \varphi)(1 + \varphi')}{N} \right)\cos\theta \qquad (3.40)$$

$$N = \left(\sqrt{L^2(1 + \varphi')^2 + S^2 - 2LS\cos(\varphi_o + \varphi)(1 + \varphi')} \right) \qquad (3.41)$$

EXAMPLE 3.2

Draw the profile of disk cam actuating an oscillating roller follower with the following data:

- Base circle radius $r_o = 50$ mm.
- The rise angle $\beta_r = 120°$.
- The upper dwell angle $\beta_{ud} = 20°$.
- The return angle $\beta_t = 120°$.
- The lift is 20°.
- The motion of the follower is parabolic during the rise and the return.
- The roller radius $R = 10$ mm.
- The length of the follower $L = 120$ mm.
- The distance between the pivots $S = 120$ mm.
- The cam rotates counterclockwise.

ANALYSIS

$$\varphi_o = \cos^{-1}\frac{120^2 + 120^2 - (40+10)^2}{2 \times 120 \times 120} = 24°$$

For the rise, $0 \le \theta \le 60°$.

$$\varphi = 40\left(\frac{\theta}{120}\right)^2$$

$$\varphi' = \frac{80}{120}\left(\frac{\theta}{120}\right)$$

For the flank, $60° \le \theta \le 120°$.

$$\varphi = 20\left[1 - 2\left(1 - \frac{\theta}{120}\right)^2\right]$$

$$\varphi' = \frac{80}{120}\left[1 - \frac{\theta}{120}\right]$$

For the upper dwell, $120° \le \theta \le 150°$.

$$\varphi = 20°$$
$$\varphi' = 0$$

For the return,
The first half, $150° \le \theta \le 210°$

$$\varphi = 20\left[1 - 2\left(1 - \frac{270 - \theta}{120}\right)^2\right]$$

$$\varphi' = -\frac{80}{120}\left[1 - \frac{270 - \theta}{120}\right]$$

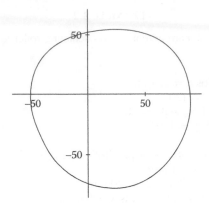

FIGURE 3.39 Cam contour for EXAMPLE 3.2.

The second half, $210° \leq \theta \leq 270°$

$$\varphi = 40\left(\frac{270-\theta}{120}\right)^2$$

$$\varphi' = -\frac{80}{120}\left(\frac{270-\theta}{120}\right)$$

For the lower dwell, $270° \leq \theta \leq 360°$.

$$\varphi = 0, \varphi' = 0$$

Equations 3.39 through 3.41 are used to obtain the profile (Figure 3.39).

3.5.1.6.3 Translating Flat-Faced Follower

Figure 3.40 shows a cam with translating flat-faced follower. The cam has a base circle radius r_o and the follower has a flat face. The centerline of the follower is offset from the center of the cam at distance h. The follower at its lowest position is tangent to the base circle.

$$y_o = r_o$$
$$\eta = r_o + y$$

The unit tangent vector to the cam profile in the plane rotating with the cam is given by Equation 3.30. This tangent, in the plane of the follower, is given by

$$\boldsymbol{u}_t = \frac{\left[(\xi' + \eta) + i(\eta' - \xi)\right]}{N}$$

Since for any cam position the follower face is always horizontal, the imaginary part of $\boldsymbol{u}_t$ is zero. Thus,

$$\eta' - \xi = 0$$

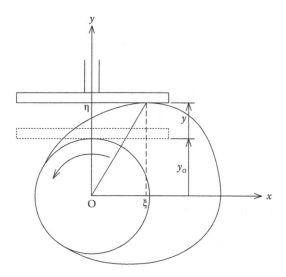

FIGURE 3.40 Translating flat-faced follower.

or,

$$\xi = \eta' = y' = v$$

where v is the reduced velocity. This type of follower does not have a pitch contour. Thus,

$$R = 0$$

Applying to Equations 3.34 and 3.35,

$$x_C = v\cos\theta + (r_o + y)\sin\theta \tag{3.42}$$

$$y_C = -v\sin\theta + (r_o + y)\cos\theta \tag{3.43}$$

EXAMPLE 3.3

Use the data of Example 3.1 to draw the profile of a disk cam actuating a translating flat-faced follower. Assume that the follower motion is cycloid during the rise and the return strokes.

ANALYSIS

For the rise, $0 \le \theta \le 120°$.

$$y = 30\left(\frac{\theta}{120} - \frac{1}{2\pi}\sin\frac{2\pi\theta}{120}\right)$$

$$v = \frac{3 \times 30}{2\pi}\left(1 - \cos\frac{2\pi\theta}{120}\right)$$

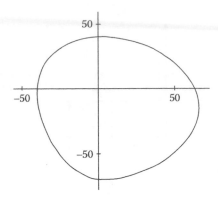

FIGURE 3.41   Contour for EXAMPLE 3.3.

For the upper dwell, $120° \le \theta \le 150°$.

$$y = 30$$
$$v = 0$$

For the return, $150° \le \theta \le 270°$, one replaces θ by $(\beta_r - \beta_{ud} \ \beta_t - \theta)$ in Equations 3.42 and 3.43

$$y = 30\left(\frac{(270 - \theta)}{120} - \frac{1}{2\pi}\sin\frac{2\pi(270 - \theta)}{120}\right)$$

$$v = \frac{3 \times 30}{2\pi}\left(-1 + \cos\frac{2\pi\theta}{120}\right)$$

For the lower dwell, $270° \le \theta \le 360°$.

$$y = 0$$
$$y' = 0$$

Equations 3.42 and 3.43 are used to obtain the profile. Math software is used to plot the profile as shown in Figure 3.41.

3.5.1.6.4   Oscillating Flat-Faced Follower

Figure 3.42 shows a cam with an oscillating flat-faced follower. The cam has a base circle radius r_o. The face of the follower is at a distance R from its pivot. The center distance between the pivots is S on the x-axis.

Consider line QA, which is parallel to the face of the follower; call it the follower pitch line. This line is in contact with the pitch contour of the cam. When the follower is at its lowest position, it makes an angle φ_o with the x-axis such that,

$$\varphi_o = \sin^{-1}\left(\frac{r_o + R}{S}\right)$$

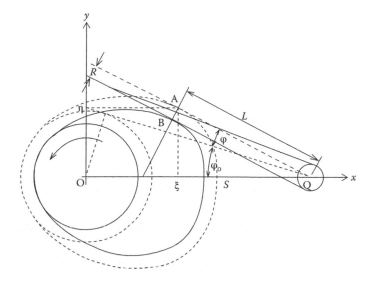

FIGURE 3.42 Oscillating flat-faced follower.

When the follower rotates an angle φ from its lowest position, the follower pitch line is in contact with the pitch contour at point B, which is at a distance l from Q; L changes with φ. Line DΓ is the common normal to the pitch contour and the follower. Thus,

$$L = QP \cos(\varphi_o + \varphi)\infty$$

As described in Section 3.5.1.5.4, it is possible to show that

$$\varphi' = \frac{OP}{PQ}$$

$$1 + \varphi' = \frac{OP + PQ}{PQ} = \frac{S}{PQ}$$

or,

$$\varphi' = \frac{OP}{PQ}$$

$$PQ = \frac{S}{1 + \varphi'}$$

Hence,

$$L = \frac{S}{1 + \varphi'} \cos(\varphi_o + \varphi) \tag{3.44}$$

Therefore,

$$\xi = S - L\cos(\varphi_0 + \varphi) \tag{3.45}$$

$$\eta = L\sin(\varphi_0 + \varphi) \tag{3.46}$$

$$\xi' = L\sin(\varphi_0 + \varphi) - \varphi'L'\cos(\varphi_0 + \varphi) \tag{3.47}$$

$$\eta' = L\cos(\varphi_0 + \varphi)\,\varphi' + L'\sin(\varphi_0 + \varphi) \tag{3.48}$$

The components of the normal vector in the fixed plane are

$$X = \eta'\xi = L\cos(\varphi_0 + \varphi)\,\varphi' + L'\sin(\varphi_0 + \varphi) - [S - L\cos(\varphi_0 + \varphi)]$$

$$Y = L\sin(\varphi_0 + \varphi)\,\varphi' - L'\cos(\varphi_0 + \varphi) + L\sin(\varphi_0 + \varphi)$$

Using Equation 3.44 and after some simplifications, we get,

$$X = -\sin(\varphi_0 + \varphi)\,[S\sin(\varphi_0 + \varphi) - L']$$

$$Y = \cos(\varphi_0 + \varphi)\,[S\sin(\varphi_0 + \varphi) - L']$$

The length of the normal vector is $[N = S\sin(\varphi_0 + \varphi) - L']$. If we consider a unit normal, its components are given by

$$u_x = -\sin(\varphi_0 + \varphi) \tag{3.49}$$

$$u_y = \cos(\varphi_0 + \varphi) \tag{3.50}$$

Substituting in Equations 3.34 and 3.35 and using Equations 3.49 and 3.50

$$x_C = -(S - L\cos(\varphi_0 + \varphi) - R\sin(\varphi_0 + \varphi))\cos\theta \\ +(L\sin(\varphi_0 + \varphi) - R\cos(\varphi_0 + \varphi))\sin\theta \tag{3.51}$$

$$y_C = -(S - L\cos(\varphi_0 + \varphi) - R\sin(\varphi_0 + \varphi))\sin\theta \\ +(L\sin(\varphi_0 + \varphi) - R\cos(\varphi_0 + \varphi))\cos\theta \tag{3.52}$$

EXAMPLE 3.4

Use the data of Example 3.2 to draw the profile of disk cam actuating an oscillating flat-faced follower with lift 10°. The motion of the follower is simple harmonic and the face of the follower is at a distance 10 mm from the pivot.

ANALYSIS

$$\varphi_o = \sin^{-1}\frac{40+10}{120} = 17°$$

For the rise, $0 \le \theta \le 120°$

$$\varphi = \frac{10}{2}\left[1-\cos\frac{3\theta}{2}\right]$$

$$\varphi' = \frac{10\times3\times\pi}{2\times2\times180}\sin\frac{3\theta}{2}$$

For the upper dwell, $120° \le \theta \le 150°$

$$\varphi = 10°$$

$$\varphi' = 0$$

For the return, $150° \le \theta \le 270°$, one replaces θ by $(\beta_r + \beta_{ud} - \beta_t.-\theta)$ in the above equations.

$$\varphi = \frac{10}{2}\left[1-\cos\frac{3(270-\theta)}{2}\right]$$

$$\varphi' = -\frac{10\times3\times\pi}{2\times2\times180}\sin\frac{3(270-\theta)}{?}$$

For the lower dwell, $270° \le \theta \le 360°$

$$\varphi = 0$$
$$\varphi' = 0$$

Equations 3.51 and 3.52 are used to obtain the profile. Math software is used to plot the profile as shown in Figure 3.43.

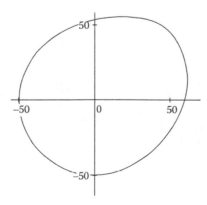

FIGURE 3.43   Contour for EXAMPLE 3.4.

3.5.1.7 Minimum Cam Size

Usually, the cam size is determined according to the available space. However, the size affects certain parameters to a great extent, which affects the cam performance. Such parameters are the pressure angle and the curvature of the cam. Smaller cam size causes large pressure angle, which, in turn, causes large undesirable lateral forces. The pressure angle is also affected by the amount of offset of the follower. Also, smaller cam size reduces the radius of curvature, which may approach zero. This results in an inaccurate cam performance besides more contact stresses. So, cams should be designed to fulfill certain requirements regarding the maximum pressure and the minimum radius of curvature.

3.5.1.7.1 Pressure Angle

The pressure angle φ as defined in Section 3.5.1.1 is the angle at any point between the common normal between the cam and the surface of the follower with the instantaneous direction of the follower motion. For translating roller followers, it is the angle between the centerline of the follower and the common normal. Figure 3.44 shows the pressure angle during the rise and the return strokes.

As described in Section 3.5.1.5.1, the common normal is obtained by rotating the reduced velocity v 90° in the opposite direction of the cam rotation. Thus, referring to Figure 3.44,

$$\tan \varphi = \frac{v - h}{y_o + y} \tag{3.53}$$

The position of the maximum pressure angle is obtained by differentiating Equation 3.53 with respect to the cam rotational angle θ and equating to zero. Thus,

$$0 = \frac{(y_o + y)\,a - (v - h)v}{(y_o + y)^2}$$

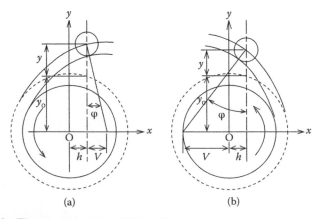

(a)   (b)

FIGURE 3.44   The pressure angle. (a) Rise (b) return.

where "a" is the derivative of v with respect to θ and is termed as the reduced acceleration. It is equal to the acceleration of the follower divided by the square of the angular velocity of the cam. Hence,

$$(y_o + y)\, a - (v - h)\, v = 0$$

The value of y where the maximum pressure angle occurs is y^* and is given by

$$(y_o + y^*) = \frac{(v - h)v}{a}$$

Substituting in Equation 3.53 the maximum pressure angle φ^* is given by

$$\tan \varphi^* = \frac{a}{v} \tag{3.54}$$

Equations 3.53 and 3.54 are shown graphically in Figure 3.45. The relations between v and y are drawn for the rise and the return strokes. Point P with coordinates (y_o, h) is located. The line from point P to any point on the curves makes an angle φ with a line parallel to the x-axis. The maximum pressure angles are obtained by drawing lines from point P tangent to the curves. Let y_r^*, v_r^*, y_t^*, and v_t^* be the values of the displacements and the reduced velocities at the positions of the maximum pressure angles. Then,

$$\tan \varphi_r^* = \frac{v_r^* - h}{y_o + y_r^*} \tag{3.55}$$

$$\tan \varphi_t^* = \frac{v_t^* - h}{y_o + y_t^*} \tag{3.56}$$

Notice that, for the return stroke, the reduced velocity is negative and so is the pressure angle.

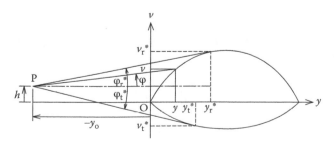

FIGURE 3.45 Graphical representation for determining the minimum cam size.

3.5.1.7.2   Minimum Cam Size and the Amount of Offset

In the design stage, one must assign the maximum values of the pressure angle in order that the cam performs properly. These values are used to obtain the minimum cam size and the associated amount of offset. This may be achieved either graphically or analytically.

1. Graphical method

   Draw the v versus y for the rise and the return strokes.

   Draw a line with an angle φ_r^* tangent to the curve for the rise stroke.

   Draw a line with an angle φ_t^* tangent to the curve for the return stroke.

   The two lines intersect at point P. The coordinates of this point determines y_0 and h. The base circle radius is obtained.

$$r_0 = \sqrt{y_0^2 + h^2} - R$$

   "R" is the radius of the roller.

2. Analytical method

   The first step is to obtain the positions at which the maximum pressure angles occur. This is achieved by solving Equation 3.54.

$$\tan \varphi^* = \frac{a}{v}$$

   The values of y_r^*, v_r^*, y_t^*, and v_t^* are obtained. Equations 3.55 and 3.56 are solved simultaneously to obtain y_0 and h, then obtain r_0.

EXAMPLE 3.5

- The rise angle $\beta_r = 120°$.
- The upper dwell angle $\beta_{ud} = 30°$.
- The return angle $\beta_t = 90°$.
- The lift is 40 mm.
- The motion of the follower is harmonic during the rise and the return.
- The roller radius $R = 10$ mm.

ANALYSIS

The motion of the follower during the rise and the return strokes is given as follows.

   For the rise,

$$y_r = 20\left[1 - \cos\frac{3\theta}{2}\right]$$

$$v_r = 30\sin\frac{3\theta}{2}$$

$$a_r = 45\cos\frac{3\theta}{2}$$

For the return,

$$y_t = 20\left[1 - \cos 2(240 - \theta)\right]$$

$$v_t = -40\sin 2(240 - \theta)$$

$$a_t = 80\cos 2(240 - \theta)$$

The positions at which maximum pressure angles occur:
For the rise,

$$\tan 30 = \frac{a_r}{v_r}$$

$$\frac{1}{\sqrt{3}} = \frac{45\cos\dfrac{3\theta}{2}}{30\sin\dfrac{3\theta}{2}}$$

This equation gives $\theta^* = 46°$. Thus,

$$y_r^* = 12.81$$
$$v_r^* = 28.99$$

For the return,

$$\tan(-30) = \frac{a_t}{v_t}$$

$$-\frac{1}{\sqrt{3}} = \frac{80\cos 2(240 - \theta)}{-40\sin 2(240 - \theta)}$$

This equation gives $\theta^* = 203°$. Thus,

$$y_t^* = 14.46$$
$$v_t^* = 38.43$$

Solving Equations 3.55 and 3.56 simultaneously,

$$v_o = 43.9\,\text{mm}$$
$$h = -4.74\,\text{mm}$$
$$r_o = 44.16\,\text{mm}$$

3.5.1.7.3 Radius of Curvature

When the cam size is reduced or the lift is increased, a certain region of the cam profile gets sharper. The radius of curvature approaches zero at certain critical cases as shown in Figure 3.46. This makes the cam loses its function. This is why it is important to design cams according to the minimum radius of curvature.

The radius of curvature ρ, in Cartesian coordinates, is given by

$$\rho = \frac{(\mathbf{r}' \cdot \mathbf{r}')^{3/2}}{\left[(\mathbf{r}' \times \mathbf{r}'') \cdot (\mathbf{r}' \times \mathbf{r}')\right]^{1/2}} \tag{3.57}$$

$$\mathbf{r} = (\xi + i\eta)\, e^{-i\theta}$$

$$\mathbf{r}' = \left[(\xi' + \eta) + i(\eta' - \xi)\right] e^{-i\theta}$$

$$\mathbf{r}'' = \left[(\xi'' + 2\eta' - \xi) + i(\eta'' - 2\xi' - \eta)\right] e^{-i\theta}$$

Substituting in Equation 3.56,

$$\rho = \frac{\left[(\xi' + \eta)^2 + (\eta' - \xi)^2\right]^{3/2}}{\left|(\xi' + \eta)(\eta'' - 2\xi' - \eta) - (\eta' - \xi)(\xi'' + 2\eta' - \xi)\right|} \tag{3.58}$$

It can be proven that $e^{-i\theta}$ does not affect Equation 3.58. The minimum radius of curvature depends on the cam geometry, the follower geometry, besides the type of motion.

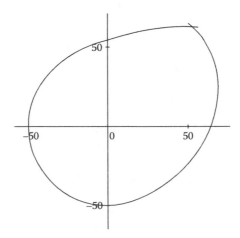

FIGURE 3.46 Interference in cams.

3.5.1.7.3.1 Translating Roller Followers   According to Section 3.5.1.6.1, for translating roller follower,

$$\xi = h, \xi' = 0, \xi'' = 0$$
$$\eta = y_o + y, \eta' = y', \eta'' = y''$$

Therefore,

$$\rho = \frac{\left[(y_o + y)^2 + (y' - h)^2\right]^{3/2}}{\left|(y_o + y)[y'' - (y_o + y)] - (y' - h)(2\,y' - h)\right|}$$

3.5.1.7.3.2 Oscillating Roller Follower   According to Section 3.5.1.6.2, for oscillating roller follower,

$$\xi = S - L\cos(\varphi_o + \varphi), \xi' = L\sin(\varphi_o + \varphi)\varphi',$$
$$\xi'' = L\left[\cos(\varphi_o + \varphi)\,\varphi' + \sin(\varphi_o + \varphi)\,\varphi''\right]$$
$$\eta = L\sin(\varphi_o + \varphi), \eta' = L\cos(\varphi_o + \varphi)\,\varphi'$$
$$\eta'' = L\left[\cos(\varphi_o + \varphi)\,\varphi'' - \sin(\varphi_o + \varphi)\right]$$

$$\rho = \frac{\left[L^2(1+\varphi')^2 + S^2 - 2LS(1+\varphi')\varphi''\right]^{3/2}}{L^2(1+\varphi')(1+3\varphi') - S^2 + SL\left[\sin(\varphi_o + \varphi)\varphi'' + \cos(\varphi_o + \varphi)\varphi' + 2\cos(\varphi_o + \varphi)\right]}$$

3.5.1.7.3.3 Translating Flat-Faced Follower   According to Section 3.5.1.6.3, for translating flat-faced follower,

$$\eta = r_o + y, \eta' = y', \eta'' = y''$$
$$\xi = y', \xi' = y'', \xi'' = y'''$$

$$\rho = r_o + y + y''$$

The proper cam size is obtained by minimizing the values of the radii of curvatures and then determining the value of the base circle, which satisfies the design requirements.

EXAMPLE 3.6

Find the base circle radius for a cam actuating a translating flat-faced follower with a simple harmonic motion. The rise angle is 120°, the return angle is 90°, and the rise is 50 mm if the minimum radius of curvature is 30 mm.

ANALYSIS

For the translating flat-faced follower,

$$p = r_o + y + y''$$

For the rise,

$$y_r = 25\left[1 - \cos\frac{3\theta}{2}\right]$$

$$y_r'' = 53.25\cos\frac{3\theta}{2}$$

Thus,

$$p = r_o + 25\left[1 - \cos\frac{3\theta}{2}\right] + 53.25\cos\frac{3\theta}{2}$$

The minimum value of p occurs when $\dfrac{3\theta}{2} = \pi$. Thus,

$$30 = r_o + 50 - 53.52$$
$$r_o = 33.52$$

For the return,

$$y_t = 25\left[1 - \cos\left(2(240 - \theta)\right)\right]$$

$$y_t'' = 100\cos\left(2(240 - \theta)\right)$$

$$p = r_o + 25\left[1 - \cos\left(2(240 - \theta)\right)\right] + 100\cos\left(2(240 - \theta)\right)$$

The minimum value of p occurs when $\dfrac{3\theta}{2} = \pi$. Thus,

$$30 = r_o + 50 - 100$$
$$r_o = 80$$

Therefore, the minimum base circle radius is 80 mm.

3.5.1.7.3.4 Oscillating Flat-Faced Follower   For the oscillating flat-faced follower, the analysis is quite lengthy. The most appropriate way is to use a trial-and-error method.

3.5.2 Specified Contour Cams

As mentioned earlier in Section 3.5, the cam is made of simple geometrical curves such as circular arcs and straight lines that are easy to produce.

3.5.2.1 Circular Cams

It is simply a circular disk that rotates around a pivot, which is eccentric from the geometrical center. The cam may actuate different types of followers.

3.5.2.1.1 Oscillating Roller Follower

For this type of cam follower system,
- The position analysis is presented in Section 1.9.4.1.
- The velocity and acceleration analysis is presented in Section 2.6.8.

3.5.2.1.2 Translating Roller Follower
- The position analysis is presented in Section 1.9.4.2.
- The velocity and acceleration analysis is presented in Section 2.6.9.

3.5.2.1.3 Oscillating Flat-Faced Follower
- The position analysis is presented in Section 1.9.4.3.
- The velocity and acceleration analysis is presented in Section 2.6.10.

3.5.2.1.4 Translating Flat-Faced Follower
- The position analysis is presented in Section 1.9.4.4.
- The velocity and acceleration analysis is presented in Section 2.6.11.

EXAMPLE 3.7

A circular cam with radius $R = 40$ mm (Figure 3.47) is actuating a translating roller follower with radius $r = 20$ mm. The contact is pure rolling. The center of rotation of the cam is 20 mm from its center. The cam rotates with a speed of 600 rpm. Plot the displacement, velocity, and the acceleration of the follower with the cam's rotational angle. Also, plot the angular velocity and angular acceleration of the roller.

ANALYSIS

For the position of point C, we use Equations 1.6 and 1.7

$$e = 20$$
$$x_C = e \cos\theta$$
$$y_C = e \sin\theta$$

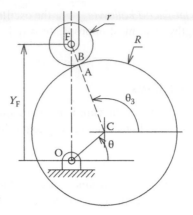

FIGURE 3.47   Circular cam with translating roller follower.

For the position of the follower, use equations given in Section 1.9.4.2.

$$Y_F = y_C + \sqrt{(R+r)^2 - y_C^2}$$

$$y_F = Y_F - (R - e)$$

$$\sin\theta_3 = -\frac{x_C}{R+r}$$

$$\cos\theta_3 = \frac{Y_F - y_C}{R+r}$$

For the velocity of point F,

$$\mathbf{V_A} = i\,\omega(e\,e^{i\theta} + R\,e^{i\theta_3})$$

Thus,

$$V_A^x = \omega\,e\sin\theta + \omega\,R\sin\theta_3$$
$$V_A^y = \omega\,e\cos\theta + \omega\,R\cos\theta_3$$

For point B, since the contact is pure rolling, then $\mathbf{V_B} = \mathbf{V_A}$. For point F,

$$\mathbf{V_F} = \mathbf{V_B} + \mathbf{V_{FB}}$$
$$i\,V_F = V_A^x + i\,V_A^y - \omega_r\,r\sin\theta_3 + i\,\omega_r\,r\cos\theta_3$$

where ω_r is the angular velocity of the roller. Considering the real parts, we get,

$$\omega_r = \frac{V_A^x}{\sin\theta_3}$$

From the imaginary parts,

$$V_F = V_A^y + r \, \omega_r \cos\theta_3 = V_A^y + V_A^x \cot\theta_3$$

For the acceleration,

$$\mathbf{A}_A = -\omega^2 (e \, e^{i\theta} + R \, e^{i\theta_3})$$
$$A_A^x = -\omega^2 e \cos\theta - \omega^2 R \cos\theta_3$$
$$A_A^y = -\omega^2 e \sin\theta - \omega^2 R \sin\theta_3$$

For point B,

$$\mathbf{A}_B = \mathbf{A}_A + \mathbf{A}_{BA}^n + \mathbf{A}_{BA}^{SL}$$

For pure rolling, $\mathbf{A}_{BA}^{SL} = 0$

$$\mathbf{A}_{BA}^n = \frac{R r (\omega - \omega_r)^2}{R+r} e^{i\theta_3}$$

For point F,

$$\mathbf{A}_F = \mathbf{A}_B + \mathbf{A}_{FB}$$

$$iA_F = A_A^x + iA_A^y + \frac{R r (\omega - \omega_r)^2}{R+r} e^{i\theta_3} + r(-\omega_r^2 + i\alpha_r) e^{i\theta_3}$$

From the real parts,

$$0 = A_A^x + \left(\frac{R r (\omega - \omega_r)^2}{R+r} - r\omega_r^2 \right) \cos\theta_3 - r\alpha_r \sin\theta_3$$

$$\alpha_r = \frac{A_A^x}{r \sin\theta_3} + \left(\frac{R r (\omega^2 - 2\omega\omega_r - r^2\omega_r^2)}{r(R+r)} \right) \cot\theta_3$$

From the imaginary parts,

$$A_F = A_A^y + \left(\frac{R r (\omega^2 - 2\omega\omega_r - r^2\omega_r^2)}{r(R+r)} \right) \sin\theta_3 + r\alpha_r \cos\theta_3$$

The plots of the position, velocity, and the angular acceleration of the follower are shown in Figure 3.48.

The plots of the angular velocity and the angular acceleration of the roller are shown in Figure 3.49.

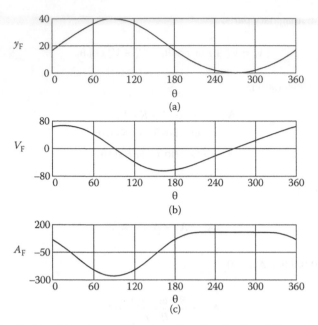

FIGURE 3.48 Position (a), velocity (b), and acceleration (c) of the follower.

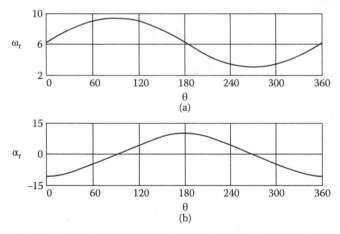

FIGURE 3.49 Angular velocity (a) and angular acceleration (b) of the roller.

3.5.2.2 Circular Arc Cams

The cam is formed from four arcs of circles (Figure 3.50):

1. The base circle with center at O and with radius r_o.
2. The flank circle with center at F and with radius R.
3. The nose circle with center at N and with radius n.
4. The upper dwell circle with center at O and with radius $r_o + s$; s is the lift.
5. The roller radius r.

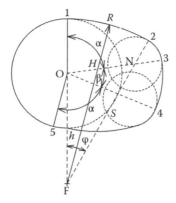

FIGURE 3.50 Circular arc cam.

3.5.2.2.1 Geometry of the Circular Arc Cam

Let h be the distance between O and F, H be the distance between O and N, and S be the distance between F and N. The geometrical relations of the cam are given by,

$$S = R - n$$
$$h = R - r_0$$
$$H = s + r_0 - n$$
$$S^2 = H^2 + h^2 + 2Hh\cos\alpha$$

$$\frac{H}{\sin\varphi} = \frac{S}{\sin\alpha} = \frac{h}{\sin(\alpha - \varphi)}$$

3.5.2.2.2 Action of the Circular Arc Cam

At the beginning of motion, the follower is considered to be in contact with the cam at point 1 which is the tangent point of the base and the flank circles. The follower is in contact with the flank until point 2, which represents the end of the flank contact and the beginning of the nose contact. At point 2, the cam has rotated through an angle φ. The period at which the follower is contact with the flank is called the acceleration period. The follower is in contact with the nose until point 3, which represents the end of the rise for an angle equal to $\alpha - \varphi$; α is the rise angle. This is called the deceleration period. The follower is in contact with the upper dwell until point 4 for an angle β. The follower returns back from point 4 to point 5. The angle $\sigma = 2\alpha + \beta$ is the called the angle of action.

3.5.2.2.3 Kinematics of the Circular Arc Cam

The circular arc cam operates with all types of followers.

3.5.2.2.3.1 Translating Roller Follower For this type of cam follower system, the position analysis, the velocity, and the acceleration analyses are presented in Example 3.7. However, there is some change in the equations, which must be considered.

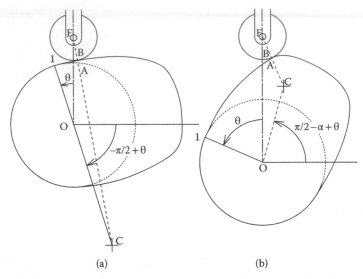

FIGURE 3.51 Contacts for the circular arc cam with translating roller follower. (a) Flank contact (b) nose contact.

For the flank contact, $0 \leq \theta \leq \varphi$ (Figure 3.51a),

- R is replaced by the flank radius.
- e is replaced h.
- θ is replaced by $-\pi/2 + \theta$.

For the nose contact, $\varphi \leq \theta \leq \alpha$ (Figure 3.51b),

- R is replaced by the nose radius n.
- e is replaced H.
- θ is replaced by $\pi/2 - \alpha + \theta$.

For the return stroke, $\alpha + \beta \leq \theta \leq \sigma$, use the equations for the rise and change θ by $\sigma - \theta$.

3.5.2.2.3.2 Translating Flat-Faced Follower   For this type of cam follower system, the position analysis, the velocity, and the acceleration analyses are presented in Sections 1.9.4.4 and 2.6.11. Considerations should be taken for base circle, flank, and nose contacts. This demonstrated by the following example.

EXAMPLE 3.8

A circular arc cam actuates a translating flat-faced follower. The radius of the base circle is $r_o = 40$ mm, the radius of the nose circle is $n = 20$ mm, the lift is $s = 30$ mm, and the rise angle is $\alpha = 90°$. The upper dwell angle is $0°$. Plot the displacement, velocity, and acceleration of the follower against the cam angle if the cam rotates at 600 rpm.

SOLUTION

$$H = s - n + r_o$$
$$= 30 - 20 + 40 = 50\,\text{mm}$$
$$S^2 = H^2 + h^2 + 2H\,h\cos\alpha$$

But,

$$S = R - n = R - 20$$
$$h = R - r_o = R - 40$$

Thus,

$$(R - 20)^2 = 50^2 + (R - 40)^2$$
$$R = 92.5\,\text{mm}$$
$$h = 52.5\,\text{mm}$$
$$S = 72.5\,\text{mm}$$

$$\frac{50}{\sin\varphi} = \frac{72.5}{\sin 90}$$

$$\varphi = 43.6°$$
$$\sigma = 210°$$

Referring to Sections 1.9.4.4 and 2.6.11, the motion of the follower is represented by,

$$y_F = R + e\sin\theta - r_o$$
$$V_F = \omega\,e\cos\theta$$
$$A_F = -\omega^2\,e\sin\theta$$

For $0 \le \theta \le 43.6°$, e is replaced by h and θ is replaced by $-\pi/2 + \theta$. Thus,

$$y_F = 52.5\sin(\theta - \pi/2) = 52.5(1 - \cos\theta)$$
$$V_F = 52.5\,\omega\sin\theta$$
$$A_F = 52.5\,\omega^2\cos\theta$$

For $43.6° \le \theta \le 90.0°$, R is replaced by n, e is replaced by h, and θ is replaced by $\pi/2 + \theta - \alpha$. Thus,

$$x_F = 50.0\sin(90.0 + \theta - 90.0) + 30.0 - 40.0$$
$$= 50.0\sin\theta - 10.0$$
$$V_F = 50.0\,\omega\cos\theta$$
$$A_F = -50.0\,\omega^2\sin\theta$$

For $90.0° \leq \theta \leq 120.0°$,

$$x_F = 30$$
$$V_F = 0$$
$$A_F = 0$$

For the return stroke, we replace θ in the rise by $\sigma - \theta$. Thus,
For $120.0° \leq \theta \leq 166.4°$,

$$x_F = 50.0\sin(210.0 - \theta) - 10.0$$
$$V_F = -50.0\,\omega\cos(210.0 - \theta)$$
$$A_F = -50.0\,\omega^2\sin(210.0 - \theta)$$

For $166.4° \leq \theta \leq 210.0°$,

$$y_F = 52.5\,[1 - \cos(210.0 - \theta)]$$
$$V_F = -52.5\,\omega\sin(210.0 - \theta)$$
$$A_F = 52.5\,\omega^2\cos(210.0 - \theta)$$

The displacement, velocity, and acceleration of the follower are shown in Figure 3.52.

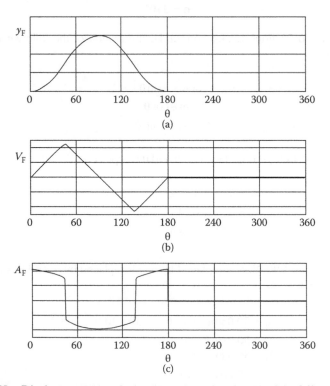

FIGURE 3.52   Displacement (a), velocity (b), and acceleration (c) of the follower.

3.5.2.3 Tangent Cams

The tangent cam is sometimes called the straight-sided cam because its flank is straight as shown in Figure 3.53. It is exactly similar to the circular arc cam discussed in Section 3.5.2.2 except that the radius of the flank is infinite.

The terminology of the cam is listed as follows:

1. H is the distance between centers of the base and the nose circle.
2. r_o is the radius of the base circle.
3. n is the radius of the nose circle.
4. r is the radius of the roller.
5. φ is the cam angle when the roller is in contact with the end of the flank and the beginning of the nose.
6. α is the rise angle.
7. β is the upper dwell angle.
8. σ is the angle of action.

The geometrical relations of the cam are

$$\cos\alpha = \frac{r_o - n}{h}$$

$$\tan\varphi = \frac{H\sin\alpha}{r_o + r}$$

$$\text{lift} = H + r + n - r_o$$

At the beginning of the rise, the roller center is at point 1. When the cam rotates an angle φ, the center is at 2 and the follower is in contact with the end of the flank and the beginning of the nose. When the cam rotates an angle α, the follower is at the end of the rise and the roller center is at point 3. The follower motion is described by:

For $0 \le \theta \le \varphi$, for position analysis, consider the vector loop,

$$[s + i(r_o + r)]\, e^{i\theta} = i(r_o + r + y_F)$$

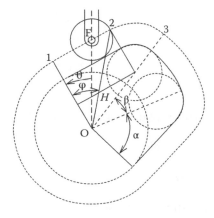

FIGURE 3.53 Tangent cam.

From the real parts and the imaginary parts, one gets,

$$s = (r_o + r) \tan \theta$$
$$y_F = (r_o + r)(\sec \theta - 1)$$

For velocity analysis, consider the vector loop,

$$i\omega [s + i(r_o + r)] e^{i\theta} - \omega_r r e^{i\theta} = iV_F$$

$$V_F = \omega (r_o + r)(\sec \theta \tan \theta) \tag{3.59a}$$

$$\omega_r = -\frac{V_F \sin \theta + r_0 \omega}{r} \tag{3.59b}$$

For acceleration analysis, consider the vector loop,

$$(-\omega^2)(s + i\, r_o)\, e^{i\theta} + i\, r\, (\omega - \omega_r)^2\, e^{i\theta} + i\, r\, (-\omega_r^2 + i\, \alpha_r)\, e^{i\theta} = iA_F$$

Multiplying both sides by $e^{i\theta}$ and considering the imaginary parts,

$$A_F = \frac{-r_o\, \omega^2 + r\, (\omega - \omega_r)^2 - r\, \omega_r^2}{\cos \theta}$$

From the real parts,

$$\alpha_r = \frac{-s\, \omega^2 - A_F \sin \theta}{r}$$

An alternative method for obtaining the acceleration of the follower and the angular acceleration of the roller can be obtained by differentiating Equations 3.59a and b with respect to time.

$$A_F = \omega^2 (r_o + r) \frac{1 + \sin^2 \theta}{\cos^3 \theta}$$

$$\alpha_r = \omega^2 (r_o + r) \frac{2 \sin \theta}{r \cos^3 \theta}$$

For $\varphi \leq \theta \leq \alpha$, the equations of the cam in Example 3.7 are used.

- R is replaced by the nose radius n.
- e is replaced H.
- θ is replaced by $\pi/2 - \alpha + \theta$.

For the return stroke, $\alpha + \beta \leq \theta \leq \sigma$, use the equations for the rise and replace θ by $\sigma - \theta$.

EXAMPLE 3.9

A tangent cam actuates a translating roller follower. The radius of the base circle is $r_o = 40$ mm, the radius of the nose circle is $n = 20$ mm, the roller radius is 20 mm, and the lift is $s = 30$ mm. Plot the displacement, velocity, and acceleration of the follower against the cam angle if the cam rotates at 600 rpm. Also, plot the angular velocity and the angular acceleration of the roller. There is no upper dwell.

Analysis

$$H = 30 + 40 - 20 = 50\,\text{mm}$$

$$\cos\alpha = \frac{r_o - n}{h} = 33.4°$$

$$\tan\varphi = \frac{H\sin\alpha}{r_o + r} = 66.4°$$

The displacement, velocity, and acceleration of the follower are shown in Figure 3.54. The angular velocity and the angular acceleration of the roller are shown in Figure 3.55.

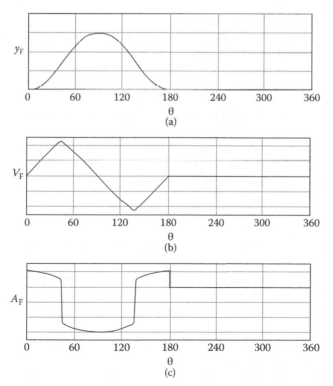

FIGURE 3.54 Displacement (a), velocity (b), and acceleration (c) of the translating follower of the tangent cam.

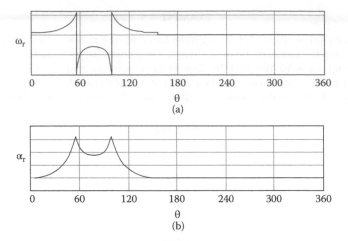

FIGURE 3.55   Angular velocity (a) and angular acceleration (b) of the roller.

PROBLEMS

3.1 The following data is provided for a disk cam.
   - The rise, upper dwell, and return angles are 90°, 30°, and 120° respectively.
   - The base circle radius is 30 mm.
   - The motion during the rise is cycloid and during the return is parabolic. The amount of lift is 50 mm.

   Draw the cam profile for the following translating followers:
   a. Radial knife edge.
   b   Radial roller; the roller radius is 50 mm.
   c. Offset roller; the roller radius is 1 in., and the amount of offset is 10 mm.
   d. Flat-faced.

3.2 A disk cam actuates an oscillating follower. The maximum swinging angle is 15°. The follower motion during the rise is sample harmonic and during the return is parabolic. The following data is given.
   - The rise, upper dwell, and return angles are 120°, 0°, and 120° respectively.
   - The radius of the base circle is 50 mm.
   - The distance between the cam center and the follower pivot is 150 mm.

   Draw the cam profile for the following cases:
   a. The follower has a roller at the end with radius 30 mm and the length of the follower is 180 mm.
   b. The follower has a flat face; the pivot center is 15 mm away from the face.

3.3 In a cam with a reciprocating roller follower, the base circle diameter is 80 mm, the roller diameter is 40 mm, the amount of offset is 15 mm (positive for the rise), the lift is 30 mm, and the rise and the return angles are 120° each. Determine the values of the maximum pressure angles during the rise and the return for the following follower motions.

Rise	Return
a. Modified uniform motion	Modified uniform motion
b. Simple harmonic	Simple harmonic
e. Parabolic	Parabolic
d. Cycloid	Cycloid
e. Simple harmonic	Parabolic
f. Simple harmonic	Cycloid
g. Parabolic	Simple harmonic
h. Parabolic	Cycloid
i. Cycloid	Simple harmonic
j. Cycloid	Parabolic

3.4 For Problem 3.3, plot curves to show the effect of the base circle diameter on the maximum pressure angles (vary the bass circle diameter and fix the other parameters).

3.5 For Problem 3.3, plot curves to show the effect of the rise and the return angles on the maximum pressure angles.

3.6 For Problem 3.3, show the effect of the lift on the maximum pressure angles.

3.7 For Problem 3.3, find the values of the maximum pressure angles when the mount of offset is zero.

3.8 A disk cam displaces a translating roller follower. The rise and the return angles are 90° each, the lift is 30 mm, and the roller radius is 15 mm. If the maximum value of the pressure angle is 30°, determine the minimum, base circle diameter, and the associated amount of offset for the follower motions listed in Problem 3.3.

3.9 Use the data given in Problem 3.8 to plot curves to shown the effect of the rise and the return angles on the minimum diameter of the base circle.

3.10 Use the data given in Problem 3.8; plot a curve to show the effect of the lift on the minimum base circle diameter.

3.11 A disk cam displaces a translating roller follower. The rise and the return angles are 90° each, the lift is 30 mm, and the roller radius is 15 mm. If the minimum value of the radius of curvature is 30 mm, determine the minimum value of the base circle radius for the follower motions listed in Problem 3.3.

3.12 A disk cam displaces a translating flat-faced follower. The rise and the return angles are 90° each and the lift is 30 mm. If the minimum value of the radius of curvature is 30 mm, determine the minimum value of the base circle radius for the three basic follower motions (simple harmonic, parabolic, and cycloid).

3.13 A disk cam actuates an oscillating follower. The maximum swinging angle is 15°. The following data is given:
- The rise and the return angles are 120°.
- The distance between the cam center and the follower pivot is 150 mm.

Find the minimum value of the base circle radius if the minimum radius of curvature is not less than 30 mm for the three basic follower motions (simple harmonic, parabolic, and cycloid).

a. The follower has a roller at the end with radius 30 mm and the length of the follower is 180 mm.

b. The follower has a flat face; the pivot center is 15 mm away from the face.

3.14 A circular cam with 150 mm diameter actuates a translating flat-faced follower. The amount of eccentricity is 75.0 mm. Plot the follower displacement, reduced velocity, and reduced acceleration against the cam rotational angle.

3.15 A circular cam with 120 mm diameter actuates a translating roller follower; the diameter of the roller is 30 mm. The amount of eccentricity is 50.0 mm Plot the follower displacement, reduced velocity, and reduced acceleration against the cam rotational angle if the amount of offset of the follower is 10 mm.

3.16 For a circular arc cam with a flat reciprocating follower, the base circle diameter is 50 mm, the nose radius is 20 mm, the lift is 50 mm, and the angle of action is 180°. There is no upper dwell. Plot the displacement, the reduced velocity, and the reduced acceleration and with the cam rotational angle.

3.17 For a circular arc cam with a flat-faced reciprocating follower, the angle of action is 150°, the lift is 25 mm, the base circle diameter is 125 mm, and the period of acceleration is one half that of the retardation. The upper dwell angle is 30.0°. Determine the flank and nose radii. Also, plot the displacement, the reduced velocity, and the reduced acceleration and with the cam rotational angle.

3.18 A tangent cam has a radial roller follower of radius 175 mm. The radius of the base circle is 150 mm, the radius of the nose circle is 50 mm, and the lift is 90 mm. There is no upper dwell. If the speed of the cam is 750 rpm, plot the displacement, the reduced velocity, and the reduced acceleration and with the cam rotational angle.

3.19 For a tangent cam with a radial roller follower, the diameter of the base circle is 175 mm, the lift is 25 mm, the angle of action is 120°, and the roller radius is 50 mm. The upper dwell angle is 30.0°. Plot the follower velocity and acceleration with the cam angle if the cam speed is 120 rpm.

4 Spur Gears

4.1 INTRODUCTION

The transmission of motion and/or power from one shaft to another is dealt with in the design of almost every machine or instrument. It is required that the angular velocities of the two shafts should remain constant.

The simplest mechanism that fulfills this requirement is a pair of cylinders in pure rolling contact as shown in Figure 4.1. As long as there is no slip between the cylinders, the drive will be satisfactory. However, this is not always possible, and the combinations of speeds, loads, and friction forces may demand an absolutely uniform speed ratio that requires a positive type of contact instead of depending on the friction alone. The motion is transmitted by gears, as the teeth on one gear push the teeth on the other.

A photo for commonly used gears is shown in Figure 4.2 (Courtesy of PCS Education Systems, Inc).

4.2 GEAR CLASSIFICATION

Gears may be classified according to the relative position of the axes of revolution. The axes may be as follows:

1. Parallel
2. Intersecting
3. Neither parallel nor intersecting

4.2.1 GEARS CONNECTING PARALLEL SHAFTS

1. *Spur gears.* Spur gears are used to transmit the motion between two parallel shafts. The teeth are parallel to the axes of rotation and the speed ratio is limited. However, high speed reduction can be obtained by using gear trains. They are not recommended for high-speed applications due to noise. The smaller gear is called a pinion, whereas the larger gear is simply called a gear. They are available as external gears, Figure 4.3a, and internal gears, Figure 4.3b.
2. *Helical gears.* A helical gear, Figure 4.4, is a cylindrical-shaped gear with helicoids teeth. Helical gears operate with less noise and vibration than spur gears. At any time, the load on the helical gears is distributed over several teeth, resulting in reduced wear. Due to their angular cut, teeth meshing results in thrust loads along the gear shaft. This action requires thrust bearings to absorb the thrust load and maintain gear alignment. They are widely

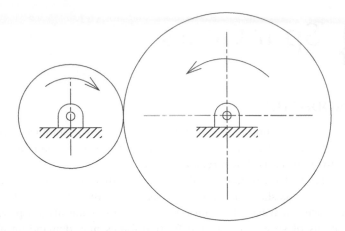

FIGURE 4.1   A pair of rolling cylinders.

FIGURE 4.2   Commonly used gears. (a) Spur gear (b) helical or spiral gears (c) worm and worm gear (d) spiral bevel gears (e) bevel gear.

used in industry, and the only drawback is the axial thrust force due to the helix form of the teeth.
3. *Herringbone gears.* To eliminate the axial thrust, Herringbone gears are manufactured as two helical gears with opposite helix angles, Figure 4.5.
4. *Rack* and *pinion gears.* In order to eliminate the axial thrust, Herringbone gears are manufactured as two helical gears with opposite helix angles, Figure 4.6.

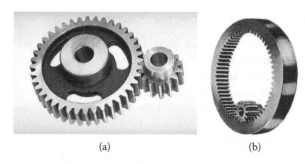

(a)   (b)

FIGURE 4.3   Spur gears. (a) External (b) internal.

FIGURE 4.4   A pair of helical gears.

FIGURE 4.5   Herringbone gears.

FIGURE 4.6   Pinion and rack set.

4.2.2 GEARS CONNECTING INTERSECTING SHAFTS

1. *Straight bevel gears.* They are used, in most applications, to transmit the motion between two perpendicular shafts, Figure 4.7. However, the angle between the shafts could vary between 90° and 180°. The teeth of each gear are parallel to its axis of rotation.
2. *Spiral bevel gears.* They are essentially bevel gears with spiral teeth form, Figure 4.8.
3. *Face gears.* In this type, the teeth of the gear are cut in the face of a cylinder, Figure 4.9. Their teeth may be straight or tapered toward the center of the gear.

FIGURE 4.7   Straight bevel gear set.

FIGURE 4.8   Spiral bevel gear set.

4.2.3 GEARS CONNECTING NONPARALLEL, NONINTERSECTING SHAFTS

1. *Crossed-helical gears.* They are called skew gears, Figure 4.10. They are used to transmit the motion between nonparallel and nonintersecting shafts.
2. *Hypoid gears.* This type is actually helical bevel gears mounted on two perpendicular shafts, but the axes of the shafts do not intersect, Figure 4.11. They are widely used in the differential system of automobile drives. Special attention should be paid to stand high tooth pressures and the rubbing action between the mating teeth.
3. *Worm and worm gear.* This type is used to obtain a high speed reduction, Figure 4.12. However, a considerable amount of power is lost due to the friction between the teeth. For this reason, the materials of the worm and the gear should be selected to be as low coefficient of friction as possible. Besides, heavy oil should be used in lubrication to prevent metal to metal contact. The tooth of the worm is like ACME screw thread, whereas the mating gear is helical.

FIGURE 4.9 Face gears.

FIGURE 4.10 Crossed helical gear set.

FIGURE 4.11   Hypoid gear set.

FIGURE 4.12   Worm and worm gear set.

4.3   GEAR-TOOTH ACTION

4.3.1   FUNDAMENTAL LAW OF GEARING

A pair of mating gear teeth acting against each other to produce rotary motion may be represented by the mechanism shown in Figure 4.13. Point C is the point of contact of the pair of teeth. Line N_2N_3 is the common normal to the surfaces and intersects the line of centers OQ at point P. Lines ON_2 and QN_3 are perpendicular to the line N_2N_3. Gear (2) rotates with angular velocity ω_2, whereas gear (3) rotates with angular velocity ω_3. The velocity of point C on gear 2 is V_2 and its velocity on gear 3 is V_3, such that,

$$V_2 = OC * \omega_2$$

$$V_3 = QC * \omega_3$$

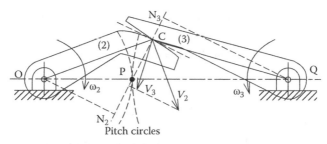

Pitch circles

FIGURE 4.13 Representation of teeth contact.

In order to ensure that the two teeth do not separate, the components of the two velocities along the common normal are the same. Thus,

$$ON_2 * \omega_2 = QN_3 * \omega_3$$

or,

$$\frac{\omega_3}{\omega_2} = \frac{ON_2}{QN_3}$$

Triangles ON_2P and QN_3P are similar. Therefore,

$$\frac{\omega_3}{\omega_2} = \frac{ON_2}{QN_3} = \frac{OP}{QP} \qquad (4.1)$$

Point P is very important to the velocity ratio, and it is called the pitch point. For a constant velocity ratio, the position of P should remain unchanged. In this case, the motion transmission between two gears is equivalent to the motion transmission between two imagined slipless cylinders with radii OP and QP. These two circles are termed as the pitch circles and are tangent at the pitch point P. The velocity ratio is equal to the inverse ratio of the diameters of pitch circles.

The fundamental law of gearing may now be stated as:

For gears with fixed center distance, the common normal to the tooth profiles at the point of contact must always pass through a fixed point (the pitch point) on the line of centers to get a constant velocity ratio.

The components of V_2 and V_3 along the common tangent are given by,

$$V_2^t = \omega_2 * N_2C$$
$$V_3^t = \omega_2 * N_3C$$

The sliding velocity between the two teeth is the difference between the two components along the common tangent, and it is given by,

$$V^s = V_2^t - V_3^t$$
$$= \omega_2 * N_2C - \omega_3 * N_3C$$
$$= \omega_2 * (N_2P + PC) - \omega_3 * (N_3P - PC)$$

But according to Equation 4.1 and the similarity of the two triangles ON_2P and QN_3P,

$$\omega_2 * N_2P = \omega_3 * N_3P$$

Therefore,

$$V^s = PC * (\omega_2 + \omega_3)$$

The analysis considers that ω_2 and ω_3 are in opposite directions.

4.3.2 CONJUGATE PROFILES

To obtain constant speed ratio of two tooth profiles, their common normal must pass through the corresponding pitch point, which is decided by the velocity ratio. The two profiles that satisfy this requirement are called conjugate profiles.

Although many tooth shapes are possible, for which a mating tooth could be designed to satisfy the fundamental law of gearing, only two are in general use, which are the involute and the cycloid profiles. The involute has important advantages. It is easy to manufacture and the center distance between a pair of involute gears can be varied without changing the velocity ratio. Thus, close tolerances between shaft locations are not required when using the involute profile. For these reasons, the most commonly used conjugate tooth profile is the involute.

4.3.3 INVOLUTES PROFILE

The involute of a circle is the curve described by the end of a taut string, as it is unwrapped from a stationary cylinder, Figure 4.14.

4.3.3.1 Properties of the Involute Curve

1. The distance BC on the string is equal to the arc AB.
2. At any instant, point B represents an instant center of rotation for the string. The path of point C is perpendicular to BC, that is, BC is normal to the tangent of the involute at point C.

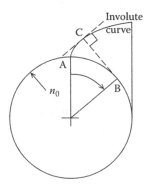

FIGURE 4.14   Involute curve.

3. The normal at any point of the involute is tangent to the circle representing the cylinder. This is because the string is tangent to the circle. This circle is called the base circle.
4. There is no involute curve within the base circle.

4.3.3.2 Involute Tooth Profile Satisfies the Fundamental Law of Gearing

Figure 4.15 shows an inversion of the system presented in Figure 4.14. When the tip of the taut string is pulled as it passes over a fixed pin while the cylinder is allowed to rotate. Point C on the string traces an involute on a plane attached to the cylinder.

Suppose that the string in Figure 4.15, instead of passing over a fixed pin, is wound over another rotating cylinder in the reverse direction, Figure 4.16. When cylinder (2) rotates, it pulls the string, causing cylinder (3) to rotate in the opposite direction. The string is always tangent to the two cylinders and intersects the line of centers OQ at a fixed point P.

If a steel plate is attached to each cylinder, and as movement takes place, any point on the string will trace simultaneously an involute on each plate. Suppose that the plates are cut along the curves, the involutes can be brought into contact as shown in Figure 4.16. No matter what points of the involutes are in contact the normal to the curves is tangent to the base circles.

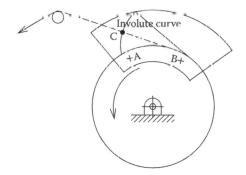

FIGURE 4.15 Tracing an involute on a rotating plane.

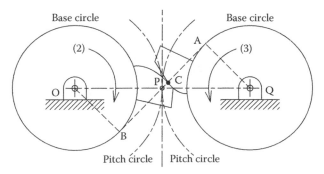

FIGURE 4.16 Formation of involute gear teeth.

It should be noted that a change in the center distance will not have any effect on the shape of the involute. Moreover, the pitch point is still fixed, and thus the law of gearing is satisfied. This is an important advantage of the involute, as stated earlier.

4.3.3.3   Construction of the Involute

Let the requirement be to construct an involute starting at point A on a base circle. The circumference of the circle is divided, as shown in Figure 4.17, into equal arc lengths AB, BC, CD, DE, and so on. The tangents to the circle are drawn at B, C, D, E, and so on. The lengths B1, C2, D3, and E4 are drawn equal to the corresponding arc lengths. A curve passing through the points A, 1, 2, 3, 4, and so on, is the desired involute.

4.3.3.4   Equation of the Involute

The involute function is widely used and is very convenient in gearing calculations. The tangent of the involute is called the pressure line.

Figure 4.18 shows an involute that has been generated from a base circle of radius R_b. The involute contains two points A and B with corresponding radii R_A and R_B and involute pressure angles φ_A and φ_B, respectively.

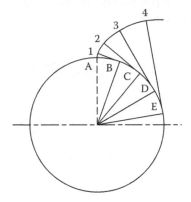

FIGURE 4.17   Construction of involute curve.

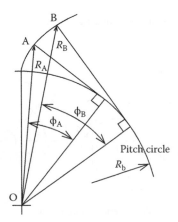

FIGURE 4.18   Representation of involute curve.

$$R_b = R_A \cos \varphi_A$$
$$R_b = R_B \cos \varphi_B$$

Thus,

$$\cos \varphi_B = \frac{R_A}{R_B} \cos \varphi_B \qquad (4.2)$$

From Equation 4.2, it is possible to determine the pressure angle for any point of known radius on the involute.

Figure 4.19 shows a tooth formed from involute curves. From this figure, it is possible to calculate the tooth thickness at any point B if the thickness at a point A is known. This is demonstrated as follows:

Arc DG is equal to the tangent GB. Hence,

$$\hat{DOG} = \frac{arc\ DG}{OG} = \frac{BG}{OG}$$
$$\tan \varphi_B = \frac{BG}{OG}$$

Thus,

$$\hat{DOG} = \tan \varphi_B$$

Also,

$$\hat{DOB} = \hat{DOG} - \varphi_B$$
$$= \tan \varphi_B - \varphi_B$$

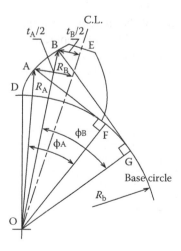

FIGURE 4.19   Thickness of an involute tooth.

Similarly,

$$D\hat{O}A = \tan\varphi_A - \varphi_A$$

The expression $(\tan\varphi-\varphi)$ is called involute function, and sometimes it is written as inv φ. Referring to Figure 4.19,

$$D\hat{O}E = D\hat{O}B + \frac{t_B}{2R_B}$$

$$= \text{inv}\,\varphi_B + \frac{t_B}{2R_B}$$

Also,

$$D\hat{O}E = D\hat{O}A + \frac{t_A}{2R_A}$$

$$= \text{inv}\,\varphi_A + \frac{t_A}{2R_A}$$

Therefore,

$$t_B = 2R_B\left[\frac{t_A}{2R_A} + \text{inv}\,\varphi_A - \text{inv}\,\varphi_B\right]$$

The following chart, Figure 4.20, is used to get the values of inv φ function and the value of φ.

4.4 TERMINOLOGY FOR SPUR GEARS

Figure 4.21 shows some of the terms for gears. The following is a list of the terms used in the analysis of spur gears:

Pitch surface: The surface of the imaginary rolling cylinder (cone, etc.) that the toothed gear may be considered to replace.

Pitch circle: A right section of the pitch surface.

Addendum circle: A circle bounding the ends of the teeth in the right section of the gear.

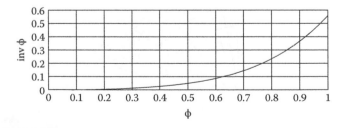

FIGURE 4.20 Determination of the involute function.

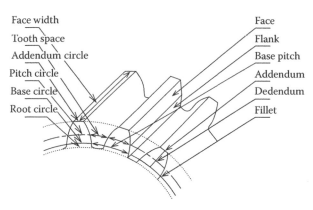

Face width

Tooth space

Addendum circle

Pitch circle

Base circle

Root circle

Face

Flank

Base pitch

Addendum

Dedendum

Fillet

FIGURE 4.21   Terminology of gear teeth.

Root (or dedendum) circle: The circle bounding the spaces between the teeth in the right section of the gear.

Addendum (a): The radial distance between the pitch circle and the addendum circle.

Dedendum (b): The radial distance between the pitch circle and the root circle.

Clearance: The difference between the dedendum of one gear and the addendum of the mating gear.

Face of a tooth: The part of the tooth surface lying outside the pitch surface.

Flank of a tooth: The part of the tooth surface lying inside the pitch surface.

Circular thickness (also called the tooth thickness): The thickness of the tooth measured on the pitch circle. It is the length of an arc and not the length of a straight line.

Tooth space: The distance between the adjacent teeth measured on the pitch circle.

Circular pitch (p): The width of a tooth and a space measured on the pitch circle. It is equal to the circular thickness and the tooth space. It is the length of the arc on the pitch circle between two adjacent teeth.

Diametral pitch (P): The number of teeth of a gear per inch of its pitch diameter. A toothed gear must have an integral number of teeth. The circular pitch, therefore, is equal to the pitch circumference divided by the number of teeth. The diametral pitch is, by definition, the number of teeth divided by the pitch diameter. That is,

$$p = \frac{\pi D}{N} \tag{4.3}$$

$$P = \frac{N}{D} \tag{4.4}$$

$$pP = \pi \tag{4.5}$$

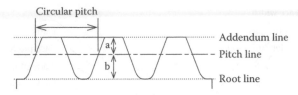

FIGURE 4.22   A rack.

where,

    p = circular pitch
    P = diametral pitch
    N = number of teeth
    D = pitch diameter

> *Module* (m): Pitch diameter divided by the number of teeth. The pitch diameter
> is usually specified in inches or millimeters; in the former case, the module
> is the inverse of diametral pitch.
> *Fillet*: The small radius that connects the profile of a tooth to the root circle.
> *Base circle*: An imaginary circle used in involute gearing to generate the invo-
> lutes that form the tooth profiles.
> *Base pitch*: The pitch of the teeth measured on the pitch circle.

4.4.1   RACK

The rack, Figure 4.22, is a gear with an infinite radius. The circles of an ordinary
gear are lines in the rack. The base circle is a base line parallel to the pitch line
and could be anywhere. The involute, in this case, is a line with any inclination to
match the mating gear. Also, the circular pitch, the addendum, and the dedendum are
decided according to the matching gear.

4.5   ENGAGEMENT ACTION

4.5.1   DEFINITIONS

> *Pinion*: The smaller gear of any pair of mating gears. The larger of the pair is
> simply called the gear.
> *Velocity ratio*: The ratio of the number of revolutions of the driving (or input)
> gear to the number of revolutions of the driven (or output) gear per time.
> *Pitch point*: The point of tangency of the pitch circles a pair of mating gears.
> *Common tangent*: The line tangent to the pitch circles at the pitch point.
> *Line of action*: A line normal to a pair of mating tooth profiles at their point of
> contact. It is tangent to the base circles of the two gears.
> *Path of contact*: The path traced by the contact point of a pair of tooth profiles.

Pressure angle (φ): The angle between the common normal at the point of tooth contact and the common tangent to the pitch circles. It is also the angle between the line of action and the common tangent.

Backlash: The difference between the circle thickness of one gear and the tooth space of the mating gear.

4.5.2 Interchangeability of Gears

A group of gears is said to be interchangeable when any two gears of the group mesh and fulfill the fundamental law of gearing. In such a group, the following conditions must be satisfied:

1. All gears must have the same diametral pitch or module.
2. All gears must have the same pressure angle.
3. The addendum "*a*" and the dedendum "*b*" must be equal.
4. The thickness of the teeth must be same and is equal to one half of the circular pitch.

4.5.3 Standard Tooth System of Spur Gears

The gears are classified according to the pressure angle. The standard values of the pressure angles are 14.5° and 20°. However, in some cases, 22.5° and 25° pressure angles are used to allow using smaller number of teeth that cause problems, which will be explained later. Tables 4.1 and 4.2 list the standard tooth system for spur gears according to American Gear Manufacturing Association (AGMA).

TABLE 4.1
Tooth Proportions

Tooth System	Pressure Angle φ (degrees)	Addendum, *a*	Dedendum, *b*
Full depth	20	$\dfrac{1}{P}$ or 1.0 m	$\dfrac{1.25}{P}$ or 1.25 m
			$\dfrac{1.35}{P}$ or 1.35 m
	22.5	$\dfrac{1}{P}$ or 1.0 m	$\dfrac{1.25}{P}$ or 1.25 m
			$\dfrac{1.35}{P}$ or 1.35 m
	25	$\dfrac{1}{P}$ or 1.0 m	$\dfrac{1.25}{P}$ or 1.25 m
			$\dfrac{1.35}{P}$ or 1.35 m
Stub	20	$\dfrac{0.8}{P}$ or 0.8 m	$\dfrac{1}{P}$ or 1.0 m

TABLE 4.2
Commonly Used Diametral Pitches

Coarse Pitch	2	2.25	2.5	3	4	6	8	10	12	16
Fine pitch	20	24	32	40	48	64	96	120	150	200

TABLE 4.3
Symbols of Spur Gears

Number of teeth	N
Pitch radius	R
Pitch diameter	D
Outside radius	R_o
Outside diameter	D_o
Base radius	R_b
Face width	F
Addendum	a
Dedendum	b
Circular pitch	p
Base pitch	p_p
Pressure angle	φ
Path of contact	Z
Contact ratio	m_p
Center distance	C
Working depth	h_k
Whole depth	h_t
Tooth thickness	t
Clearance	c
Backlash	B

For the British metric standard, the tooth proportions are the same. The recommended modules in millimeters are 1, 1.25, 1.5, 2, 2.5, 3, 4, 5, 6, 8, 10, 12, 16, 20, 25, 32, 40, and 50.

The symbols used for spur gears are listed in Table 4.3.

4.5.4 NATURE OF CONTACT

Consider a pair of meshing gears, a pinion with radius r is driving a gear with radius R, Figure 4.23. The two pitch circles are tangent at the pitch point P. Point O_2 is the center of rotation of the driving gear, and point O_3 is the center of rotation of the driven gear. Because the profiles of the teeth are involute, their common normal is tangent to the base circles at points A and B. Line AB is the line of action and makes angle φ, the pressure angle, with the tangent of the pitch circles. All contact points between the two teeth are

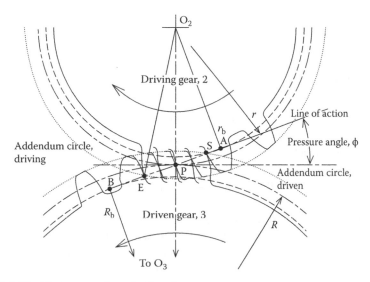

FIGURE 4.23   The nature of a pair of teeth.

along this line, as explained in Section 4.3.3.2. In this case, the first point of contact is point S, Figure 4.23. It is the point of intersection of the addendum circle of the driven gear and the line of action. The contact between the two teeth proceeds diagonally downward to the left along the line of action until the two teeth separate. Point E is the last point of contact, and it is at the intersection of the addendum circle of the driving gear and the line of action. Line SE is called the path of contact. The path of contact is divided into two parts. The first is the path of approach, line SP, and the second is the path of recess, line PE.

The base circle radii are given by,

$$r_b = r \cos \varphi \tag{4.6a}$$

$$R_b = R \cos \varphi \tag{4.6b}$$

The base pitch p_b is defined as the arc on the base circle between two adjacent teeth. The relation between the base pitch p_b and the circular pitch is given by,

$$p_b = p_p \cos \varphi$$

The path of approach, the path of recess, and the path of contact are calculated as follows:

For the path of approach,

$$SP = SB - PB$$

In triangle O_3SB,

$$SP = \sqrt{(R+a)^2 - R_b^2} - R \sin \varphi \tag{4.7}$$

For the path of recess,

$$PE = AE - AP$$

In triangle O_2AE,

$$PE = \sqrt{(r+a)^2 - r_b^2} - r\sin\varphi \qquad (4.8)$$

The path of contact Z is given by,

$$Z = \sqrt{(R+a)^2 - R_b^2} + \sqrt{(r+a)^2 - r_b^2} - C\sin\varphi, \qquad (4.9)$$

where C is the center distance $= R + r$.

Figure 4.24 shows the tooth at the beginning and at the end of contact. Line SE is the path of contact. The arc of contact with length Z_p is the arc on the pitch circle through which a tooth profile moves from the beginning to the end of contact. The arc of contact is the same for both gears. It is divided into two parts. The first is the arc of approach, arc S_pP, from the beginning of contact until the pitch point, and the second is the arc of recess, arc PE_p, from the pitch point till the end of contact. Also, the base arc of contact is the arc on the base circle through which a tooth profile moves from the beginning to the end of contact. The arc of contact and the base arc of contact are the same for both gears.

When the line of action is laid on the base circle, point S moves along the tooth profile passing through point S_p on the pitch circle and ending at point S_b on the base circle. Similarly, Point E traces an involute with point E_p on the pitch circle and point S_b on the base circle. Thus, arc S_bE_b on the base circle with length Z_b is equal to the path of contact SE with length Z. Arc S_pE_p on the pitch circle is the arc of contact with length Z_p. Both arcs include the same radial angle. Thus,

$$\frac{Z_p}{R} = \frac{Z_b}{R_b}$$

The relation between R and R_b is given by Equation 4.6b. Hence,

$$Z_b = Z_p \cos\varphi \qquad (4.10)$$

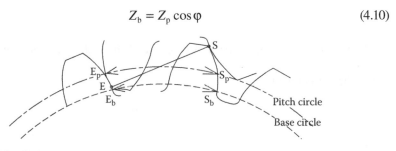

FIGURE 4.24   Path and arc of contact.

Also,

$$Z = Z_p \cos \varphi \tag{4.11}$$

4.5.5 Contact Ratio

The ratio m_c of the arc of contact to the circular pitch is known as the contact ratio. It is also equal to the path of contact divided by the base pitch.

$$m_c = \frac{Z_p}{p} = \frac{Z}{p_b}$$

Physically, the contact ratio is considered to be an indication of the number of teeth in contact. The contact ratio is never less than one. If it is equal to one, it means that a pair of meshing teeth is at the beginning of contact and the leading pair is at the end of contact. This is a critical situation. The contact ratio must be greater than one. If the contact ratio is 2, it means that a pair is at the beginning of contact, a pair is at the end of contact, and one pair is at halfway. Now, if the contact ratio contains a fraction, say 1.5, it does not mean that 1.5 pairs are in contact. It means that the number of pairs in contact is sometimes one and other times two. In general, suppose that the contact ratio is a number with an integer n and a fraction f. Figure 4.25 shows the arc of contact of a gear, arc S_pE_p. Consider the situation when one tooth is at the beginning of contact, point S_p. At the same time, another tooth is at the end of contact, point N. The region S_pN contains $n + 1$ tooth in contact, whereas in the region NE_p, there are no teeth in contact. When the gear rotates, point N moves to E_p. During this interval, $n + 1$ are in contact. After that, n teeth are only in contact. This means that the number of teeth in contact is n or $n + 1$.

$$NE_p = Z_p - S_pN$$
$$= m_p * p - n * p$$

Because $m_p = n + f$, then,

$$NE_p = fp$$

FIGURE 4.25 Arc of contact and contact ratio.

This means that during a period equivalent to "fp" there are $n + 1$ teeth in contact, whereas for a period equivalent to "$(1 - f)p$" there are only "n" teeth in contact.

Evidently, it is advantageous to have a contact ratio as large as possible. It is an indication for the number of teeth sharing the transmitted load. A gear set with a high contact ratio can be used for transmitting more power. Moreover, a large contact ratio may result in less noise when gears are operated at high speeds.

EXAMPLE 4.1

A pinion of 24 teeth drives a gear with 60 teeth. The module is 3 mm and the pressure angle is 20°. Determine the contact ratio, the number of teeth sharing the load, and the periods for sharing the load.

SOLUTION

The diameters of the gears are

$$d = 3 * 24 = 72 \, \text{mm}$$
$$D = 3 * 60 = 180 \, \text{mm}$$

The radii of the base circles are

$$r_p = 36 * \cos 20 = 33.83 \, \text{mm}$$
$$R_p = 90 * \cos 20 = 84.57 \, \text{mm}$$

The path of contact is given by Equation 4.9,

$$Z = \sqrt{(R+a)^2 - R_b^2} + \sqrt{(r+a)^2 - r_b^2} - C\sin\varphi$$

$$Z = \sqrt{(90+3)^2 - 84.57^2} + \sqrt{(36+3)^2 - 33.83^2} - (36+90) * \cos 20$$
$$= 15.0 \, \text{mm}$$

The arc of contact is equal to

$$Z_p = \frac{15.0}{\cos 20} = 15.96 \, \text{mm}$$

The circular pitch is given by

$$p = \frac{15.96}{60} = 9.42 \, \text{mm}$$

The contact ratio is

$$m_p = \frac{15.96}{9.42} = 1.69$$

The number of teeth in contact is 2 for 0.69 p and one for 0.31 p.

4.5.6 Interference in Involute Gears

Point A in Figure 4.26 is the tangent point of the line of action with the base circle. It is called the interference point. If the root circle of the gear is smaller than the base, the part of the flank between the two circles is noninvolute because an involute tooth profile cannot exist inside the base circle. It is usually made radial, Figure 4.26. Point S is the beginning of contact of the mating gears. There is no problem as long as the point S is inside PA. If the beginning of contact is beyond PA, contact occurs in the noninvolute part of the flank. In this case, the fundamental law of gearing is violated. The velocity ratio is not any more constant that is not desirable. This act is called interference. In order to avoid interference, there are three solutions as follows:

1. Undercutting: The part of the flank inside the base circle is undercut with some arc, arc AC, to avoid contact at this part. On the other hand, this undercutting reduces the arc of contact, and, hence, the contact ratio; besides, it results in a weaker tooth. In general, undercutting should be avoided as possible except for small quantities.
2. Shorter addendum: The part of the face of the mating gear is cut with an amount called correction.
3. Controlling the minimum number of teeth: In general, this is a better solution for the interference problem.

4.5.7 Addendum Correction

To prevent interference, the addendum circle of a gear must intersect the line of action within the interference points, points A and B in Figure 4.27. These points are the tangent points of the line of action with the base circles as explained earlier. It is clear from the drawing that the addendum of the larger gear is more effective. r and R are the radii of pitch circles of the pinion and the gear, respectively. To calculate the correct addendum a_c, consider triangle O_gAP. Thus,

$$(R + a_c)^2 = R^2 + (r \sin \varphi)^2 - 2Rr \sin \varphi \cos(\varphi + 90)$$
$$= R^2 + (r \sin \varphi)^2 + 2Rr \sin^2 \varphi$$

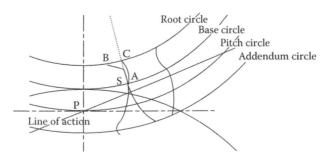

FIGURE 4.26 Undercutting the gear tooth to prevent interference.

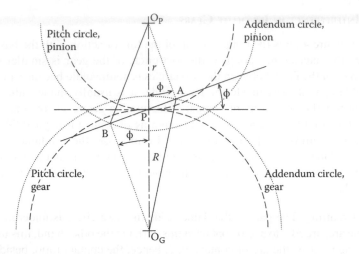

FIGURE 4.27 Correction of the addendum to prevent interference.

Therefore,

$$a_c = \sqrt{R^2 + (r\sin\varphi)^2 + 2Rr\sin^2\varphi} - R \qquad (4.12)$$

The correction c is given by,

$$c = a - a_c$$

4.5.8 Minimum Number of Teeth to Prevent Interference

It is possible to adjust the number of teeth of pinion so that the addendum does not exceed the correct addendum a_c. Let

N_P = the number of teeth of the pinion
N_G = the number of teeth of the gear

$$\lambda = \text{the speed ratio} = \frac{N_G}{N_P} = \frac{R}{r}$$

$$a_c = \frac{K}{P} = \frac{2rK}{N_P}$$

where K is a constant. For a full-depth system, $K = 1$, and for the sub system, $K = 0.8$. P is the diametral pitch, R is the radius of the gear, and r is the radius of the pinion. Substitute in Equation 4.12, then,

$$N_P = \frac{2rK}{\sqrt{R^2 + (r \sin \varphi)^2 + 2Rr \sin^2 \varphi} - R}$$

$$= \frac{2K}{\lambda \left[\sqrt{1 + \left(\dfrac{1}{\lambda} + 2 \right) \dfrac{\sin^2 \varphi}{\lambda}} - 1 \right]} \tag{4.13}$$

where λ is the gear ratio and is equal to $\dfrac{N_G}{N_P}$. Let

$$m = \frac{1}{\lambda} \left(\frac{1}{\lambda} + 2 \right)$$

Therefore,

$$N_P = \frac{2K}{\lambda(\sqrt{1 + m \sin^2 \varphi} - 1)} \tag{4.14}$$

The number of teeth of the gear is given by,

$$N_G = \frac{2K}{\sqrt{1 + m \sin^2 \varphi} - 1} \tag{4.15}$$

For equal gears, $N_P = N_G$ and $\lambda = 1$, thus,

$$N_P = N_G = \frac{2K}{\sqrt{1 + 3 \sin^2 \varphi} - 1}$$

For a pinion meshing with a rack, λ is equal to infinity. In this case, N_P is equal to zero divided by zero. To determine the value of N_P, let $\lambda = \dfrac{1}{\varepsilon}$ and applying the limits to Equation 4.13 as ε tends to zero. Therefore,

$$N_P = \frac{2K}{\sin^2 \varphi} \tag{4.16}$$

EXAMPLE 4.2

A full depth 40 mm pinion having 10 teeth with standard addendum is to be in mesh without interference with (a) 30.tooth gear and with (b) a rack. The pressure angle is 20°. Find the amount of tooth correction.

The module of the pinion is

$$m = \frac{40}{10} = 4 \text{ mm}$$

For a standard gear, the addendum is equal to the module. For the pinion meshing with the gear, the correct addendum is obtained by using Equation 4.12. Thus,

$$a_c = \sqrt{R^2 + (r\sin\varphi)^2 + 2Rr\sin^2\varphi} - R$$

The gear ratio is 3. Then

$$R = 60 \text{ mm}$$

$$a_c = \sqrt{60^2 + (20\sin 20)^2 + 2*60*20\sin^2 20} - 60$$
$$= 2.67 \text{ mm}$$

The amount of correction c is

$$c = 4 - 2.67 = 1.33 \text{ mm}$$

For the pinion meshing with the rack, the correct addendum is obtained from Figure 4.28. Thus

$$a_c = r\sin^2\varphi$$
$$= 2.34 \text{ mm}$$

Therefore,

$$c = 4 - 2.34 = 1.66 \text{ mm}$$

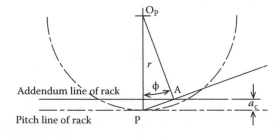

FIGURE 4.28   Correct addendum for a pinion meshing with a rack.

EXAMPLE 4.3

For Example 4.2, find the minimum number of teeth to prevent interference.

SOLUTION

The minimum number of teeth can be obtained from Equations 4.14 through 4.16. They can be obtained from the correct amount of addendums obtained.

For the pinion meshing with the gear, $a_c = 2.67$. Thus

$$2.67 = \frac{2*20}{N_P}$$

$$N_P = \frac{40}{2.67} = 14.98$$

Therefore, the minimum number of teeth of the pinion is 15.
For the pinion meshing with the rack, $a_c = 2.34$. Thus

$$N_P = \frac{40}{2.34} = 17.09$$

Therefore, the minimum number of teeth of the pinion is 17 (the fraction 0.09 is small and can be tolerated).

4.6 INTERNAL GEARS

In many applications, an internal involute gear is meshed with a pinion instead of using two external gears to achieve certain advantages (Figure 4.29). Perhaps, the most important advantage is the compactness of the drive. Also, for the same tooth proportions, internal gears will have greater length of contact, greater tooth strength, and lower relative sliding between meshing teeth than external gears.

In an internal gear, the tooth profiles are concave instead of convex as in an external gear. The number of teeth of the internal gear must be considerably larger than the pinion to prevent interference of the teeth.

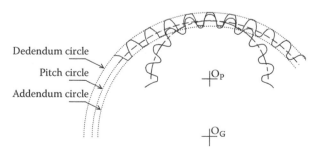

FIGURE 4.29 A pair of internal gears.

4.6.1 NATURE OF CONTACT

Referring to Figure 4.30, the beginning of contact is at point S, which is the intersection of the addendum circle of the internal gear with the line of action. The end of contact is at point E, which is the intersection of the addendum circle of the pinion with the line of action. The line of action is tangent to the base circle at point A. To avoid interference, point S must coincide with point A. The path of contact is SE = Z and is obtained by evaluating the path of approach SP and the path of recess PE.

The path of approach SP is given by

$$SP = R_G \sin \varphi - \sqrt{(R_G - a)^2 - (R_G \cos \varphi)^2}$$

The path of recess PE is given by

$$PE = \sqrt{(R_p + a)^2 - (R_P \cos \varphi)^2} - R_p \sin \varphi$$

The path of contact SE is given by

$$SE = R_G \sin \varphi + \sqrt{(R_p + a)^2 - (R_P \cos \varphi)^2} - \sqrt{(R_G - a)^2 - (R_G \cos \varphi)^2} - R_p \sin \varphi$$

When the start of contact is at point A,

$$SE = \sqrt{(R_p + a)^2 - (R_P \cos \varphi)^2} + R_P \sin \varphi$$

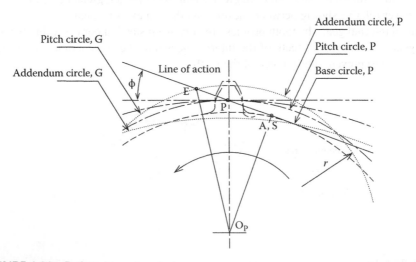

FIGURE 4.30   Path and arc of contact.

where,

> R is the radius of the pitch circle of the internal gear
> R_p is the radius of the pitch circle of the pinion
> R_b is the radius of the base circle of the pinion
> "a" is the addendum

The arc of contact and the contact ratio are the same for external gears.

$$Z = Z_p \cos \varphi$$

$$m_c = \frac{Z_p}{p}$$

4.7 THE CYCLOIDAL SYSTEM

Although the cycloid tooth form is seldom used today in the production of modern gears, still a study of mechanisms is incomplete without referring to the cycloidal system. Historically, it is the first theoretically correct form to be used for satisfying the law of gearing. A cycloid is generated by a point on a circle, called the generating circle, as it rolls on the outside of another circle. The generated curve is called epicycloid. When the generating circle rolls inside the other circle, the point describes a hypocycloid, Figure 4.31.

The teeth of a gear are formed from a face above the pitch circle and a flank below the pitch circle. The face of the gears is formed from epicycloids, whereas the flank is formed from a hypocycloid, Figure 4.32. The teeth profiles of the mating gears are formed from the same generating circles.

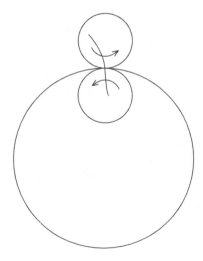

FIGURE 4.31   Cycloidal tooth form.

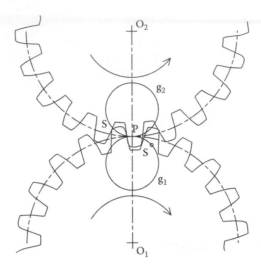

FIGURE 4.32   Contact of a pair of cycloidal gears.

The starting point of contact S is at the intersection of the addendum circle of the driven gear and the generating circle g_1. The end point of contact E is the intersection of the addendum circle of the driven gear and the generating circle g_2. Thus, the path of approach is the arc SP. It should be noticed that the pressure line will not have a constant inclination. As the point of contact comes closer to the pitch point, the pressure line approaches a perpendicular direction with the line of centers. The path of contact during recess is the arc PE. Thus, the path of contact or the line of action is SPE.

Ultimately, it will be interesting to point out the following major differences between the cycloid and the involute forms:

1. The path of contact or line of action is curved in the cycloid system, whereas it is straight in the involute system.
2. The pressure angle varies during contact in the cycloidal system. This is a major disadvantage when compared with the involute system in which the pressure angle is constant. As a result, there exists a varying separating force (the component of the tooth load along the line of centers), which is periodic and therefore causes vibration, noise, wear, and a change in bearing reactions at the shaft supports.
3. Cycloid gears must be operated at exactly the correct center distance. If for example, the distance between the centers O_1 and O_2 in Figure 4.32 is either increased or decreased, the point of contact will no longer intersect the line of centers O_1O_2 at a fixed point and the motion transmitted will no longer be uniform. Because the deflections due to transmission of load are bound to occur, it seems impossible to keep an exact center distance under all loading conditions. The center distance of involute gears can be changed without destroying the conjugate action.

4. The faces and flanks in contact must be generated by the same-size generating circle. Thus, for an interchangeable system, all faces and flanks must be generated by the same-size generating circle.

However, there is an important advantage of cycloid teeth. There is no danger of interference no matter how many teeth are used. Therefore, cycloid profiles may be used with gears having, say, three or four teeth, if necessary.

If the diameter of the generating circle is made equal to the radius of the pitch circle of the pinion to be used in the system, the flanks of the teeth will be radial. The pinion is the gear that establishes the generating circle diameters of each system. In general, the systems for industrial gears were based on pinions of 12 or 15 teeth.

PROBLEMS

4.1 An involute is generated on a base circle radius R_b of 102 mm. As the involute is generated, the angle that corresponds to inv φ varies from $0°$ to $15°$ for points on the involute. For increments of $3°$ for this angle, calculate the corresponding pressure angle φ and the radius of the points on the involute. Plot this series of points in polar coordinates and connect them with a smooth curve to give the involute.

4.2 Write a computer program for Problem 4.1, for $R_b = 76.2$, 102, and 127 mm. Determine the corresponding values of pressure angle φ and radius R for each value of R_b.

4.3 The thickness of an involute gear tooth is 7.98 mm at a radius of 88.9 mm and a pressure angle of $14\frac{1}{2}°$. Calculate the tooth thickness and radius at a point on the involute that has a pressure angle of $25°$.

4.4 If the involutes that form the outline of a gear tooth are extended, they will intersect and the tooth becomes pointed. Determine the radius at which this occurs for a tooth that has a thickness of 6.65 mm at a radius of 102 mm and a pressure angle of $20°$.

4.5 The thickness of an involute gear tooth is 4.98 mm at a radius of 50.8 mm and a pressure angle of $20°$. Calculate the tooth thickness on the base circle.

4.6 The pitch radii of two spur gears in mesh are 51.2 mm and 63.9 mm, and the outside radii are 57.2 mm and 69.9 mm, respectively. The pressure angle is $20°$. Make a full-size layout of these gears as shown in Figure 4.23, and label the beginning and end of contact. The pinion is the driver and rotates clockwise. Determine and label the angles of approach and recess for both gears. The angle is equal to the arc divided by the radius.

4.7 A pinion of 50.0 mm pitch radius rotates clockwise and drives a rack. The pressure angle is $20°$ and the addendum of the pinion and of the rack is 5.0 mm. Make a full-size layout of these gears and label the beginning and end of contact. Determine and label the angle of approach and recess for the pinion.

4.8 Two equal spur gears of 48 teeth mesh together with pitch radii of 96.0 mm and addendums of 4.0 mm. If the pressure angle is $20°$, calculate the path of contact Z and the contact ratio m_p.

4.9 The contact ratio is defined either as the arc of action divided by the circular pitch, or, as the ratio of the path of contact to the base pitches. Prove that

$$\frac{\text{Arc of contact}}{\text{Circular pitch}} = \frac{\text{Path of a contact}}{\text{Base pitch}}$$

4.10 Verify the equation for the path of contact Z for a pinion driving a rack in terms of the pitch radius R, the base radius R_b, the addendum a, and the pressure angle φ.

4.11 A pinion with a pitch radius of 38.0 mm drives a rack. The pressure angle is $20°$. Calculate the maximum addendum possible for the rack without having involute interference on the pinion.

4.12 A 2.module, $20°$ pinion of 24 teeth drives a 40.tooth gear. Calculate the pitch radii, base radii, addendum, dedendum, the tooth thickness on the pitch circle, and the contact ratio.

4.13 A 3.module, $20°$ pinion of 18 teeth drives a 45.tooth gear. Calculate the pitch radii, base radii, addendum, dedendum, tooth thickness on the pitch circle, and the contact ratio.

4.14 A 2.module, $20°$ pinion of 42 teeth drives a gear of 90 teeth. Calculate the contact ratio.

4.15 If the radii of a pinion and gear are increased so that each becomes a rack, the arc of contact theoretically becomes a maximum. Determine the equation for the path of contact under these conditions and calculate the maximum contact ratio for $14½°$, $20°$, and $25°$ full-depth systems.

4.16 A 6.module, $20°$ pinion of 20 teeth drives a rack. Calculate the pitch radius, the base radius, the working depth, the whole depth, and the tooth thickness on the pitch line.

4.17 Determine the approximate number of teeth in a $20°$ involute spur gear so that the base circle diameter is equal to the dedendum circle diameter.

4.18 Determine the following for a pair of standard spur gears in mesh:
   • An equation for the center distance C as a function of the number of teeth and module.
   • The various combinations of $20°$ gears that can be used to operate at a center distance of 120 mm with an angular velocity ratio of 3:1.
   The module is not to be less than 2, and the gears are not to be undercut.

4.19 A 4.module, $20°$ pinion with 24 teeth drives a rack. Calculate the path of action and the contact ratio.

4.20 A 12.module, $20°$ pinion with 24 teeth drives a rack. If the pinion rotates counterclockwise at 360 rpm, determine graphically the sliding velocity between the pinion tooth and the rack at the beginning of contact, at the pitch point, and at the end of contact.

4.21 Two shafts whose axes are 216 mm apart are to be coupled together by standard spur gears with an angular velocity ratio 1.5:1. Using a module of 4, select two pairs of gears to best fit the above requirements. What change in the given data would have to be allowed if each set were to be used?

4.22 For a pressure angle of 22.5°, calculate the minimum number of teeth in a pinion to mesh with a rack without interference. Also, calculate the number of teeth in a pinion to mesh with a gear of equal size without interference. The addendum is equal to the module.

4.23 A 3.module, 20° pinion with 24 teeth drives a 56 teeth gear. Determine the outside radii so that the addendum circle of each gear passes through the interference point of the other. Calculate the value of K for each gear.

4.24 Two equal 5.modules, 20° gears mesh together such that the addendum circle of each gear passes through the interference point of the other. If the contact ratio is 1.622, calculate the number of teeth and the outside radius of each gear.

4.25 Two equal 20° gears are in mesh in the standard center distance. The addendum circle of each gear passes the interference point of the other. Derive an equation for K as a function of the number of teeth.

4.26 The pitch diameter of a gear is 120 mm and the module is 5. Find the radius of a pin that contacts the profile at the pitch point.

4.27 A 2.5.module, 20° pinion with 40 teeth meshes with a rack with no backlash. If the rack is pulled out at 1.27 mm, calculate the amount of backlash produced.

4.28 A 2.module, 20° pinion of 18 teeth drives a gear of 54 teeth. If the center distance at which the gears operate is 73.27 mm, calculate the operating pressure angle.

4.29 A 2.5.module, 20° pinion with 36 teeth drives a gear with 60 teeth. If the center distance is increased by 1.650 mm, calculate:
   • The radii of the operating pitch circles
   • The operating pressure angle
   • The backlash produced

4.30 A 6.module, 20° pinion of 24 teeth drives a gear of 40 teeth. Calculate:
   • The maximum theoretical distance that these gears can be drawn apart and still mesh together with continuous driving
   • The amount of backlash on the new pitch circles when the gears are drawn apart the amount calculated

4.31 A pinion with 25 teeth has a tooth thickness of 6.477 mm at a pitch radius of 37.50 mm and a pressure angle of 20°. A gear having 42 teeth has a tooth thickness 5.842 mm at a pitch radius of 63.00 mm and a pressure angle of 20°. Calculate the pressure angle and center distance if these gears are meshed together without backlash.

4.32 Two meshing spur gears have 19 and 36 teeth. The diametral pitch is 6. Determine:
   • Pitch diameters
   • The center distance
   • The base circle diameters for a pressure angle of 20°
   • The pitch line velocity if the pinion rotates at 600 rpm
   • The speed of the gear

4.33 Two parallel shafts 7.4 in. apart are to be connected by spur gears having 18 and 56 teeth. Calculate the pitch diameters, the circular pitch, and the diametral pitch.

4.34 A 4 diametral pitch, 20 tooth pinion is to drive a 30 tooth gear. The teeth are 14½ full-depth involute. Make a neat drawing of the gears showing a profile of one tooth on the gear making contact with a profile of a mating tooth on the pinion and at the pitch point. Find and tabulate the following results: the addendum, dedendum, clearance, circular pitch, tooth thickness and the base circle diameters, the arcs of approach, recess and action, and the contact ratio.

4.35 A pair of spur gears have 16 and 22 teeth, 2 diametral pitch, ½ in. addendum, 9/16 dedendum, and 20° pressure angle. The direction of rotation of the pinion is counterclockwise. Determine the following:
• The pitch circle radii
• The base circle radii
• The circular pitch
• The base pitch
• The length of the path of contact
• The contact ratio
• The angles of approach and recess for the driver and for the follower
Indicate the pitch point and the first and last points of contact.

4.36 Same as Problem 4.35 except that the pressure angle is 14½°.

4.37 Same as Problem 4.35 except that the gear is replaced by a rack.

4.38 A pair of gears has 14 and 16 teeth, the diametral pitch is 2, the addendum is 3/4 in., and the pressure angle is 14½°. Check whether or not the gears have interference.

4.39 A pinion having a pitch circle diameter of 10 in. drives a rack. The addendum for the pinion and the rack is 0.5 in. and the teeth have involute form with a pressure angle of 20°. Show that interference does not occur and find the minimum number of teeth on the pinion.

4.40 What are the requirements of the mating profiles of two gear teeth for a constant velocity ratio between the two gears? Show how they may be satisfied by an involute toothed pinion meshing with an internally toothed gear. Find graphically or analytically the length of the path of contact when a pinion with 17 teeth meshes with an internally toothed gear with 68 teeth with a pressure angle of 20° and a diametral pitch of 4. The addendum for the pinion is 0.35 in. and for the gear is 0.15 in.

4.41 A pinion of 16 involute teeth and diametral pitch of 4 is driving a rack. The pinion has a standard addendum of ¼ in. What is the least pressure angle to avoid undercutting? Determine the contact ratio.

4.42 Calculate the tip thickness of a tooth in a standard gear of 20 teeth and diametral pitch 2.

4.43 A standard 32 tooth, 20° pressure angle, full depth, 4 diametral pitch gear is driven by a 20 tooth pinion that rotates at 1200 rpm. Using the length of the path of contact as abscissa, plot a curve showing the sliding velocity at all points of contact. Note that the sliding velocity changes sign when the point of contact passes through the pitch point.

4.44 Two meshing gears having 16 and 38 teeth of involute form and a diametral
      pitch of 4 are to be proportioned to satisfy the following condition: the
      addendum on each gear is to be made of such a length that the line of con-
      tact on each side of the "itch point" has half the maximum possible length
      of the line of contact.
      If the pinion rotates at 400 rpm, find:
      • The velocity of the point of contact along the surface of each tooth
      • The velocity of sliding
4.45 Show that, if the diameter of the generating circle is equal to the radius of
      the directing circle, then the flank of a cycloidal tooth will be radial.
4.46 An 18 tooth 6 in. pinion drives a 9 in. gear. If the cycloidal teeth are based
      on the 15 tooth system, determine the initial and final point of contact, the
      line of action, and the pressure angles at the initial and final points.

5 Helical, Worm, and Bevel Gears

5.1 INTRODUCTION

In Chapter 4, attention has been focused on the tooth form in the plane of motion, that is, the tooth form of spur gears. There are many other forms of gears. In this chapter, we shall present the principles of other types: helical, worm, and bevel gears.

5.2 PARALLEL HELICAL GEARS

In spur gears, a line in plane parallel to the axis of a cylinder forms the involute surface of the gear (Figure 5.1a). In this case, the teeth are parallel to the axis. Since the teeth of the helical gear are inclined to the axis of the gear, involute teeth form can be obtained by an inclined line in the plane. The line is inclined with an angle, which is known as the helix angle ψ. Each point on the generating line traces an involute (Figure 5.1b). The surface generated by the line is known as involute helicoids.

The involute tooth form for the mating gear is obtained by wrapping the plane over another cylinder as was done for spur gears. Actually, the initial contact of helical–gear teeth is a point that changes gradually into a diagonal line across the face of the tooth. This is a great advantage when compared with the action of spur gears. This property allows smooth engagement and more teeth in contact. For this reason, helical gears are used to transmit heavy load at high speeds. Helical gears can be considered as a great number of infinitesimally thin spur gears attached to each other and are rotated relative to one another. There are two types of helical gears. The first is used to transmit the motion between parallel shafts and is called parallel helical gears (Figure 5.2). The second is used to transmit the motion between nonparallel nonintersecting shafts and is called crossed helical gears (Figure 5.3).

5.2.1 PROPERTIES OF HELICAL GEARS

Figure 5.4 shows a top view of a helical gear. The gear has a helix angle ψ. The helix angle is the angle of inclination of the teeth with the axis of the gear. It is also the angle of inclination of the line forming the teeth profile. The gear has a circular pitch p, which is in a plane normal to the axis. The normal circular pitch p_n is measured in a plane normal to the teeth. The relation between the two pitches is given by,

$$p_n = p \cos \psi \qquad (5.1)$$

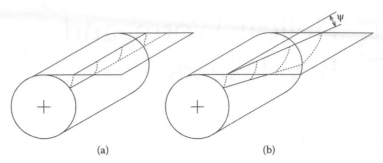

(a)   (b)

FIGURE 5.1 Formation of involute surface for helical gears. (a) Teeth parallel to the axis (b) helical teeth.

FIGURE 5.2 A pair of helical gears.

FIGURE 5.3 A pair of crossed helical gears.

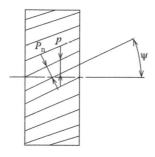

FIGURE 5.4   Top view of a helical gear.

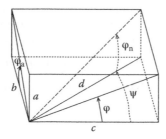

FIGURE 5.5   Angles incorporated in helical gears.

Since $p \times P = \pi$, the normal diametral pitch P_n is given by,

$$P_n = \frac{P}{\cos \psi} \qquad (5.2)$$

Because of the angularity of the teeth, there are two pressure angles: the transverse pressure angle φ and the normal pressure angle φ_n. The relation between the two angles can be obtained by the help of the parallelogram OABC (Figure 5.5).

$$\tan \varphi = \frac{a}{c}$$

$$\tan \varphi_n = \frac{a}{d}$$

Dividing the two equations, then

$$\frac{\tan \varphi_n}{\tan \varphi} = \frac{c}{d}$$

But

$$\frac{c}{d} = \cos \psi$$

Therefore,

$$\tan \varphi_n = \tan \varphi \cos \psi \qquad (5.3)$$

There is also an axial pressure angle φ_a and it occurs in the plane along the tooth and is given by

$$\tan \varphi_a = \frac{a}{b} = \frac{a}{d} \times \frac{d}{b} = \frac{\tan \varphi_n}{\sin \psi} \tag{5.4}$$

5.2.2 EQUIVALENT SPUR GEAR

The equivalent spur gear concept is used widely for design purposes in the determination of the approximate strength of the helical gear strength. In Figure 5.6, the radius of the pitch cylinder of the gear is r. The normal plane AA intersects the pitch cylinder in an ellipse that has a radius of curvature r_e at the pitch point P. The radius r_e is called the equivalent pitch radius. It is equal to the radius of curvature at the minor axis of the ellipse. The minor axis of the ellipse is equal to r and the major axis is equal to $\dfrac{r}{\cos \psi}$. Note that when $\psi = 0$, this radius of curvature is r. If ψ is increased from 0 to 90°, r_e will begin at a value of r and increase until infinity when ψ becomes 90°. The radius of curvature at the minor axis is given by

$$r_e = \frac{r}{\cos^2 \psi} \tag{5.5}$$

The equivalent pitch radius r_e can be thought of as the pitch of a spur gear whose teeth has a shape that approximates the shape of the helical gear teeth in the normal plane. To obtain the equivalent number of teeth N_e, we should remember that r and r_e describe circles corresponding to gears in different planes.

$$pN = 2\pi r$$

$$p_n N_e = 2\pi r_e$$

Using Equations 5.1 and 5.5,

$$N_e = \frac{N}{\cos^3 \psi} \tag{5.6}$$

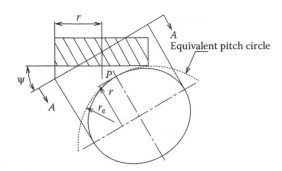

FIGURE 5.6 The equivalent spur gear.

5.2.3 Helical Gear Tooth Proportion

In determining the tooth proportions of a helical gear, it is necessary to consider the manner in which the teeth are to be cut, whether rack cut or to be hobbed. There is a considerable variety in the tooth proportions used for the helical gears. The proportions depend to a great extent on the tools available for gear manufacturing.

As a general guide, Table 5.1 lists tooth proportions for various helix angles based on a normal diametral pitch $P_n = 1$. If P_n takes values other than 1, all the linear values in the table should be divided by the specified normal pitch.

In Table 5.2, tooth proportions for different helix angles are based on a transverse diametral pitch of 1. At helix angles of 30° or more, this series provides better tooth contact ratios than the series in Table 5.1. The proportions shown are full depth in the normal section. In the transverse section, they appear stubbed. The 30° and 45° angles are usually used for double helical gears. The series in Table 5.1 is recommended where the noise level must be kept low.

TABLE 5.1
Tooth Proportions for Helical Gears

Helix Angle ψ	Diametral Pitch P	Circular Pitch p	Axial Pitch p_a	Pressure Angle φ	Working Depth (in.)	Whole Depth (in.)
0	1.000	3.14159	—	20° 0.0' 00.0"	2.000	2.250
5	0.996195	3.15359	36.04560	20° 0.4' 13.1"	2.000	2.250
8	0.990268	3.12747	22.57327	20° 10' 50.6"	2.000	2.250
10	0.984808	3.19006	18.09171	20° 17' 00.7"	2.000	2.250
12	0.978148	3.21178	15.11019	20° 24' 37.1"	2.000	2.250
15	0.965926	3.25242	12.13817	20° 38' 48.8"	2.000	2.250
18	0.951057	3.30326	10.16640	20° 56' 30.7"	2.000	2.250
20	0.939693	3.34321	9.18540	21° 10' 22.0"	2.000	2.250
21	0.933580	3.36510	8.76638	21° 17' 50.4"	2.000	2.250
22	0.927184	3.38832	8.38636	21° 25' 57.7"	2.000	2.250
23	0.920505	3.41290	8.04029	21° 34' 26.3"	2.000	2.250
24	0.913545	3.43890	7.72389	21° 43' 22.9"	2.000	2.250
25	0.906308	3.46636	7.43364	21° 52' 58.7"	2.000	2.250
26	0.898794	3.49539	7.16651	22° 02' 44.2"	2.000	2.250
27	0.891007	3.52589	6.91994	22° 13' 10.6"	2.000	2.250
28	0.882948	3.55807	6.69175	22° 24' 0.9.0"	2.000	2.250
29	0.874620	3.59195	6.48004	22° 35' 40.0"	2.000	2.250
30	0.866025	3.62762	6.28318	22° 47' 45.1"	2.000	2.250

Normal diametral pitch $P_n = 1$.
Normal pressure angle $\varphi_n = 20°$.
Normal circular pitch $p_n = 3.14159$.
Edge radius of generating rack = 0.300 in.

TABLE 5.2
Tooth Proportions for Different Helix Angles

Helix Angle $y°$	Normal Diametral Pitch P_n	Normal Circular Pitch p	Axial Pitch p_a	Normal Pressure Angle f_n	Working Depth (in.)	Whole Depth (in.)	Edge Radius Rack (in.)
15	1.03528	3.03154	11.7245	19° 22' 12.2"	2.00	2.35	0.35
23	1.0836	2.89185	7.40113	18° 31' 21.6"	1.84	2.20	0.35
30	1.1547	2.72070	5.44140	17° 29' 42.7"	1.74	2.05	0.3
45	1.41421	2.22144	3.14159	14° 25' 57.9"	1.42	1.70	0.25

Normal diametral pitch $P = 1$.
Normal pressure angle $\varphi = 20°$.
Normal circular pitch $p_n = 3.14159$.

5.2.4 UNDERCUTTING OF HELICAL GEARS

In Section 4.5.8, we have found that for any standard spur gear system with an addendum constant K and a pressure angle φ, the smallest number of teeth N for a gear to prevent interference is when it is engaged with a rack, which is given by Equation 4.16.

$$N_P = \frac{2K}{\sin^2 \varphi} \tag{5.7}$$

For helical gears, the corresponding equation in terms of dimensions in the plane of rotation is the same.

$$N_P = \frac{2K}{\sin^2 \varphi} \tag{5.8}$$

We shall now compare the results when a spur gear generator with $K = 1$ and $\varphi = 20°$ is used to cut spur gears and helical gears having $\psi = 30°$. For the spur gears, Equation 5.7 gives $N = 17.1$.

For the helical gears, $K_n = 1$. Then,

$$K = K_n \cos \psi = 0.866$$

$\varphi_n = 20°$, then

$$\tan \varphi_n = \tan \varphi \cos \psi$$

Thus,

$$\varphi = 22° 47' 45.1"$$

Substitute in Equation 5.7, therefore $N_P = 11.6$. This proves that helical gears can use less number of teeth than spur gears without interference when both are cut with the same rack.

5.2.5 CONTACT ACTION OF HELICAL GEARS

The study of the action between helical gears can be simplified by considering them either as twisted spur gears or as a series of infinitesimally thin spur gears rotated uniformly relative to one another. In Figure 5.7, AB is a projection of a plane tangent to the base cylinders and represents the plane of contact. F is the face width of the gear and Z is the length of the path of contact of any tooth. SE is the path of contact of the tooth at the face. For a detailed study, we have to use an extra view. The rectangle $S_F E_F E_B S_B$ is a projected view of the part of the tangent plane where action between the tooth helicoids takes place. Let $S_F E_F$ represent the front transverse plane and $E_B S_B$ represent the back transverse plane. Let us define the axial pitch p_a as the distance on the line of action between the tooth at the face and the same tooth at the back. Assume, for example, that the face of the gear is such that the face of the tooth is the beginning of contact and the back of the tooth is at the end of contact, indicated by F' as shown in the Figure 5.7. This means that the back of the tooth has described a contact of length Z and the front of the tooth will describe a path of contact of length Z. Therefore, the total path of contact of one tooth is 2 Z. In general, the path of contact Z_t of any tooth is given by

$$Z_t = Z + F \tan \psi \tag{5.9}$$

The contact ratio is given by

$$m_t = \frac{Z_t}{p}$$

$$= m_e + m_F$$

The quantity m_F is called the overlap or axial contact ratio and is given by

$$m_F = \frac{F \tan \psi}{p} = \frac{F}{p_a} \tag{5.10}$$

where p_a is known as the axial pitch and is given by

$$p_a = \frac{p}{\tan \psi} \tag{5.11}$$

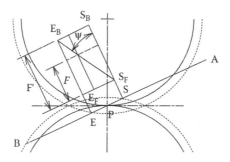

FIGURE 5.7 Contact action in helical gears.

5.2.6 Force Analysis in Helical Gears

Usually gears are designed to transmit certain torque T. This torque is equal to the tangential force F_t times the radius of the pitch circle R of the gear. Thus,

$$F_t = \frac{T}{R} \tag{5.12}$$

F_n is the normal force transmitted from one gear to its mating gear. Due to the inclinations of the face of the tooth, other forces are developed. They are the radial force F_r and the axial force F_a. Referring to Figure 5.8, the relations between the forces are given by

$$F_r = F_t \tan \varphi \tag{5.13}$$

$$F_a = F_t \tan \psi \tag{5.14}$$

$$F_n = F_t \sqrt{1 + \tan^2 \varphi + \tan^2 \psi} \tag{5.15}$$

For a spur gear, $\psi = 0$. The axial force F_a is equal to zero.

5.2.7 Herringbone Gears

The disadvantage of the helical gears is that they produce axial loads, which cause axial thrust to the bearings. If this axial thrust causes problems to the bearing, Herringbone gears are the proper solution. They are actually two helical gears with the same helix angle but opposite in directions attached to each other (Figure 5.9). The axial force of one gear is eliminated by the other. Herringbone gears can be formed from two separate helical gears (Figure 5.10a) or cut from one cylinder. In this case, it is better to leave a space between the gears (Figure 5.10b).

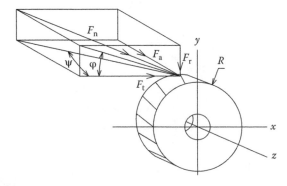

FIGURE 5.8   Forces in helical gears.

FIGURE 5.9 Herringbone gear.

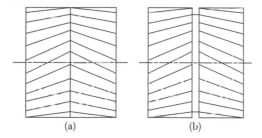

(a)   (b)

FIGURE 5.10 (a) Two separate gears (b) one unit with two gears.

5.3 CROSSED HELICAL GEARS

Helical gears on nonparallel shafts (Figure 5.3) are called crossed helical gears. The tooth action of these gears is different from that when helical gears are on parallel shafts. Here, contact occurs at a point that changes to line contact as the gears wear in. Crossed helical gears are identical to helical gears until they are mounted in mesh with each other. A pair of meshed crossed helical gears must have the same normal pitch or module. Their pitches in the plane of rotation are not necessarily and not usually equal. Their helix angles may or may not be equal. The gears may be of the same or of opposite hand. The directions of the hand are shown in Figure 5.11. The gear in Figure 5.11a is left hand while the gear in Figure 5.11b is right hand.

Suppose that gears (1) and (2) are mesh and are mounted on two shafts. The angle between the two shafts is σ, the helix angle of gear (1) is ψ_1, and the helix angle of gear (2) is ψ_2. When the two gears have opposite hands (Figure 5.12a), then

$$\sigma = \psi_1 . \psi_2 \tag{5.16}$$

When the two gears have the same hand (Figure 5.12b), then

$$\sigma = \psi_1 + \psi_2 \tag{5.17}$$

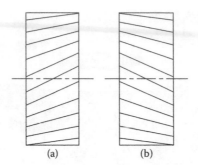

FIGURE 5.11   Teeth directions. (a) Left hand (b) right hand.

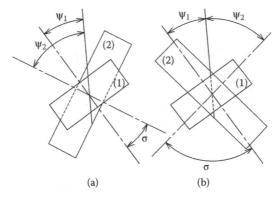

FIGURE 5.12   Orientation of the crossed helical gears. (a) Gears with opposite hands (b) gears with the same hands.

Opposite hand crossed helical gears are used when the shaft angle is small. The pitch diameters are obtained from the following equations:

$$D_1 = \frac{N_1}{P_n \cos \psi_1} \qquad (5.18)$$

$$D_2 = \frac{N_2}{P_n \cos \psi_2} \qquad (5.19)$$

where N_1 and N_2 are the number of teeth and P_n is the normal diametral pitch. The center distance C is given by

$$C = \frac{D_1 + D_2}{2}$$

It can be written as

$$C = \frac{1}{2 P_n} \left[\frac{N_1}{\cos \psi_1} + \frac{N_2}{\cos \psi_2} \right] \qquad (5.20)$$

For a given gear ratio $m_G = N_2/N_1$, we can put $N_2 = m_G N_1$ and Equation 5.20 can take the form:

$$C = \frac{N_1}{2 P_n} \left[\frac{1}{\cos \psi_1} + \frac{m_G}{\cos \psi_2} \right] \qquad (5.21)$$

It should be noted that the gear ratio m_G is not inversely proportional to the diameters unless the two helix angles are the same. From Equations 5.18 and 5.19, we get

$$m_G = \frac{N_2}{N_1} = \frac{D_2 \cos \psi_2}{D_1 \cos \psi_1} \qquad (5.22)$$

EXAMPLE 5.1

A pair of gears is used to connect two shafts at an angle of 60° with a velocity ratio 1.5:1. The pinion has a normal diametral pitch of 6, a pitch diameter of 7.75 in., and a helix angle of 35°. Determine the helix and the pitch diameter of the gear and the number of teeth of both the pinion and the gear.

SOLUTION

$$\sigma = \psi_1 + \psi_2$$

$\sigma = 60°$ and $\psi_1 = 35°$. Therefore,

$$\psi_2 = 25°$$

From Equation 5.22,

$$m_G = \frac{N_2}{N_1} = \frac{D_2 \cos \psi_2}{D_1 \cos \psi_1}$$

$$D_2 = m_G \frac{D_1 \cos \psi_1}{\cos \psi_2} = 1.5 \frac{7.75 \times \cos 35°}{\cos 25°}$$

$$D_2 = 10.5 \text{ in}$$

The number of teeth on the pinion and the gear is

$$N_1 = P_n D_1 \cos \psi_1 = 6 \times 7.75 \times \cos 35°$$
$$N_1 = 38$$

$$N_2 = N_1 m_G$$
$$N_2 = 57$$

5.3.1 Sliding Velocity in Crossed Helical Gears

Figure 5.13 shows a pair of helical gears in mesh. The driven gear (2) is below gear (1), which is the driver. Point P is the point of contact. The shaft angle σ is equal to $\psi_1 + \psi_2$ in the plane of projection as shown. Line PT is tangent to the helix on the upper surface of the lower gear. Line PN is normal to the helix at the same point. The absolute velocity of the point of contact P on the pitch cylinder of each gear is tangent to the pitch circle (perpendicular to the axis of rotation), as shown in the Figure 5.13. The relative velocity V_{12} is the velocity of sliding. It is to be noted that the components of two absolute velocities in a direction normal to the helices at P, that is, V_{13}, should be the same so that the teeth may not separate. Now it is clear that

$$V_{12} = V_1 \sin \psi_1 + V_2 \sin \psi_2 \qquad (5.23a)$$

This can be put in a different form

$$V_{12} = r_1 \omega_1 [\sin \psi_1 + \cos \psi_1 \sin \psi_2] \qquad (5.23b)$$

EXAMPLE 5.2

Two shafts are to be connected by a pair of crossed helical gears with a velocity ratio of 3:1 (Figure 5.13). The angle between the shafts is 45° and the shortest distance between the two skew lines is 9 in. The normal diametral pitch is 5 and the pinion has 20 teeth. Determine the pitch circle diameters and the helix angles when they have the same hand. If the pinion rotates at 300 rpm, find the sliding speed between the teeth.

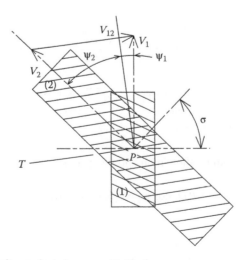

FIGURE 5.13 Velocity analysis in crossed helical gears.

SOLUTION

$$C = 9$$

From Equation 5.21,

$$C = \frac{N_1}{2 \times 5}\left[\frac{1}{\cos\psi_1} + \frac{m_G}{\cos\psi_2}\right]$$

$$9 = \frac{20}{10}\left[\frac{1}{\cos\psi_1} + \frac{3}{\cos\psi_2}\right]$$

$$\frac{1}{\cos\psi_1} + \frac{3}{\cos\psi_2} = 4.5 \qquad\qquad (a)$$

Since the two helices have the same hand,

$$\sigma = \psi_1 + \psi_2 = 45° \qquad\qquad (b)$$

From Equations a and b,

$$\psi_1 = 15°$$

$$\psi_2 = 30°$$

From Equations 5.18 and 5.19

$$\frac{D_2}{D_1} = \frac{N_2 \cos\psi_1}{N_1 \cos\psi_2} = 0 \times \cos 15° \cos 30° = 3.346$$

But

$$D_1 + D_2 = 2 \times 9 = 18$$

Thus,

$$D_1 = 4.141\,\text{in}$$

$$D_2 = 13.859\,\text{in}$$

$$V_{12} = r_1\omega_1[\sin\psi_1 + \cos\psi_1\sin\psi_2]$$

To obtain the sliding velocity, we use Equation 5.23a.

$$V_1 = 2\pi\frac{N_1}{60} \times \frac{D_1}{2} = 2\pi \times \frac{300}{60} \times \frac{4.141}{2}$$

$$= 65.1\,\text{in/s}$$

$$V_2 = 2\pi \frac{N_2}{60} \times \frac{D_2}{2} = 2\pi \times \frac{100}{60} \times \frac{13.859}{2}$$
$$= 73.0 \text{ in/s}$$

The sliding velocity is given by

$$V_{12} = V_1 \sin\psi_1 + V_2 \sin\psi_2$$
$$= 65.1 \times \sin 15° + 73.0 \times \sin 30°$$
$$= 52.92 \text{ in/s}$$

5.4   WORM GEARING

The terminology of the worm gearing is shown in Figures 5.14 through 5.16.

- D_2 is the pitch diameter of the wheel.
- D_{2m} is the maximum diameter of the wheel.
- D_{2t} is the throat diameter of the wheel.
- p is the circular pitch of the wheel.
- N_2 is the number of teeth of the wheel.
- D_1 is the pitch diameter of the worm.
- D_{1o} is the outside diameter of the worm.
- D_{1r} is the root diameter of the worm.
- N_1 is the number of teeth of the worm.
- a is the addendum.
- b is the dedendum.
- p_x is the axial pitch of the worm. It is the axial distance between two adjacent teeth.
- L is the lead. It is the axial distance that a point on the helix of the worm will move in one revolution of the worm.
- $L = N_1 p_x$.
- λ is the lead angle of the worm.

$$\tan\lambda = \frac{L}{\pi D_1} = \cot\psi \tag{5.24}$$

- ψ is the helix angle of the worm teeth. It is the angle of inclination of the teeth with the axis of the worm.

Because of the appearance of the worm and its resemblance to a screw, its teeth are often called threads. Two worms are shown in Figure 5.17; Figure 5.17a has one thread and is called a single thread worm, while Figure 5.17b is a double thread worm.

It is a usual practice to use a normal pressure angle of 20° for lead angles less than 30° and a normal pressure angle of 25° for lead angles up to 45°. The number of threads on a worm may range from 1 to 10. The pitch of worm gear sets is not standardized as in the case of spur gears.

FIGURE 5.14   A set of worm and worm gear.

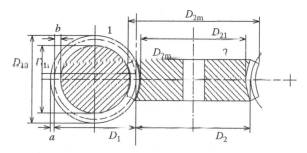

FIGURE 5.15   Notations of the worm gear set, cross-section view.

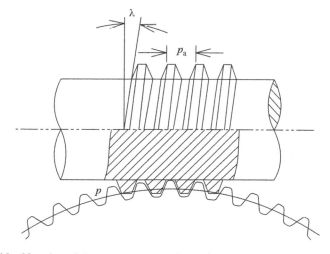

FIGURE 5.16   Notation of the worm gear set, front view.

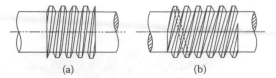

(a)   (b)

FIGURE 5.17   (a) Single thread worm (b) double thread worm.

The pitch is usually specified by the axial pitch of the worm. Some common pitches are $\dfrac{3}{16}, \dfrac{1}{14}, \dfrac{5}{16}, \dfrac{3}{8}, \dfrac{1}{2}, \dfrac{5}{16}, \dfrac{3}{8}, 1$ in, and so on.

When the shortest distance between two perpendicular shafts and the speed ratio is known, the pitch diameters are not unique. As a guide, the following empirical formula for the worm pitch diameter D_1 may be used

$$D_1 = \frac{C^{0.875}}{K} \tag{5.25}$$

where C is the center distance and K varies in the range $1.7 \le K \le 3.0$. The corresponding gear pitch diameter D_2 may now be approximated from

$$C = \frac{D_1 + D_2}{2} \tag{5.26}$$

When a suitable axial pitch p_a is chosen, D_2 must be adjusted to satisfy the equation,

$$D_2 = \frac{N_2 p_a}{\pi} \tag{5.27}$$

If C cannot be changed, D_1 will be adjusted to satisfy the equation

$$D_1 = 2\, C \cdot D_2$$

For most purposes, we can use the following proportions based on an addendum of $a = \dfrac{1}{P}$,

$$a = 0.3183\, p_n$$
$$\text{Whole depth} = 0.6366\, p_n \tag{5.28}$$
$$\text{Clearance} = 0.05\, p_n$$

5.4.1 Sliding Velocity of Worm Gears

The formula for the sliding speed on worm gears may be obtained from Equation 5.23 for crossed helical together with Equation 5.24. If D_1 is the pitch diameter, ω_w is the angular velocity of the worm, and λ is the lead angle of the worm. When the shaft angle $\sigma = 90°$, the sliding velocity V_{12} is given by

$$V_{12} = \left(D_1/2 \right) \omega_w [\cos\lambda + \sin\lambda \times \tan\lambda]$$

or

$$V_{12} = \frac{D_1 \omega_w}{2\cos\lambda} \tag{5.29}$$

Due to this sliding, which takes place along the teeth, worm gears are less efficient than both spur gears and helical gears.

5.5 BEVEL GEARS

Bevel gears (Figure 5.18) are used to connect two intersecting shafts. The pitch surfaces are frustums of right circular cones (Figure 5.19). The angle between the two shafts is σ, while the pitch angles of the gear and the pinion are γ_G and γ_P respectively. The angular velocities of the gear and pinion are ω_G and ω_P respectively.

If D_1 and D_2 are the pitch diameters of the pinion and the gear respectively and assuming no slipping, then

$$\omega_P D_1 = \omega_G D_2$$

or

$$\frac{\omega_P}{\omega_G} = \frac{D_2}{D_1} \tag{5.30}$$

From the figure,

$$\sin\gamma_P = \frac{D_1}{2\,OA}$$

$$\sin\gamma_G = \frac{D_2}{OA}$$

FIGURE 5.18   A set of bevel gears.

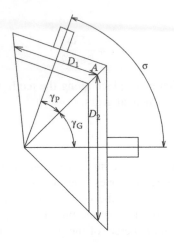

FIGURE 5.19   Pitch cones of the bevel gear set.

Therefore,

$$\frac{\sin \gamma_G}{\sin \gamma_P} = \frac{D_2}{D_1} \tag{5.31}$$

Also,

$$\sigma = \gamma_P + \gamma_G \tag{5.32}$$

$$\sin \gamma_P = \sin (\sigma \cdot \gamma_G)$$

$$= \sin \sigma \cos\gamma_G - \cos \sigma \sin \gamma_G$$

$$\frac{\sin \gamma_P}{\sin \sigma \sin \gamma_G} = \frac{\cos \gamma_G}{\sin \gamma_G} - \frac{\cos \sigma}{\sin \sigma}$$

$$\frac{1}{\sin \sigma}\left[\frac{\sin \gamma_P}{\sin \gamma_G} + \cos \sigma\right] = \frac{1}{\tan \gamma_G}$$

From Equation 5.32, we get

$$\tan \gamma_G = \frac{\sin \sigma}{\cos \sigma + \dfrac{D_1}{D_2}}$$

$$= \frac{\sin \sigma}{\cos \sigma + \dfrac{N_1}{N_2}} \tag{5.33}$$

$$= \frac{\sin \sigma}{\cos \sigma + \dfrac{\omega_G}{\omega_P}}$$

Also,

$$\tan \gamma_P = \cfrac{\sin \sigma}{\cos \sigma + \cfrac{D_2}{D_1}}$$

$$= \cfrac{\sin \sigma}{\cos \sigma + \cfrac{N_2}{N_1}} \qquad (5.34)$$

$$= \cfrac{\sin \sigma}{\cos \sigma + \cfrac{\omega_P}{\omega_G}}$$

5.5.1 TYPES OF BEVEL GEARS

If σ lies between 90° and 180°, then cos σ will be negative. Since $\dfrac{\omega_G}{\omega_P}$ is less than 1, the denominator of Equation 5.33 could be zero or negative. If it is zero, tan γ_G is infinity and $\gamma_G = 90°$. The pitch surface of the gear becomes a plane as shown in Figure 5.20. Such a gear is known as a crown gear and it is the rack among bevel gears.

If tan γ_G is negative, γ_G will be greater than 90°. Such a gear is an internal gear (Figure 5.21). When $\sigma = 90°$ and the speed ratio is 1, the gears are called miter gears.

5.5.2 TOOTH PROFILE OF BEVEL GEARS

Referring to Figure 5.19, we realize that all points on the large end of the pinion pitch cone move on a circle in a plane perpendicular to the axis of the pinion, while all points on the large end of the gear pitch cone move on a circle in a plane perpendicular to the axis of the gear. The two circles are in different planes but the points

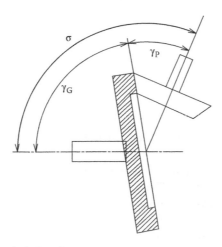

FIGURE 5.20 Pitch angle in bevel gears.

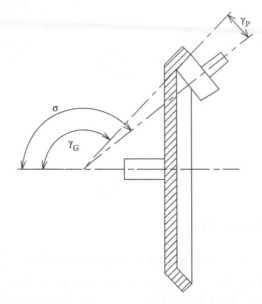

FIGURE 5.21   Miter gear set.

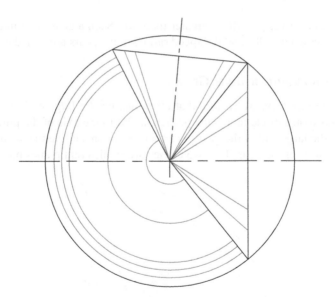

FIGURE 5.22   Formation of the gear teeth in bevel gears.

always remain at a fixed distance from the apex O. This implies that the two points move on the surface of a sphere. Thus, the pitch cones of the pinion and the gear may be considered to be cut from the same sphere as shown in Figure 5.22.

If the pitch cones are to be replaced by actual gears, the action between the teeth must be considered on the surface of a sphere. We are interested in determining a suitable tooth form.

An involute crown gear tooth is shown in Figure 5.23. The base cone angle is less than the 90° pitch angle. When the generating plane is rolled on the base cone, the generating line OA forms the spherical involute tooth shown. This crown gear having involute teeth would be difficult to cut. For this reason, the teeth of bevel gears are not of involute form. A more satisfactory tooth form and a means for producing it were invented by Hugo Bilgram in 1884. This tooth form is called the octoid, and because of its resemblance to the involute, the two forms are often confused.

θ_{bc} is the base cone angle.
θ_p is the pitch angle.

Octoid bevel gears are conjugate to a crown gear having teeth made with straight sides. An octoid crown gear is shown in Figure 5.24. The sides lie in planes that pass through the center of the sphere. The complete path of contact of the teeth on the surface of a sphere is in the form of number 8 and that is why the name octoid. Only a portion of the path is used, APB or A'PB'. These portions are nearly straight. The curve is symmetrical about point P, and for a short tooth height, it is practically on a plane perpendicular to line DE. Therefore, the octoid tooth form satisfies the condition for interchangeability, and when the tooth height is small compared to the diameter of the sphere, the involute line of action is practically indistinguishable from the involute.

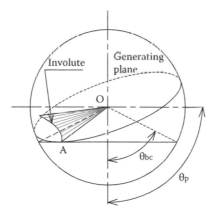

FIGURE 5.23 Formation of the involute crown gear.

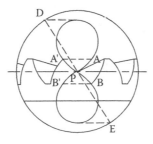

FIGURE 5.24 Octoid crown gear.

Strictly speaking, the tooth profile should be drawn on the surface of a sphere. For practical purposes, this is not necessary and an approximation is used.

5.5.3 BEVEL GEARS DETAILS

For considering the details of a bevel gear, an axial section of a pair of Gleason straight–tooth bevel gears is shown in Figure 5.25a. The Gleason system has been adopted as the standard for bevel gears. As seen in the figure, the dedendum elements are drawn toward the apex of the pitch cones. The addendum elements, however, are drawn parallel to the dedendum elements of the mating member, thus giving a constant clearance and eliminating possible fillet interference at the small ends of the teeth. Elimination of this possible interference allows larger edge radii to be used on the generating tools, which will increase tooth strength through increased fillets. The large ends of the teeth are proportioned according to the long and short addendum system discussed in Chapter 4 so that the addendum on the pinion will be greater than that on the gear. Long addendums are used on the pinion primarily to avoid undercutting, to balance tooth wear, and to increase tooth strength. Figure 5.25b is the transverse section A–A, which shows the tooth profiles.

The addendum and dedendum are measured perpendicularly to the pitch cone element at the outside of the gear; therefore, the dedendum angle is given by

$$\tan \delta = \frac{b}{A_O} \tag{5.35}$$

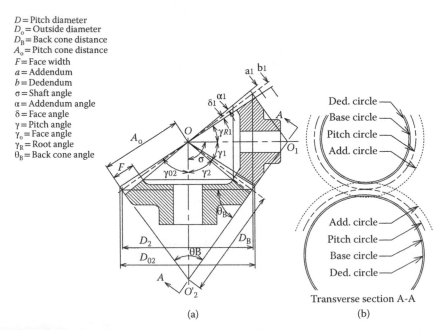

D = Pitch diameter
D_o = Outside diameter
D_B = Back cone distance
A_o = Pitch cone distance
F = Face width
a = Addendum
b = Dedendum
σ = Shaft angle
α = Addendum angle
δ = Face angle
γ = Pitch angle
γ_o = Face angle
γ_R = Root angle
θ_B = Back cone angle

(a)

Transverse section A-A

(b)

FIGURE 5.25   Bevel gear details. (a) Axial section (b) transverse section.

Because the addendum element is not drawn toward the apex of the pitch cones, the addendum angle α must be determined indirectly. It can be shown that the addendum angle of the pinion is equal to the dedendum angle of the gear. Likewise, the addendum angle of the gear is equal to the dedendum angle of the pinion. The face angle and the root angle are therefore

$$\gamma_o = \sigma + \alpha \tag{5.36}$$

$$\gamma_R = \sigma - \delta \tag{5.37}$$

Because the back angle is equal to the pitch angle, the outside diameter of a bevel gear is

$$D_o = D + 2a \cos \sigma \tag{5.38}$$

The face width of a bevel gear is not determined by the kinematics of tooth action but by requirements of manufacture and load capacity. Manufacturing difficulties are encountered if the face width of the gear is too large a proportion of the cone distance A_O. Therefore, the face width is limited as follows:

$$F \leq 0.3\, A_O \tag{5.39a}$$

or

$$F \leq \frac{10.0}{P} = 10.0 \text{ m} \tag{5.39b}$$

The smaller value is chosen.

5.5.4 BEVEL GEAR TOOTH PROPORTION

1. Number of teeth
   16 teeth or more teeth in the pinion.
   15 teeth in pinion and 17 teeth or more in gear.
   14 teeth in pinion and 20 teeth or more in gear.
   13 teeth in pinion and 30 teeth or more in gear.
2. Pressure angle, $\varphi = 20°$
3. Working depth:

$$h_w = \frac{2.000}{P}$$
$$= 2.000 \text{ m}$$

4. Whole depth:

$$h_t = \frac{2.188}{P} + 0.002$$
$$= 2.188 \text{ m} + 0.05$$

5. Addendum
   Gear:

$$a_G = \frac{0.540}{P} + \frac{0.460}{P\left(\dfrac{N_2}{N_1}\right)^2}$$

$$= 0.540 \text{ m} + \frac{0.460}{\left(\dfrac{N_2}{N_1}\right)^2}$$

Pinion:

$$a_G = \frac{2.000}{P} - a_G$$

$$= 2.000 \text{ m} - a_G$$

6. Dedendum
   Gear:

$$b_G = \frac{2.188}{P} - a_G$$

$$= 2.188 \text{ m} - a_G$$

Pinion:

$$b_P = \frac{2.188}{P} - a_P$$

$$= 2.188 \text{ m} - a_P$$

7. Addendum

Gear: $t_G = \dfrac{p}{2} - (a_P - a_G) \tan \varphi$ (approximately)

Pinion: $t_P = p - t_G$

where p is the circular pitch.

5.5.5 ANGULAR STRAIGHT BEVEL GEARS

The proportions of angular straight bevel gears can be determined from the same relations as given for bevel gears at right angles with the following exceptions:

1. The limiting number of teeth cannot be taken from item one in Section 5.5.4. Each application must be examined separately for undercutting with the aid of a chart available from the manufacturer manual. This chart shows a plot of maximum pinion dedendum angle for no undercut versus pitch angle. Curves are given for several pressure angles.
2. The pressure angle is determined in conjunction with the preceding item.

3. In determining the gear addendum from item five in Section 5.5.4, it is necessary to use an equivalent 90° bevel gear ratio for the ratio N_2/N_1.

$$\text{Equivalent } 90° \text{ ratio} = \frac{N_2 \cos\sigma_2}{N_1 \cos\sigma_1}$$

For a crown gear ($\sigma = 90°$), this ratio equals infinity. For angular bevel gears where the shaft angle is greater than 90° and the pitch angle of the gear is also greater than 90°, an internal bevel gear results. In this case, the calculations should be referred to the manufacturer whether the gears can be cut or not.

5.5.6 Zerol Bevel Gears

The teeth of Zerol bevel gears (Figure 5.26) are curved but lie in the same general direction as teeth of straight bevel gears. They may be thought of as spiral bevel gears of zero spiral angle and are manufactured on the same machines as spiral bevel gears. The face cone elements of Zerol bevel gears do not pass through the pitch cone apex but instead are approximately parallel to the root cone elements of the mating gear to provide uniform tooth clearance. The root cone elements also do not pass through the pitch cone apex because of the manner in which these gears are cut. Zerol bevel gears are used in place of straight bevel gears when generating equipment of the spiral type but the straight type is not available and may be used when hardened bevel gears of high accuracy (produced by grinding) are required.

5.5.7 Spiral Bevel Gears

Spiral bevel gears (Figures 5.26 and 5.27) have curved oblique teeth on which contact begins gradually and continues smoothly from end to end. They mesh with a rolling contact similar to straight bevel gears. As a result of their overlapping tooth action, however, spiral bevel gears will transmit motion more smoothly than straight bevel or Zerol bevel gears, reducing noise and vibration, which become especially noticeable at high speeds.

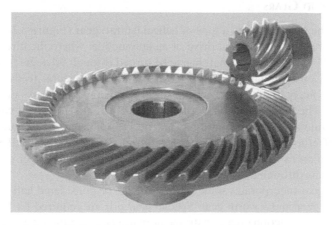

FIGURE 5.26 A form of spiral bevel gear set.

FIGURE 5.27   Spiral bevel gear set.

One of the advantages associated with spiral bevel gears is the complete control of the localized tooth contact. By making a slight change in the radii of curvature of the mating tooth surfaces, the amount of surface over which tooth contact takes place can be changed to suit the specific requirements of each job. Localized tooth contact promotes smooth, quiet running spiral bevel gears and permits some mounting deflections without concentrating the load dangerously near either end of the tooth.

5.5.8   HYPOID GEARS

A hypoid is the name given to a type of helical (spiral) gear (Figure 5.28). The main application of this is in the final drive of an automobile, where the direction of the drive carried by the propeller shaft has to be turned through 90 degrees to drive the rear wheels. Conventional straight cut gears, with perpendicular teeth, are considered to be too noisy in use, and a normal spiral bevel does not always give sufficient contact area. The hypoid gear places the pinion off-axis to the crown wheel, which allows the pinion to be larger in diameter. In a normal passenger car, the pinion is always offset to the bottom of the crown wheel. This allows propeller shaft that drives the pinion to be lowered so that the "hump" in the passenger compartment floor that it runs through does not intrude too much.

A hypoid gear incorporates some sliding and can be considered halfway between a straight cut gear and a worm gear. Special gear oils are required for hypoid gears because the sliding action creates extreme pressure between the teeth.

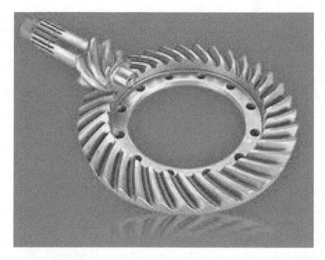

FIGURE 5.28 Hypoid gear set.

PROBLEMS

5.1 A helical gear with a helix angle of 23° that has 24 teeth is cut with a standard 14½° full-depth hob of diametral pitch 2.5. Find the circular pitch in the normal and transverse planes, the transverse pressure angle, and the face width if the tooth in one end is rotated one p_c on the other end.

5.2 The same as Problem 5.1 except that the gear has 36 teeth, a helix angle of 18°, and it was generated with a full-depth, 6-pitch, and 25° pressure angle hob.

5.3 The same as Problem 5.1 except that the gear has 48 teeth, a helix angle of 12°, and was generated with a full-depth, 8-pitch, and 14½° pressure angle hob.

5.4 Find the equivalent pitch radius and the equivalent number of teeth for the gear in Problem 5.1.

5.5 Find the equivalent pitch radius and the equivalent number of teeth for the gear in Problem 5.2.

5.6 A pair of helical gears with a velocity ratio of 2 is required. The pinion should not have less than 15 teeth cut with a standard 20° full-depth hob of 5-pitch. The center distance is 5 in. Determine the helix angle, the diametral pitch, the transverse pressure angle, and the minimum face width.

5.7 A 16-teeth helical pinion to run at 2000 rpm and drive a helical gear on a parallel shaft at 500 rpm. The centers of the shafts are 12 in. apart. Using a helix angle at 24° and a pressure angle at 20°, determine the number of teeth, pitch diameters, normal circular pitch, and the diametral pitch.

5.8 For an overlap of 1.5, calculate the required face width of the gear in Problem 5.1.

5.9 For an axial contact ratio of 1.4, calculate the required face width of the gear of Problem 5.2.

5.10 A pair of helical gears having 15 and 24 teeth is to be mounted on parallel shafts 2.5 in. apart. They are to be cut with a standard hob and the helix angle is to be a minimum. Find the diametral pitch of a suitable hob.

5.11 A 14-teeth helical gear is to be cut by a 2.5-module, 20° hob. Calculate the minimum helix angle for this gear without undercutting.

5.12 Two equal spur gears of 48 teeth, 25.4-mm face width, and a 4-module mesh together in the drive of a fatigue tester. Calculate the helix angle of a pair of helical gears to replace the spur gears if the face width, center distance, and velocity ratio are to remain the same. Use the following cutters: (a) 4-module (b) 4-normal-module hob.

5.13 Two standard spur gears were cut with a 2.5-module, 20° hob to give a velocity ratio of 3.5:1 and center distance of 168.75 mm. Helical gears are to be cut with the same hob to replace the spur gears keeping the center distance and angular velocity ratio the same. Determine the helix angle, number of teeth, and face width of the new gears keeping the helix angle to a minimum.

5.14 Two standard spur gears are to be replaced by helical gears. The spur gears were cut by a 3-module, 20° hob, the velocity ratio is 1.75:1, and the center distance is 132 mm. The helical gears are to be cut with the same hob and maintain the same center distance. The helix angle is to be between 15° and 20° and the velocity ratio between 1.70 and 1.75. Find the number of teeth, helix angle, and velocity ratio.

5.15 In a proposed gear drive, two standard spur gears (1.5 module) with 36 and 100 teeth respectively are meshed at the standard center distance. It is decided to replace spur gears with helical gears having a helix angle of 22° and the same number of teeth. Determine the change in center distance required if the helical gears are cut with 1.5-module, 20° hob.

5.16 A pair of helical gears for parallel shafts is to be cut with a 3-module hob. The helix is to be 20° and the center distance between 152.40 and 158.75 mm. The angular velocity ratio is to approach 2:1 as closely as possible. Calculate the circular pitch and the module in the plane of rotation. Determine the number of teeth, pitch diameters, and the center distance to satisfy the above conditions.

5.17 A 2.5-module, 20-teeth spur pinion drives two gears, one of 36 teeth and the other of 48 teeth. It is desired to replace all three gears with helical gears and to change the velocity ratio between the 20-teeth gear shaft and the 48-teeth gear shaft to 2:1. The velocity ratio and the center distance between the 20-teeth gear shaft and the 36-teeth remain the same. Using a 3-module, 20° hob and keeping the helix angle as low as possible, determine the number of teeth, helix angle and hand, face width, and the outside diameter for each gear. Calculate the change in center distance between the shafts that originally mounted the 20- and the 48-teeth gears.

5.18 A 2-module, 24-teeth spur pinion drives two gears, one of 36 teeth and the other of 60 teeth. It is necessary to replace all three gears with helical gears keeping the same velocity ratios and center distances. Using a 1.5-module, 20° hob and keeping the helix angle as low as possible, determine the number of teeth, helix angle and hand, face width, and the outside diameter for each gear.

5.19 Two parallel shafts are to be connected by a pair of helical gears (gears 1 and 2). The angular velocity ratio is to be 1.25:1 and the center distance 114.3 mm. In addition, gear 2 is to drive a helical gear 3 whose shaft is at right angles to shaft 2. The angular velocity ratio between gears 2 and 3 is to be 2:1. Using a 2.75-module, 20° hobs, determine the number of teeth, helix angle, and pitch diameter of each gear and find center distance C_{23}.

5.20 Two parallel shafts are to be connected by a pair of helical gears (gears 1 and 2). The angular velocity ratio is to be 1.75:1 and the center distance 69.85 mm. In addition, gear 2 is to drive a third helical gear (gear 3) with an angular velocity ratio of 2:1. Three hobs are available for cutting the gears: hob A (3.5 module), hob B (2.75 module), and hob C (2 module). (a) Choose the hob that will result in the smallest helix angle ψ. (b) Which hob will permit the shortest center distance C_{23} while maintaining a helix angle less than 35°?

5.21 If a pair of crossed helical gears, A and B, rotates at rates D_A and D_B respectively and helix angles ψ_A and ψ_B respectively, show that the sliding velocity along the tooth helices at the pitch point is equal to

$$\pi(n_A D_A \sin\psi_A + n_B D_B \sin\psi_B)$$

where n_A and n_B are the speeds. Hence, show that if the gear ratio m_G, the center distance C, the normal diametral pitch P_n, and the shaft angle θ are fixed, this sliding velocity will be a minimum, when

$$\cot\psi_A = m_G + \frac{\cos\theta}{\sin\theta}$$

5.22 Two crossed helical gears 1 and 2 have their axes inclined at an angle θ. Show that the teeth can be designed to give either clockwise or counterclockwise rotation to 2 while 1 is rotating in a clockwise direction. If the center distance is 15 in., the velocity ratio $N_1/N_2 = 2$, the normal pitch is 0.75 in., the angle θ is 50°, and the helix angle is the same for both gears, find for both directions of rotation of 2:
- The number of teeth on each gear.
- The helix angle.
- The circular pitch.
- The exact center distance.

5.23 Two nonintersecting shafts at an angle of 60° are to be connected by crossed helical gears of the smallest possible size. The shortest distance between their axes is 3 in. The teeth are cut with a standard 14½° hob of 4-pitch. The angular speed of the driver is 10 rad/s and the speed ratio is 2. Find the tooth numbers, helix angles, diametral pitches, diameters, pressure angles, and sliding speed (graphically and analytically).

5.24 Two shafts crossed at right angles are connected by helical gears (gears 1 and 2) cut with a 2-module, 20° hob. Both gears are right handed and the angular velocity is 15:1; $D_2 = 131.64$ mm and $\psi = 60°$. A design modification requires a reduction of the outside diameter (o.d.) of gear 1 by 6.35 mm to provide clearance for a new component. Assuming that the same hob must be used for cutting any new gears, show that o.d. of gear 1 can be reduced without changing the velocity ratio, the shaft angle, and the number of gear teeth N_1 and N_2. The o.d. of gear 2 and the center distance may be altered if necessary. In the analysis, calculate and compare the following data for both the original and the new gears: C_{12}, D_1, D_2, N_1, N_2, ψ_1, ψ_2.

5.25 Two crossed shafts are connected by helical gears. The velocity ratio is 18.1:1 and the shaft angle 45°. If $D_1 = 57.735$ mm and $D_2 = 93.175$ mm, calculate the helix if both gears have the same hand.

5.26 Two crossed shafts are connected by helical gears. The velocity ratio is 3:1, the shaft angle is 60°, and the center distance is 254 mm. If the pinion has 35 teeth and a normal module of 3, calculate the helix angles and the pitch diameters if the gears are of the same hand.

5.27 A double-thread worm drives a gear having 60 teeth. The axes are at 90°. The axial pitch of the worm is 1¼ in. and the pitch diameter is 3 in. Determine the helix angle of the worm, the lead, and the center distance of the gears.

5.28 A worm having 4 teeth and a lead of 1 in. drives a gear at a velocity ratio of 8. Determine the pitch diameters of the worm and worm gear for a center distance of 2 in.

5.29 Solve Problem 5.28 with the following changes: Number of teeth on the worm is 3, the speed ratio is 36, axial pitch is 3/8 in., center distance is 10 in., and the angle between the shafts is 90°.

5.30 A worm having 4 threads and rotating at 2400 rpm drives a gear at a velocity ratio of 15. The axial pitch of the worm is ½ in., the center distance is 6 in., and the shaft angle 90°. Calculate the gear pitch diameter, the worm pitch diameter, the outside diameter of the worm, the lead angle of the worm, and the sliding speed.

5.31 It is required to design a worm and worm gear set. Let the circular pitch be 1.5163 in. and the pressure angle be 26°. The lead angle is 40.5°, and the center distance is close to 16 in. Use worms with 3, 4, or 5 teeth. Make your choice based on the merits of the proportions of each. Then, calculate the normal circular and diametral pitches and the pressure angle.

5.32 A double-threaded worm having a lead of 64.292 mm drives a worm gear with a velocity ratio of 19.5:1; the angle between the shafts is 90°. If the center distance is 235.0, determine the pitch diameter of the worm and worm gear.

5.33 A worm and worm gear with shafts at 90° and a center distance of 178.0 mm have a velocity ratio of 17.5:1. If the axial pitch of the worm is to be 26.192 mm, determine the maximum number of teeth in the worm and worm gear that can be used for the drive and determine their corresponding pitch diameters.

5.34 A worm and worm gear connect shafts at 90°. Derive equations for the diameters of the worm and the worm gear in terms of the center distance C, velocity ratio ω_1/ω_2, and the lead angle λ.

5.35 A worm and worm gear with shafts at 90° and a center distance of 152.0 mm have a velocity ratio of 20:1. If the axial pitch of the worm is to be 17.463 mm, determine the smallest diameter worm that can be used for the drive.

5.36 A double-threaded worm drives a 31-teeth worm gear with shafts at 90°. If the center distance is 210.0 mm and the lead angle of the worm 18.83°, calculate the axial pitch of the worm and the pitch diameters of the two gears.

5.37 A three-threaded worm drives a 35-teeth worm gear having a pitch diameter of 207.8 mm and a helix angle of 21.08°. If the shafts are at right angles, calculate the lead and the pitch diameter of the worm.

5.38 A four-threaded worm drives a worm gear with an angular velocity ratio of 8.75:1 and a shaft angle of 90°. The axial pitch of the worm is 18.654 mm and the lead angle 27.22°. Calculate the pitch diameters of the worm and worm gear.

5.39 A six-threaded worm drives a 41-teeth worm gear with a shaft angle of 90°. The center distance is 88.90 mm and the lead angle 26.98°. Calculate the pitch diameter, the lead, and the axial pitch of the worm.

5.40 A worm and worm gear with shafts at 90° and a center distance of 76.20 mm have a velocity ratio of 71:1. Using a lead angle of 28.88°, determine the pitch diameters. Select numbers of teeth for the gears considering worms with 1–10 threads.

5.41 A worm and worm gear with shafts at 90° and a center distance of 102.0 mm have a velocity ratio of 16.5:1 and a lead angle of the worm of 13.63°. Determine the various pairs of gears that can be used considering worms with 1–10 threads. Specify the numbers of teeth and pitch diameters.

5.42 On Gleason straight bevel gears, the working depth is $2/P$ and the clearance is $0.188/P + 0.002$ in. For ratios between 1.7 and 1.76, the gear addendum is $0.79/P$. Calculate the root and face angles of a bevel gear pair having 24 and 36 teeth and a diametral pitch of 10. The two shafts are perpendicular.

5.43 A pair of 2-diametral-pitch straight bevel gears has 18 teeth and 29 teeth. The two axes are perpendicular. Determine the pitch diameters, pitch angles, addendum, dedendum, face width, and the pitch diameters of the equivalent spur gears (add. $= 0.17/P$, working depth $= 2/P$, whole depth $= 2.188/P_{eq}$).

5.44 A pair of bevel gears has a velocity ratio ω_1/ω_2, and the shaft centerlines intersect at an angle σ. If the distances x and y are laid off from the intersection point along the shaft axes in the ratio ω_1/ω_2, prove that the diagonal of the parallelogram with sides x and y will be the common pitch cone element of the bevel gears.

5.45 A Gleason crown bevel gear of 24 teeth and a module of 5.08 is driven by a 16-teeth pinion. Calculate the pitch diameter and the pitch angle of the pinion, the addendum and dedendum, the face width, and the pitch diameter of the gear.

5.46 A Gleason crown bevel gear of 48 teeth and a module of 2.12 is driven by 24-teeth pinion.
   • Calculate the pitch angle of the pinion and the shaft angle.
   • Make a sketch (to scale) of the pitch cones of the two gears in mesh. Show the back cones of each gear and label the pitch cones and the back cones.

5.47 A pair of Gleason miter gears has 20 teeth and a module of 6.35. Calculate the pitch diameter, the addendum and dedendum, the face width, the pitch cone face angle, the root angle, and the outside diameter. Make a full-size axial sketch of the gears in mesh using reasonable proportions for the hub and web. Dimension the drawing with the values calculated.

5.48 A Gleason 4.23-module, straight bevel pinion of 21 teeth drives a gear of 27. The shaft angle is 90°. Calculate the pitch angles, the addendums, and the face width of each gear. Make a full-size axial sketch of the gears in mesh with reasonable proportions for the hubs.

5.49 A Gleason 6.35-module, straight bevel pinion of 14 teeth drives a gear of 20 teeth. The shaft angle is 90°. Calculate the addendum and dedendum, circular tooth thickness for each gear, and the pitch and base radii of the equivalent spur gears.

5.50 A pair of Gleason bevel gears meshes with a shaft angle of 75°. The module is 2.54 and the number of teeth in the pinion and gear are 30 and 40 respectively. Calculate the pitch angles and the addendum and dedendum of the pinion and the gear.

6 Gear Trains

6.1 INTRODUCTION

A gear train is a system of two or more meshing gears. The simplest system consists of a driver on one shaft meshing with a follower on another shaft. If both gears are external, the shafts rotate in opposite directions. If one of the pair is an internal gear, the two shafts rotate in the same direction.

Gear trains are used for transmission of power between two shafts when the distance between them is not too large, and when a certain velocity ratio between them is either necessary or desirable.

New trends are toward higher speeds for the prime movers, which necessitate in most cases a step-down in speed for the driven machines. In few cases, a step-up in speed may also be desirable.

If the number of teeth on all members of a gear train is known, the overall speed ratio between the input and output shafts can be easily determined. However, to find a gear train to produce a desired ratio is rather difficult.

Gear trains can be classified into two types, namely, ordinary gear trains and planetary gear trains. In the ordinary gear trains, all gears in a system rotate about fixed axes. In the planetary gear trains, some gears rotate about moving axes.

The train value e is defined as the output speed divided by the input speed.

6.2 ORDINARY GEAR TRAINS

Ordinary gear trains are divided into two types: simple and compound.

6.2.1 SIMPLE GEAR TRAINS

In this type of trains, all shafts and axles are fixed relative to the frame and each of the shafts carries only one gear (Figure 6.1). It is clear that all gears have the same diametral pitch or module.

The magnitude of the velocity ratio of a simple train depends only on the number of teeth on the input and output gears. The intermediate gears or idlers are used only to bridge a given center distance and to satisfy a desired direction of output rotation.

Let m_G be the gear ratio. It is equal to the angular velocity of gear 1 divided by the angular velocity of gear 3:

$$m_G = \frac{\omega_1}{\omega_3}$$

$$= \left(-\frac{N_2}{N_1}\right)\left(-\frac{N_3}{N_2}\right) = \frac{N_3}{N_1}$$

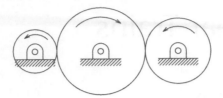

FIGURE 6.1   Simple gear train.

where N is the number of teeth of the gears. The train value e is defined as the reciprocal of the gear ratio.

$$e = \frac{1}{m_G} \tag{6.1}$$

If two idlers are used between the first and the last shafts, then

$$m_G = \left(-\frac{N_2}{N_1}\right)\left(-\frac{N_3}{N_2}\right)\left(-\frac{N_4}{N_3}\right) = -\frac{N_4}{N_1}$$

or, in general,

$$m_G = \frac{\omega_i}{\omega_o} = \frac{N_o}{N_i} \qquad \text{when the number of idlers is odd}$$

$$m_G = -\frac{\omega_i}{\omega_o} = -\frac{N_o}{N_i} \qquad \text{when the number of idlers is even} \tag{6.2}$$

where "i" stands for input and "o" stands for output.

6.2.2   Compound Gear Trains

As in simple trains, the centerlines of all gears are fixed relative to the frame. However, each intermediate shaft carries two gears. If we start from gear 1 (Figure 6.2), whose speed is n, and pass from pair to pair until we reach gear 6, whose speed is n_6, we obtain

$$n_2 = \left(-\frac{N_1}{N_2}\right)n_1 = n_3$$

$$n_4 = \left(-\frac{N_3}{N_4}\right)n_3 = \left(-\frac{N_1}{N_2}\right)\left(-\frac{N_3}{N_4}\right)n_1 = n_5 \tag{6.3}$$

$$n_6 = \left(-\frac{N_5}{N_6}\right)n_5 = \left(-\frac{N_1}{N_2}\right)\left(-\frac{N_3}{N_4}\right)\left(-\frac{N_5}{N_6}\right)n_1$$

If we had started with gear 6, it is easy to show that we would have obtained

$$n_1 = \left(-\frac{N_6}{N_5}\right)\left(-\frac{N_4}{N_3}\right)\left(-\frac{N_2}{N_1}\right)n_6 \tag{6.4}$$

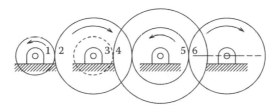

FIGURE 6.2 Compound gear train.

If we start from gear 1, then gears 1, 3, and 5 are drivers, while gears 2, 4, and 6 are followers or driven. Equation 6.3 states that the speed of the last follower is equal to the speed of the first driver multiplied by the product of the tooth numbers on the drivers divided by the product of the tooth numbers on the followers. If we start from gear 6, then gears 6, 4, and 2 are drivers, while gears 5, 3, and 1 are followers, and Equation 6.4 shows that the same rule for speed ratios applies. Thus, any gear in a train may be either a driver or a follower. It depends on which gear we start with. Equations 6.3 and 6.4 may be written as

$$m_G = \frac{\text{Speed of first driver}}{\text{Speed of last follower}} = \frac{\text{Product of teeth on followers}}{\text{Product of teeth on drivers}} \tag{6.5}$$

The sign of the direction of rotations between the first and last gears should be considered.

6.2.3 Reverted Compound Gear Trains

A reverted gear train is one in which the driving shaft is collinear with the output shaft. Figure 6.3 illustrates this type of train. The outstanding feature of this type is the compactness of the assembly. There are some imposed conditions in constructing such gears. Let D be the diameter, N be the number of teeth, P_1 be the diametral pitch of gears 1 and 2, and P_2 be the diametral pitches of gears 3 and 4. Since the axis of the input shaft is aligned with the output shaft,

$$D_1 + D_2 = D_3 + D_4$$

Multiplying the left side by $\frac{P_1}{P_1}$, and multiplying the right hand side by $\frac{P_2}{P_2}$ and expanding, then we get

$$\left(N_1 + N_2\right)P_2 = \left(N_3 + N_4\right)P_1 \tag{6.6}$$

EXAMPLE 6.1

A reverted gear train, as shown in Figure 6.3, is to have a gear ratio of 24/1. The minimum number of teeth in any gear should be 16. Gears 1 and 2 have a diametral pitch of 4; gears 3 and 4 have a diametral pitch of 3. Find the number of teeth required in each gear.

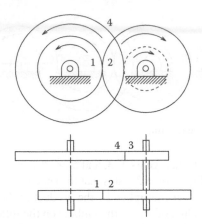

FIGURE 6.3 Reverted gear train.

SOLUTION

The ratio 24/1 is accomplished in two stages. Therefore, the ideal ratio for each stage would be $\sqrt{24}$. However, $\sqrt{24}$ does not give an integer and imposes complication to the calculation. Thus, it is best to factor 24 into two factors that are close to each other. In this case, the ratio 24/1 could be factored into a ratio of 6/1 for one set of gears, say gears 1 and 2, and 4/1 for gears 3 and 4.

$$\frac{N_2}{N_1} = 6 \quad \text{or} \quad N_2 = 6N_1 \tag{a}$$

$$\frac{N_4}{N_3} = 4 \quad \text{or} \quad N_4 = 4N_3 \tag{b}$$

Substituting Equations a and b into Equation 6.6, we arrive at

$$3(N_1 + 6N_1) = 4(N_3 + 4N_3)$$

or

$$\frac{N_1}{N_3} = \frac{20}{21}$$

Since N_1 and N_2 are both integers, the least number of teeth in gears 1 and 3 are

$$N_1 = 20$$

$$N_3 = 21$$

The number of teeth could be multiples of these numbers. However, the stated values are satisfactory since they are more than the minimum number of teeth to avoid undercutting. Therefore, the number of teeth is given by

$$N_1 = 20 \text{ teeth}$$

$$N_2 = 120 \text{ teeth}$$

$$N_3 = 21 \text{ teeth}$$

$$N_4 = 84 \text{ teeth}$$

EXAMPLE 6.2

A reverted gear train, as shown in Figure 6.3, is to have a velocity ratio of 15. Assuming that gears with as few as 12 teeth may be used, find the required tooth numbers. All gears have the same pitch.

SOLUTION

Let

$$\frac{N_2}{N_1} = 5 \quad \text{or} \quad N_2 = 5N_1 \tag{a}$$

$$\frac{N_4}{N_3} = 3 \quad \text{or} \quad N_4 = 3N_3 \tag{b}$$

Since all gears have the same diametral pitch, according to Equation 6.6,

$$N_1 + 5N_1 = N_3 + 3N_3$$

or

$$\frac{N_1}{N_3} = \frac{4}{6}$$

Since the minimum number of teeth allowed is 12,

$$N_1 = 12 \text{ teeth}$$

$$N_2 = 60 \text{ teeth}$$

$$N_3 = 18 \text{ teeth}$$

$$N_4 = 54 \text{ teeth}$$

6.3 EPICYCLIC OR PLANETARY GEAR TRAINS

In an epicyclic train, one or more of the gears rotate about a central axis in a manner similar to that of planets revolving around the sun. These gears are called planet gears or just planets.

6.3.1 ORDINARY PLANETARY GEAR TRAINS

Figure 6.4 shows a type of a planetary gear train; gear 1 is called the sun gear, gear 2 is called the planet gear, and lever A is called the planet carrier or simply the arm. The planet gear rotates freely on the arm pin. The arm rotates freely about a fixed center. Figure 6.4a and b is front and schematic section views of the system, respectively.

The speeds of the components depend on the situation. They depend on which element is the driving element and which is the driven element. There are two approaches used to determine the relations between the speeds of the elements; namely, the analytical method and the tabular method.

6.3.1.1 Analytical Method

For the train shown in Figure 6.4, let

- r_1 be the radius of gear 1
- r_2 be the radius of gear 2
- ω_1 be the angular speed of gear 1; positive in the counterclockwise direction; n_1 is the speed
- ω_2 be the angular speed of gear 2; positive in the counterclockwise direction; n_2 is the speed
- ω_A be the angular speed of the arm; positive in the counterclockwise direction; n_A is the speed

Note that the directions of the arrows in Figure 6.4b indicate the directions of rotations of the elements in the face view.

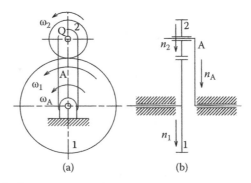

(a)   (b)

FIGURE 6.4 Simple planetary gear train.

Consider the velocity of the pitch point P. Its velocity as a point on gear 1 is given by

$$V_P = \omega_1 r_1 \tag{6.7}$$

The velocity of P as a point on gear 2 is given by

$$V_P = V_Q + V_{PQ}$$

or

$$V_P = \omega_A (r_1 + r_2) - \omega_2 r_2 \tag{6.8}$$

Notice the direction of V_{PQ} as obtained from the direction of ω_2.
Since V_P is the same,

$$\omega_1 r_1 = \omega_A (r_1 + r_2) - \omega_2 r_2$$

The radii r_1 and r_2 are proportional to the number of teeth N, and the angular velocity ω is proportional to the rotating speeds n. Thus,

$$n_1 N_1 = n_A (N_1 + N_2) - n_2 N_2$$

or

$$\frac{n_1 - n_A}{n_2 - n_A} = -\frac{N_2}{N_1} \tag{6.9}$$

When the arm is fixed, the speed ratio of gear 1 to gear 2 is given by

$$\frac{n_1}{n_2} = -\frac{N_2}{N_1} \tag{6.10}$$

Comparing Equations 6.9 and 6.10, we conclude that

$$\frac{n_1 - n_A}{n_2 - n_A} = \left.\frac{n_1}{n_2}\right|_{\text{arm is fixed}} \tag{6.11}$$

Equation 6.11 implies that the speed of any gear in the train relative to the arm is equal to the speed of the gear when the arm is fixed. Equations 6.10 and 6.11 hold for any pair of gears in the ordinary planetary gear trains. Suppose that the sun gear 2 rotates inside the internal gear 3 (Figure 6.5). Using the previous analysis, we can arrive at

$$\frac{n_3 - n_A}{n_2 - n_A} = \left.\frac{n_3}{n_2}\right|_{\text{arm is fixed}} \tag{6.12}$$

Notice that when the arm is fixed, the directions of rotation of gears 2 and 3 are the same. Suppose that an ordinary planetary gear train contains both sun and internal gears (Figure 6.6). Notice that the planet gear and the arm are duplicated in the other side for balancing (shown in dotted lines) and do not affect the speeds.

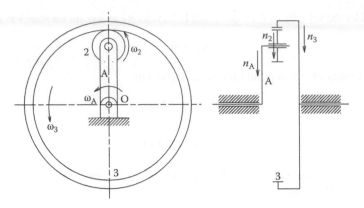

FIGURE 6.5 Planetary gear system with internal gear.

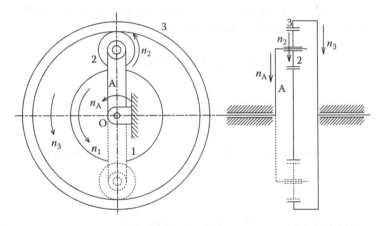

FIGURE 6.6 Balancing the planetary gear system.

We can combine Equations 6.11 and 6.12 to get,

$$\frac{n_3 - n_A}{n_1 - n_A} = -\frac{N_1}{N_3}$$

or

$$\frac{n_3 - n_A}{n_1 - n_A} = \frac{n_3}{n_1}\bigg|_{\text{arm is fixed}}$$

In general, for any two gears, i and j, in one planetary train, the speed ratios of the two gears are given by

$$\frac{n_i - n_A}{n_j - n_A} = \frac{n_i}{n_j}\bigg|_{\text{arm is fixed}} \tag{6.13}$$

EXAMPLE 6.3

In the planetary gear tram of Figure 6.5, the sun gear 1 rotates at 1000 rpm clockwise and the internal (ring) gear 3 rotates at 1000 rpm counterclockwise. Determine the speed of the arm n_A. The number of teeth is

$$N_1 = 30$$

$$N_2 = 20$$

SOLUTION

$$r_3 = r_1 + 2\,r_2$$

Since all the gears have the same diametral pitch,

$$N_3 = N_1 + 2\,N_2$$
$$= 70$$

Applying Equation 6.12, we get

$$\frac{1000 - n_A}{-1000 - n_A} = -\frac{30}{70}$$

This equation gives $n_A = 400$ rpm (400 rpm counterclockwise)

6.3.1.2 Tabular Method

Referring to Equation 6.13

$$\frac{n_i - n_A}{n_j - n_A} = \frac{n_i}{n_j}\bigg|_{\text{arm is fixed}}$$

$$n_i - n_A = (n_j - n_A)\frac{n_i}{n_j}\bigg|_{\text{arm is fixed}} = x$$

Thus,

$$n_i - n_A = x$$

$$(n_j - n_A) = x\frac{n_j}{n_i}\bigg|_{\text{arm is fixed}} \tag{6.14}$$

When the arm is fixed, $n_A = 0$, then

$$n_i\big|_{\text{arm is fixed}} = x$$

and

$$n_j = x\frac{n_j}{n_i}\bigg|_{\text{arm is fixed}}$$

Going back to Equation 6.14, and if $n_A = y$, the total speeds of the gears are given by

$$n_i = x + y$$

$$n_j = x \left. \frac{n_j}{n_i} \right|_{\text{arm is fixed}} + y \tag{6.15}$$

Equation 6.16 can be applied in the form of a table. The number of columns is equal to the number of gears plus the arm. In the first row, we write the name of the element. In the second row, we assume that the arm is fixed. Assume the speed of any gear to be x and find the speeds of the other gears accordingly. In the third row, we write the speed of the arm, say y. In the last row, we add the values in the second and third rows, which represent the total speeds of the gears. Finally, we use the given data to obtain the values of x and y. The procedure is demonstrated by the following examples.

EXAMPLE 6.4

Solve Example 3 using the tabular method (Table 6.1).

SOLUTION

$n_1 = -1000$, and $n_2 = 1000$. Thus,

$$x + y = -1000 \tag{a}$$

$$y - x \frac{30}{70} = 1000 \tag{b}$$

Solving Equations a and b yields

$$x = -1400$$

$$y = 400$$

Therefore, the speed of the arm is 400 counterclockwise.

TABLE 6.1
Solution using Tabular Method

Arm	Gear 1	Gear 2	Gear 3
0	x	$-x \dfrac{N_1}{N_2} = -x \dfrac{30}{20}$	$-x \dfrac{30}{20} \times \dfrac{N_2}{N_3} = -x \dfrac{30}{70}$
y	y	y	y
y	$x + y$	$y - x \dfrac{30}{20}$	$y - x \dfrac{30}{70}$

EXAMPLE 6.5

Automobile Differential

The gears of a differential (Figure 6.7) allow a car's powered wheels to rotate at different speeds as the car turns around corners. The car's drive shaft rotates the crown wheel, which in turn rotates the half shafts leading to the wheels. When the car is traveling straight ahead, the planet pinions do not spin, so the crown wheel rotates both wheels at the same rate. When the car turns a corner, however, the planet pinions spin in opposite directions, allowing one wheel to slip behind and forcing the other wheel to turn faster. Notice that the planet pinions can rotate freely relative to the frame attached to the crown wheel. A photo of an actual automobile differential box is shown in Figure 6.8.

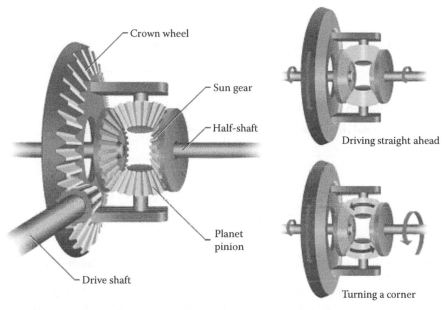

FIGURE 6.7   Automobile differential.

FIGURE 6.8   An actual automobile differential.

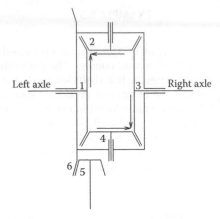

FIGURE 6.9 Schematic diagram for the automobile differential.

TABLE 6.2
Solution of the Automobile Differential System

Arm and crown gear	Gear 1	Gears 2 and 4	Gear 3
0	x	$x\dfrac{N_1}{N_2}$	$-x$
y	y		y
y	$x+y$		$y-x$

ANALYSIS

The analysis is better understood with the help of the schematic sketch shown in Figure 6.9.

Gear 1 is attached to the left axle, while gear 3 is attached to the right axle; both have the same number of teeth. Planet gears 2 and 4 have the same number of teeth and spin freely in the box attached to the crown gear. We should be aware that the plane of rotation of gears 1 and 3 is different from the plane of rotation of gears 2 and 4. We construct the table for the system (Table 6.2).

When the automobile moves on a straight road ($x = 0$), both wheels rotate with the same speed as the crown gear. When it moves on a turn, the inside wheel slows down by amount x due to friction, while the outside wheel speeds up with the same amount x. This is done automatically.

6.3.2 COMPOUND PLANETARY GEAR TRAINS

In the compound gear trains, some axes carry more than one gear.

EXAMPLE 6.6

Consider the epicyclic gear train of Figure 6.10. Gears 1 and 5 are sun gears, while the planets 2 and 3 are attached together and rotate freely on arm A. The arm rotates freely about the central axis, which coincides with the axes of the

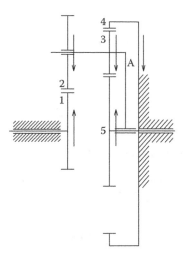

FIGURE 6.10 Compound planetary gear train system.

input and output shafts. The internal gear 4 is fixed in the frame. If the input shaft (gear 1) rotates at 700 rpm clockwise, determine the speeds of all the gears. All gears have the same diametral pitch. The number of teeth is

$$N_1 = 25, \ N_2 = 30, \ N_3 = 20$$

SOLUTION

Since all gears have the same diametral pitch,

$$N_1 + N_2 + N_3 = N_4$$

$$N_4 = 75$$

Also,

$$N_1 + N_2 = N_3 + N_5$$

$$N_5 = 35$$

1. Analytical method
   The directions of the arrows indicate the directions of rotations of the gears relative to each other when the arm is fixed.

$$\frac{n_2 - n_A}{n_1 - n_A} = -\frac{N_1}{N_2}$$

$$\frac{n_2 - n_A}{-700 - n_A} = -\frac{25}{30}$$

(a)

Also,

$$\frac{n_3 - n_A}{n_4 - n_A} = \frac{N_4}{N_3}$$

or

$$\frac{n_3 - n_A}{0 - n_A} = \frac{75}{20} \tag{b}$$

Since $n_2 = n_3$, from Equations a and b

$$\frac{-700 - n_A}{0 - n_A} = -\frac{75}{20} \times \frac{30}{25} = -\frac{9}{2}$$

$$n_A = -\frac{1400}{11}$$

We can include gears 4 and 5 in one equation. In this case, gear 3 is idler. Thus,

$$\frac{n_5 - n_A}{0 - n_A} = -\frac{75}{35} = -\frac{15}{7}$$

$$n_5 = -400$$

2. Tabular method (Table 6.3)
   From the second and fourth columns,

$$x + y = -700 \tag{a}$$

$$y - x\frac{25}{30} \times \frac{20}{75} = 0$$

or

$$9y - 2x = 0 \tag{b}$$

From Equations a and b

TABLE 6.3

Solution of EXAMPLE 6.6

Arm	Gear 1	Gears 2 and 3	Gear 4	Gear 5
0	x	$-x\frac{N_1}{N_2} = -x\frac{25}{30}$	$-x\frac{25}{30} \times \frac{20}{75}$	$x\frac{25}{30} \times \frac{20}{35}$
y	y	y	y	y
y	$x + y$	$y - x\frac{25}{30}$	$y - x\frac{25}{30} \times \frac{20}{75}$	$y + x\frac{25}{30} \times \frac{20}{35}$

$$y = -\frac{1400}{11} \quad \text{(which is the speed of the arm)}$$

$$x = -\frac{3000}{11}$$

Substituting in the last column,

$$n_5 = -400 \text{ rpm}$$

EXAMPLE 6.7

Consider the gear train shown in Figure 6.11. Gear 1 rotates at 1200 rpm counter-clockwise. Find the speed of gear 5. The number of teeth is

$$N_1 = 20, N_2 = 80, N_3 = 100, N_4 = 15, N_5 = 30$$

SOLUTION

Applying Equation 6.11 to gears 1 and 3, we arrive at

$$\frac{n_1 - n_A}{n_3 - n_A} = \frac{n_1}{n_3}\Big|_{\text{arm is fixed}}$$

$$\frac{1200 - n_A}{0 - n_A} = -\frac{100}{20}$$

$$n_A = 200 \text{ rpm}$$

Using gears 3 and 5, we get

$$\frac{n_5 - 200}{0 - 200} = \frac{100 \times 15}{80 \times 30}$$

$$n_5 = 75 \text{ rpm}$$

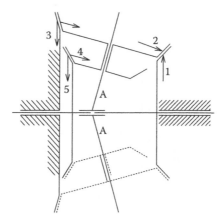

FIGURE 6.11 System including inclined arms.

Tabular method (Table 6.4)
From the second and fourth columns,

$$x + y = 1200$$

$$y - x\frac{1}{5} = 0$$

$$y = 200 \text{ (which is the speed of the arm)}$$

$$x = 1000$$

Substituting in the last column, we get

$$n_5 = 75 \text{ rpm}$$

Important Note: When the gear train includes two axes not in line, the group of gears on each axis should be treated separately.

EXAMPLE 6.8

For the system shown in Figure 6.12, shaft A rotates at 100 rpm clockwise, while shaft B rotates at 50 rpm clockwise. Determine the output speed at shaft C. The number of teeth is

$$N_1 = 24, N_2 = 30, N_3 = 60, N_4 = 50, N_5 = 52, N_6 = 58$$

TABLE 6.4
Solution of EXAMPLE 6.7

Arm	Gear 1	Gears 2 and 4	Gear 3	Gear 5
0	x	$x\dfrac{20}{80}$	$-x\dfrac{20}{100}$	$-x\dfrac{20}{80} \times \dfrac{15}{30}$
y	y			y
y	$x+y$		$y - x\dfrac{20}{100}$	$y - x\dfrac{20}{80} \times \dfrac{15}{30}$

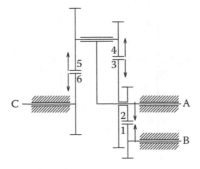

FIGURE 6.12   System of EXAMPLE 6.8.

TABLE 6.5
Table for EXAMPLE 6.8

Arm	Gear 3	Gears 4 and 5	Gear 6
0	x	$-x\dfrac{60}{50}$	$x\dfrac{60}{50}\times\dfrac{52}{58}$
y	y	y	y
y	$x+y$	$y-x\dfrac{60}{50}$	$y+x\dfrac{60}{50}\times\dfrac{52}{58}$

<small>SOLUTION</small>

Axes A and B are treated separately. Thus, the speed of gear 2 is

$$n_2 = n_3 = -50 \times \left(-\frac{24}{30}\right) = 40 \text{ rpm}$$

Axes A and C are treated separately. Thus,

$$\frac{n_6 + 100}{40 + 100} = \frac{60 \times 52}{50 \times 58}$$

$$n_C = n_6 = 50.6 \text{ rpm}$$

Tabular Method (Table 6.5)

$$y = -100$$

$$x = 140$$

$$n_6 = 50.6 \text{ rpm}$$

6.3.3 SYSTEMS WITH SEVERAL ARMS

Some gear trains contain more than one arm. Each arm carries its own planet gears. In this case, each arm is treated separately as demonstrated by the following examples.

EXAMPLE 6.9

Find the overall speed reduction ratio for the gear train shown in Figure 6.13 in terms of the number of teeth of the gears.

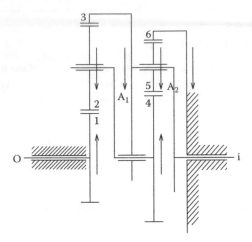

FIGURE 6.13   Gear train of EXAMPLE 6.9.

SOLUTION

The input shaft "i" is connected to gear arm A_2 and gear 3. Gear 4 is connected to arm A_1. Gear 6 is fixed. The output shaft is connected to gear 1. Assume that the output speed is one. Thus,

$$n_3 = n_{A_2} = 1$$

$$n_6 = 0$$

$$n_4 = n_{A_1}$$

For the second arm A_2,

$$\frac{n_6 - n_{A_2}}{n_4 - n_{A_2}} = -\frac{N_4}{N_6}$$

Thus,

$$n_4 = 1 + \frac{N_6}{N_4} \tag{a}$$

For the first arm A_1,

$$\frac{n_3 - n_{A_1}}{n_1 - n_{A_1}} = -\frac{N_1}{N_3}$$

$$\frac{1 - n_4}{n_1 - n_4} = -\frac{N_1}{N_3} \tag{b}$$

$$n_1 = -\frac{N_3}{N_1} + n_4\left(1 + \frac{N_3}{N_1}\right)$$

From Equations a and b

$$n_o = 1 + \frac{N_6}{N_4}\left(1 + \frac{N_3}{N_1}\right)$$

If we use the tabular method (Table 6.6), we have to construct a table for each arm and then apply the given conditions.

For arms 1 and 2 (Table 6.6), the table contains four unknowns and we have four train conditions

For $n_{A_2} = 1$,

$$v = 1$$

For $n_6 = 0$,

$$v - u\frac{N_4}{N_6} = 0$$

or

$$u = \frac{N_6}{N_4}$$

For $n_4 = n_{A_1}$,

$$y = u + v - 1 + \frac{N_6}{N_4}$$

TABLE 6.6
Table for EXAMPLE 6.9

Arm 1	Gear 1	Gear 2	Gear 3
0	x	$-x\dfrac{N_1}{N_2}$	$-x\dfrac{N_1}{N_2}\times\dfrac{N_2}{N_3}$
y	y	y	y
y	$x+y$	$y-x\dfrac{N_1}{N_2}$	$y-x\dfrac{N_1}{N_3}$

Arm 2	Gear 4	Gear 5	Gear 6
0	u	$-u\dfrac{N_4}{N_5}$	$-u\dfrac{N_4}{N_5}\times\dfrac{N_5}{N_6}$
v	v	v	v
v	$u+v$	$v-u\dfrac{N_4}{N_5}$	$v-u\dfrac{N_4}{N_6}$

For $n_3 = 1$,

$$y - x\frac{N_1}{N_3} = 1$$

$$x = \frac{N_3 N_6}{N_1 N_4}$$

The speed of the output shaft is n_3 and is given by

$$n_o = x + y$$

$$= \frac{N_3 N_6}{N_1 N_4} + 1 + \frac{N_6}{N_4}$$

$$n_o = 1 + \frac{N_6}{N_4}\left(1 + \frac{N_3}{N_1}\right)$$

EXAMPLE 6.10

Find the speed reduction for the triple-planetary gear drive shown in Figure 6.14.

SOLUTION

Gears 2, 4, and 6 are fixed to the output shaft. Gear 1 is fixed to the input shaft. Arm A_1 is fixed. Arm A_2 is attached to gear 5, while arm A_3 is attached to gear 3. Let the output speed be one. Thus,

$$n_2 = n_4 = n_6 = 1$$

Consider arm 1,

$$\frac{n_5 - 0}{n_6 - 0} = -\frac{N_6}{N_5}$$

(a)

$$n_5 = -\frac{N_6}{N_5} = n_{A_2}$$

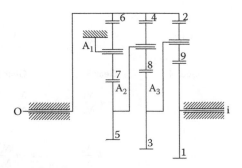

FIGURE 6.14 System for EXAMPLE 6.10.

Consider arm 2,

$$\frac{n_3 - n_5}{n_4 - n_5} = -\frac{N_4}{N_3}$$

(b)

$$n_3 = n_5 - \frac{N_4}{N_3}(1 - n_5) = n_{A_2}$$

Consider arm 3,

$$\frac{n_1 - n_3}{n_2 - n_3} = -\frac{N_2}{N_1}$$

(c)

$$n_1 = n_3 - \frac{N_2}{N_1}(1 - n_3)$$

From Equations a, b, and c

$$n_o = \left(1 + \frac{N_2}{N_1}\right)\left[\left(1 + \frac{N_4}{N_3}\right)\left(-\frac{N_6}{N_5}\right) - \frac{N_4}{N_3}\right] - \frac{N_2}{N_1}$$

EXAMPLE 6.11

Figure 6.15 shows a diagrammatic representation of the Trojan automotive gear-box that has three forward gears and one reverse. The first gear is engaged by tightening Band 2. The application of Band 3 gives the second gear, and the closure of Band 1 yields the reverse. Top gear, that is, direct coupling of the output and input shafts, is brought about by interlocking the drums of B_2 and B_3 by means of separate bands (not shown in the sketch). The numbers of teeth are as follows:

$$N_1 = 17, \ N_2 = 26, \ N_3 = 21, \ N_4 = 22, \ N_5 = 25, \ N_6 = 18, \ N_7 = 20, \ N_8 = 23$$

Determine the four resulting ratios.

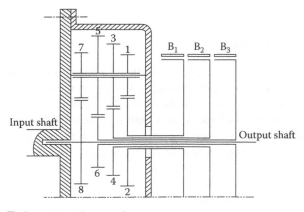

FIGURE 6.15 Trojan automotive gear box.

Solution

The speed of the input shaft is the same as the speed of the arm. Gears 1, 3, 5, and 7 are fixed together and rotate freely on the arm axle. Gear 8 is fixed to the output shaft. If Band 2 is tightened, gear 4 is fixed. Thus,

$$\frac{n_8 - n_A}{n_4 - n_A} = \frac{N_4 N_7}{N_3 N_8}$$

$$n_8 = n_A \left(1 - \frac{N_4 N_7}{N_3 N_8} \right) = n_A \left(1 - \frac{22 \times 20}{21 \times 23} \right)$$

$$n_o = \frac{1}{11.24} n_i$$

If Band 6 is tightened, gear 6 is fixed. Thus,

$$\frac{n_8 - n_A}{n_6 - n_A} = \frac{N_6 N_7}{N_5 N_8}$$

$$n_8 = n_A \left(1 - \frac{N_6 N_7}{N_5 N_8} \right) = n_A \left(1 - \frac{18 \times 20}{25 \times 23} \right)$$

$$n_o = \frac{1}{2.67} n_i$$

If Band 1 is tightened, gear 3 is fixed. Thus,

$$\frac{n_8 - n_A}{n_2 - n_A} = \frac{N_2 N_7}{N_1 N_8}$$

$$n_8 = n_A \left(1 - \frac{N_2 N_7}{N_1 N_8} \right) = n_A \left(1 - \frac{26 \times 20}{17 \times 23} \right)$$

$$n_o = -\frac{1}{3.03} n_i$$

6.4   GEAR TRAIN DESIGN

If the numbers of teeth for the gears of a train are known, it is simple to determine the speed ratio between the input and output shafts. On the other hand, finding the number of teeth for the gears of a train to satisfy a given ratio is much more difficult. Moreover, the specified speed ratio cannot be exactly obtained in some cases because the gears must have integral numbers of teeth. It may be possible to approximate the specified ratio within a certain degree of accuracy that is sufficient for the required application. This may include some trial-and-error techniques.

The general gear–ratio problem is to determine the number of teeth in a train, termed a, b, c, d, ..., such that the gear ratio m_G is equal to a specified ratio within specified limits.

$$m_G = \frac{a}{b} \times \frac{c}{d} \times \dots$$

An additional restriction is that the number of teeth in any gear has to be less than some number, generally 100, because gears with more teeth are not usually convenient, especially in complicated trains. No other restrictions are imposed on the number of teeth in the gears to be used.

In order to find a gear train ratio equal to a specified value m_{GO} within certain limits, say $\pm\varepsilon$, two requirements must be met:

1. A rational fraction must be produced that represents m_G within the specified limits.
2. The numerator and denominator of this fraction must be factorable into numbers sufficiently small to be within the limits for the minimum and maximum numbers of teeth on a gear.

There are two methods that may be used for this goal.

6.4.1 METHOD OF CONTINUED FRACTIONS

Any number that includes an integer and a fraction may be converted to a continued fraction of the following form:

$$a_0 + \cfrac{1}{a_1 + \cfrac{1}{a_2 + \cfrac{1}{a_3 + \dots}}}$$

This is usually written in the form

$$a_0 + \frac{1}{a_1 +} \ \frac{1}{a_2 +} \ \frac{1}{a_3 +} \dots$$

The term a_0 corresponds to the integer and the remainder corresponds to the fraction. A continued fraction is best understood by a numerical example. Suppose that it is required to fraction 60/127. Thus,

$$\frac{60}{127} = \cfrac{1}{\cfrac{127}{60}} = \cfrac{1}{2 + \cfrac{7}{60}} = \cfrac{1}{2 + \cfrac{1}{\cfrac{60}{7}}} = \cfrac{1}{2 + \cfrac{1}{8 + \cfrac{4}{7}}}$$

$$= \cfrac{1}{2 + \cfrac{1}{8 + \cfrac{1}{\cfrac{7}{4}}}} = \cfrac{1}{2 + \cfrac{1}{8 + \cfrac{1}{1 + \cfrac{3}{4}}}} = \cfrac{1}{2 + \cfrac{1}{8 + \cfrac{1}{1 + \cfrac{1}{\cfrac{4}{3}}}}}$$

Finally,

$$\frac{60}{127} = \cfrac{1}{2 + \cfrac{1}{8 + \cfrac{1}{1 + \cfrac{1}{1 + \cfrac{1}{3}}}}}$$

Thus, the continued fraction is formed by dividing 1 by the inverse of the fraction. This gives an integer a (always 1 or greater) plus a second fraction. This procedure is continued as long as necessary. If two such fractions are generated, corresponding to the two allowed limits of the desired ratio m_G, the first few values of a_i will be identical, after which they will diverge. Assuming that divergence occurs after a_k, the continued fraction is

$$a_0 + \frac{1}{a_1} + \frac{1}{a_2} + \frac{1}{a_3} + \cdots \frac{1}{a_3 + A}$$

where the value of A is between the two decimal remainders. Sometimes, it is convenient to let A take negative values. The first convergent is, for the given example,

$$a_0 = 0, \; a_1 = 2, \; a_2 = 8, \; a_3 = 1, \; a_4 = 1, \; a_5 = 1$$

The values of a are always 1 or more.

Now, we form a series of convergent fractions in the form

$$\frac{b_0}{c_0} = \frac{a_0}{1} \; \left(a_0 = 0 \text{ if } m_G < 1 \right) \tag{6.16}$$

The second convergence is

$$\frac{b_1}{c_1} = \frac{a_0}{1} + \frac{1}{a_0} = \frac{a_0 a_1 + 1}{a_1} \tag{6.17}$$

The remaining convergence is obtained for the recurrence formula

$$\frac{b_{i+1}}{c_{i+1}} = \frac{b_i a_{i+1} + b_{i-1}}{c_i a_{i+1} + c_{i-1}} \tag{6.18}$$

where b_i and c_i are the numerator and the denominator of the ith convergent fraction.

Returning to the numerical example, we determine the values of parameters (a). Now we obtain the values of b and c. From Equation 6.16,

$$\frac{b_0}{c_0} = \frac{a_0}{1} = \frac{0}{1}$$

Thus,

$$b_o = 0$$

$$c_o = 1$$

The values of b_1 and c_1 are obtained from Equation 6.15. $a_o = 0$, $a_1 = 2$. Thus,

$$\frac{b_1}{c_1} = \frac{a_o}{1} + \frac{1}{a_o} = \frac{a_o a_1 + 1}{a_1} = \frac{1}{2}$$

Hence,

$$b_1 = 1$$

$$c_1 = 2$$

To obtain b_2 and c_2, we use the recurrence equation, Equation 6.18:

$$\frac{b_{i+1}}{c_{i+1}} = \frac{b_1 a_{i+1} + b_{i-1}}{c_1 a_{i+1} + c_{i-1}}$$

Putting $i = 1$ and using the previous values b and c, we arrive at

$$\frac{b_2}{c_2} = \frac{b_1 a_2 + b_o}{c_1 a_2 + c_o} = \frac{(1 \times 8) + 0}{(2 \times 8) + 1} = \frac{8}{17}$$

Thus,

$$b_2 = 8$$

$$c_2 = 17$$

The iteration process is carried out until we reach a reasonable convergence.

EXAMPLE 6.12

Find a reasonable fraction for π.

SOLUTION

$$\pi = 3.14159265$$

$$= 3 + \cfrac{1}{\cfrac{1}{0.14159265}} = 3 + \cfrac{1}{7.06251348} = 3 + \cfrac{1}{7 + \cfrac{1}{15.99654986}}$$

The last fraction is almost equal to 16. Thus,

$$a_o = 3, \ a_2 = 7, \ a_3 = 16$$

Thus,

$$b_o = 3, \; b_1 = 22, \; b_2 = 355$$
$$c_o = 3, \; c_1 = 7, \; c_2 = 113$$

Therefore, the first approximation is

$$\frac{b_1}{c_1} = \frac{22}{7} = 3.1428571$$

The second approximation is

$$\frac{b_2}{c_2} = \frac{333}{106} = 3.1415292$$

Comparing with the value of π, we find that the second iteration yields a closer value. Of course, we can still get closer values but with large values of b and c. Therefore, if we want to obtain a gear ratio approximately equal to π, the best solution is to use two gears with 22 and 7 teeth. It is clear that the gear with 7 teeth causes interference. Thus, multiples of 22 and 7 may be used according to the design situation.

6.4.2 Method of Approximate Fractions

Let the speed ratio of the first and last shafts in a train be m_G. If a, b, c, and d are integers chosen arbitrarily such that a/b is slightly larger than m_G and c/d is slightly smaller than m_G, it can be shown that the fraction $\dfrac{ax + c}{bx + d}$ lies between a/b and c/d. The factor x is a rational fraction such that

$$\frac{ax + c}{bx + d} = m_G \tag{6.19}$$

Solving Equation 6.19, we get

$$x = \frac{c - m_G d}{b m_G - a} \tag{6.20}$$

It is clear that for a specified speed ratio, m_G must be presented as a common fraction whose numerator and denominator can be factored out into terms that represent the numbers of teeth on the gears. It is not expected that the value of x obtained from Equation 6.20 rarely fulfills this requirement. A certain approximation should be used to adjust its value as demonstrated by the following example.

EXAMPLE 6.13

Design a gear train for a speed ratio of π such that the error is less than 1×10^6.

SOLUTION

$$\pi = 3.14159265$$

Let

$$\frac{a}{b} = \frac{22}{7} = 3.14285714, \qquad \text{which is higher than } \pi$$

$$\frac{c}{d} = \frac{91}{29} = 3.13793103, \qquad \text{which is less than } \pi$$

According to Equation 6.20

$$x = 11.99659441$$

We put x as a whole number, $x = 12$; then, the left-hand side of Equation 6.19 gives

$$\frac{ax + c}{bx + d} = \frac{(22 \times \pi) + 91}{(7 \times \pi) + 29} = \frac{355}{113} = 3.14159292$$

Notice that the fraction is the same as obtained by the previous method. The error is

$$3.14159292 - \pi = 2.66764189 \times 10^{-7}$$

This result is less than the allowed error.

6.5   BRAKING TORQUE

Since the speed of the output shaft differs from the speed of the input shaft, the torques are not the same. The torque balance necessitates that the frame of the train should be fixed by some torque. There are two methods. The first is by considering the forces in the system, which is not the scope of this chapter. The second is by considering the power flow in the system.

Since the power input is equal to the power output, the sum of the power is equal to zero. Thus,

$$T_{in}\omega_{in} + T_{ot}\omega_{ot} = 0 \qquad (6.21)$$

Also, the sum of the torques in a system, including the braking torque T_{br}, is equal to zero. Thus,

$$T_{in} + T_{ot} + T_{br} = 0 \qquad (6.22)$$

From these two equations, the braking torque is determined. We should consider the following:

- The input torque is in the direction of the input speed.
- The output torque is opposite to the direction of the output speed.
- The output power in Equation 6.21 is negative.

EXAMPLE 6.14

For Example 6.6, if the input power is 20 kW, find the braking torque.

SOLUTION

The input speed is 700 rpm clockwise. Thus, the input torque is

$$T_{in} = \frac{20 \times 60}{2 \times \pi \times 700} = 0.273 \text{ mN clockwise}$$

The output speed is 400 rpm clockwise. Thus, from Equation 6.21

$$0.273 \times 700 = T_{ot} \times 400$$

$$T_{ot} = 0.477 \text{ mN counterclockwise}$$

Considering that the torque is positive in the counterclockwise direction, we obtain the braking torque by applying Equation 6.22:

$$0.273 + 0.477 + T_{br} = 0$$

$$T_{br} = -0.204 \text{ mN clockwise}$$

PROBLEMS

6.1 Determine the speed and the direction of rotation of gear 7 in the compound gear train shown in Figure P6.1

$$N_1 = 26, N_2 = 78, N_3 = 18, N_4 = 72, N_5 = 16, N_6 = 50, N_7 = 54$$

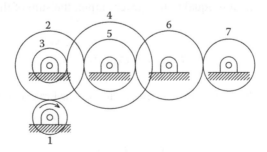

FIGURE P6.1

6.2 For the reverted gear train shown in Figure P6.2, determine the diametral pitch and the number of teeth of gears 3 and 4.

$N_1 = 16$, $N_2 = 24$, $N_3 = 18$, $n_1 = 100$ rpm, $n_4 = 25$ rpm. The center distance is 20 mm.

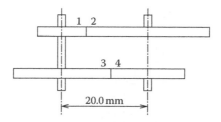

FIGURE P6.2

6.3 Solve Problem 6.2 for n4 = 36 rpm.

6.4 Determine the speed and direction of rotation of gear 8 in the gear train shown in Figure P6.4. Use arrow convention to indicate the directions of rotation.

$$N_1 = 18, N_2 = 27, N_3 = 20, N_4 = 41, N_5 = 18, N_6 = 38,$$

$$N_7 = 2\ R.H., N_8 = 24$$

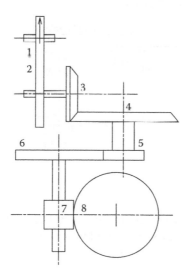

FIGURE P6.4

6.5 The motor connected to worm 1 (Figure P6.5) rotates at 1750 rpm. A wire rope is connected to the drum and is used for lifting purposes. The drum diameter is 1 m. Determine the lifting speed.

$$N_1 = 3, N_2 = 90, N_3 = 24, N_4 = 72, N_5 = 15, N_6 = 40,$$

$$N_7 = 4, N_8 = 48$$

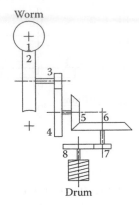

FIGURE P6.5

6.6   The spindles of the minute and hour hands of a clock mechanism are connected by means of a reverted gear train as shown in Figure P6.6. Choosing pinions from the range 9 to 16 teeth, determine the minimum number of teeth in all gears.

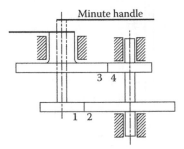

FIGURE P6.6

6.7   Figure P6.7 shows a truck transmission. It has four forward speeds and one reverse. The gears are shifted to obtain the drives shown in the figure. Determine all the speed ratios.

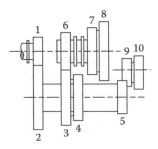

FIGURE P6.7

$$N_1 = 17, N_2 = 43, N_3 = 36, N_4 = 727, N_5 = 17, N_6 = 24,$$
$$N_7 = 33, N_8 = 43, N_9 = 110, N_{10} = 22$$

The first speed is obtained by meshing 1 with 2 and 5 with 8.
The second speed is obtained by meshing 1 with 2 and 4 with 7.
The third speed is obtained by meshing 1 with 2 and 3 with 6. The fourth speed is obtained by direct connection.
The reverse is obtained by meshing 1 with 2, 5 with 9, and 10 with 8.

6.8 Find a suitable gear train consisting of external spur gears to transmit power from a shaft rotating at 2800 rpm clockwise to another rotating at 200 rpm in the opposite direction. No idlers should be used and the ratio in any gear pair should not exceed 4.

6.9 Solve Problem 6.8 if the driven shaft is to rotate clockwise at 150 rpm.

6.10 For the gear train shown in Figure P6.10, find the speed and direction of rotation of the arm.

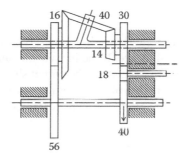

FIGURE P6.10

6.11 For the gear train shown in Figure P6.11, find the speed ratio of the input shaft and the output shaft.

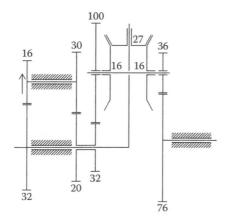

FIGURE P6.11

6.12 In the train shown in Figure P6.12, the arm rotates at 300 rpm in the counterclockwise direction.
  a. Write down an equation relating n_1, n_4, and n_A. Use given tooth numbers but do not reduce to decimals.
  b. If $n_1 = 0$, find n_4.
  c. $n_4 = 0$, find n_1.

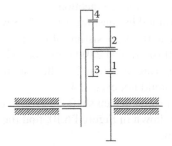

FIGURE P6.12

6.13 In the epicyclic gear train shown in Figure P6.13, the sun gears and the planet gears are equal in size. If the input shaft runs at 1000 rpm clockwise, find the alternative speeds of the output shaft that can be obtained by fixing the ring gears in turn. If the torque on the input shaft is 50 mN, find the torques required to fix the ring gears in each case.

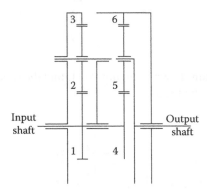

FIGURE P6.13

6.14 The train of Problem 6.13 is actually an automotive epicyclic gearbox where the ring gears 3 and 6 can be locked independently by means of band brakes. Let $N_1 = N_4 = 23$ teeth and $N_2 = N_5 = 22$ teeth (accordingly, $N_3 = N_6 = 67$ teeth). Find the output speed if the input speed is 3000 rpm clockwise and either 3 or 6 is being fixed.

6.15 In the gear train shown in Figure P6.15, find n_5 if the arm rotates at 80 rpm clockwise if gear 1 is fixed.

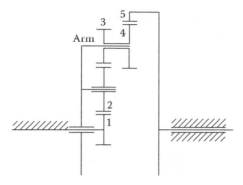

FIGURE P6.15

6.16 For the gear train shown in Figure P6.16, check the possible inversions as given by Table P6.16.

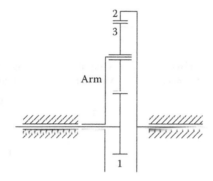

FIGURE P6.16

TABLE P6.16

Output Member	Fixed Member	Input Member	Speed Ratio Equation
2	A	1	$-\dfrac{N_2}{N_1}$
1	A	2	$-\dfrac{N_1}{N_2}$
A	2	1	$1+\dfrac{N_2}{N_1}$
A	1	2	$1+\dfrac{N_1}{N_2}$
1	2	A	$\dfrac{1}{1+\left(N_2/N_1\right)}$
2	1	A	$\dfrac{1}{1+\left(N_1/N_2\right)}$

6.17 For the gear train shown in Figure P6.17, check the possible inversions as given by Table P6.17.

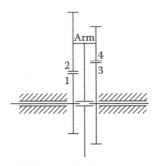

FIGURE P6.17

TABLE P6.17

Input Member	Fixed Member	Output Member	Speed Ratio Equation
1	A	3	$\dfrac{N_2 N_3}{N_1 N_4}$
1	3	A	$1 - \dfrac{N_2 N_3}{N_1 N_4}$
3	1	A	$1 - \dfrac{N_1 N_4}{N_2 N_3}$
3	A	1	$\dfrac{N_4 N_1}{N_2 N_3}$
A	1	3	$\dfrac{N_2 N_3}{N_2 N_3 - N_1 N_4}$
A	3	1	$\dfrac{N_1 N_4}{N_1 N_4 - N_2 N_3}$

6.18 For the gear train shown in Figure P6.18, show that the speed ratio is

$$\left[1 + \frac{N_1}{N_2 \left(1 + N_4/N_3 \right)} \right] \left(1 + \frac{N_6}{N_5} \right)$$

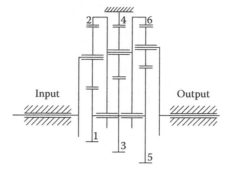

FIGURE P6.18

6.19 For the gear train shown in Figure P6.19, show that the speed ratio is

$$\frac{1+\dfrac{N_1N_4}{N_1N_3+N_2N_3}}{1+\dfrac{N_4N_5}{N_3N_5+N_3N_6}}$$

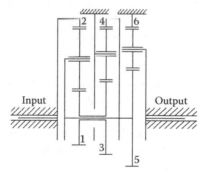

FIGURE P6.19

6.20 For the gear train shown in Figure P6.20, show that the speed ratio is

$$\frac{1}{1+\dfrac{N_1}{N_2}}\left[1+\frac{N_4}{N_3}\left(1+\frac{N_6}{N_5}\right)\right]$$

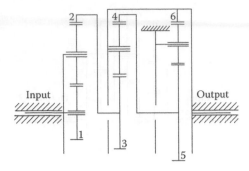

FIGURE P6.20

6.21 In a differential similar to that shown in Figures 6.9 and 6.13, the tooth numbers are as follows:

$$N_1 = N_3 = 16,\ N_2 = N_4 = 11,\ N_5 = 11,\ N_6 = 54$$

Gear number 5 turns at 1000 rpm. Determine the speed of a wheel if it is raised up while the other is resting on the road.

6.22 For the differential of Problem 6.21, $N_5 = 540$ rpm. If the engine is turning clockwise as you look at it from the front of the car, in what direction is the raised wheel turning?

6.23 A car using a differential as that of Problem 6.22 turns to the right at a speed of 40 km/h on a curve of 30-m radius. The tires are 35 cm in diameter. Use 1.8 m as the center-to-center distance between the treads. Calculate the speed of each rear wheel.

6.24 Find suitable numbers of teeth for the gears of a train ratio $m_G = 1/2.54 = 0.3937008$.

6.25 The input shaft of a gear train is to make 2.7182818 revolutions for each revolution of the output shaft. Find suitable numbers of teeth for the gears of the train.

6.26 Find suitable numbers of teeth for the gears of a train if the input shaft makes one revolution per hour and the output shaft is to make one revolution per month of one-twelfth part of the solar year of 365.24220 mean solar days.

6.27 Find a gear train to generate the sidereal ratio 1.0027379093 within one part in 10^8 (equivalent to an error of 1 second in 3 years).

6.28 The ratio $\sqrt{3} = 1.73203$ is required. Use the continued fractions to obtain the number of teeth in a train such that no gear exceeds 100 teeth. The error should not exceed one part in 15 million.

6.29 Using continued fractions, find the first 10 approximations to the fraction 0.548891. Obtain a four-gear train having tooth numbers less than 100 that will approximate the required ratio within at least one part in 150.000.

7 Force Analysis

7.1 INTRODUCTION

Force causes a body to move and accelerate. The distance moved by the body times the force is work. So, we actually get work by applying forces. Not all forces make work. Some types of forces are only constraint forces, which are used to keep a body in some state.

Forces are classified as follows:

1. Static forces. These are the forces in a system without consideration of motion.
2. Friction forces. These are the forces developed between two bodies in contact and resist relative motion.
3. Dynamic forces. These are the forces that cause acceleration of the bodies.

Forces may cause the bodies to rotate. They actually produce moments or torques, which cause rotation. So, we can consider that a force deals with translation motion, while a moment deals with rotational motion. Moment is equal to the force times a normal distance.

Force analysis in mechanisms is a very important stage for designing machines. The size of machine parts depends mainly on the applied forces and moments and the material. So, for any mechanism, it is necessary to determine the forces and moments on each part. The interaction force between the links is also important for designing the nature of the joint between them.

For example, consider the engine mechanism. The crankshaft requires a torque to overcome all the resisting forces applied on the vehicle during drive, which are mainly the drag forces, the friction forces, and the forces required for speeding up. In the mechanism itself, although the crank rotates approximately at a constant speed, other parts have accelerations. The gas produces a force on the piston that is sufficient to overcome the resisting torque on the crankshaft, the accelerating forces for the links, and the friction forces between the links. Reciprocating pumps use the same mechanism. A driving torque is applied on the crank to overcome all the other forces.

The accelerating forces may be insignificant for machines operating at a slow speed. However, they become substantially large for high-speed machines.

7.2 STATIC FORCE ANALYSIS

7.2.1 Principles

In static force analysis, the following points should be taken into consideration:

1. Forces are transmitted from one link to the other normal to the surfaces of contact, neglecting friction. Referring to Figure 7.1:
   a. For links connected by turning joints, the force transmitted from one link to the other passes through the center of the joint.
   b. For links connected by sliding joints, the force is transmitted normal to the slider.
2. Action and reaction. For links (i) and (j), which are connected together, the force exerted by link (i) on link (j) is denoted as F_{ij}. The force exerted by link (j) on link (i) is denoted as F_{ji}. From Newton's third law,

$$F_{ij} = -F_{ji}$$

3. Equilibrium. A body is considered in static equilibrium when the sum of the applied forces F_i and the sum of the applied moments M_i are zero. That is,

$$\sum F_i = 0 \qquad (7.1)$$

$$\sum M_i = 0 \qquad (7.2)$$

Let us now consider a system of forces and moments applied on a link.

Case 1: Link (i) is transmitting a force from link (j) to link (k) at joints A and B. The forces acting on link (i) are $\mathbf{F}_{ji}$ and $\mathbf{F}_{ki}$. The link is also subjected to a moment M (Figure 7.2). Thus,

$$\mathbf{F}_{ji} + \mathbf{F}_{ki} = 0$$

or

$$\mathbf{F}_{ji} = -\mathbf{F}_{ki}$$

This means that $\mathbf{F}_{ji}$ is equal in magnitude and opposite in direction to $\mathbf{F}_{ki}$. Also, $\mathbf{F}_{ji}$ is at a distance h from $\mathbf{F}_{ki}$ to balance the moment such that

$$h = \frac{M}{F_{ij}}$$

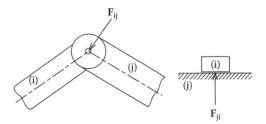

FIGURE 7.1　Forces transmitted by turning and sliding joints.

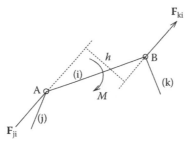

FIGURE 7.2   Forces of a three-links system.

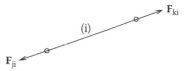

FIGURE 7.3   Case when there is no external moment.

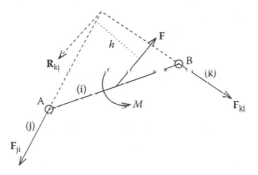

FIGURE 7.4   The middle link is subjected to a moment and an external force.

If the moment M is equal to zero, then $h = 0$. The two forces are along the center-line of the link as shown in Figure 7.3.

Case 2: The link of case 1 is also subjected to an external force **F** and a moment M (Figure 7.4).

In this case, the resultant of $\mathbf{F}_{ji}$ and $\mathbf{F}_{ki}$ is $\mathbf{R}_{kj}$ and must be equal in magnitude and opposite in direction to **F**. The distance between $\mathbf{R}_{kj}$ and **F** is h such that

$$h = \frac{M}{F}$$

If the moment M is equal to zero, then $h = 0$. **F** and $\mathbf{R}_{kj}$ must have the same line of action. This means that the three forces, **F**, $\mathbf{F}_{ji}$, and $\mathbf{F}_{ki}$, must intersect at one point.

EXAMPLE 7.1

The bell crank (2) shown in Figure 7.5a is subjected to a resisting force $\mathbf{F}_2$ at point B known in magnitude and direction. The driving force $\mathbf{F}_1$ at point A is known in

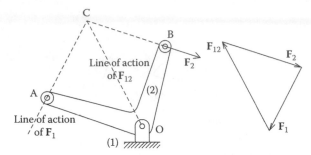

FIGURE 7.5   Forces in a bell crank.

direction. Find the magnitude of the driving force and the reaction force F_{12} at the support O.

ANALYSIS

The forces acting on the bell crank are F_1, F_2, and F_{12}. Since there is no moment applied on the bell crank, the three forces intersect at one point. The directions of F_1 and F_2 are known. They intersect at point C. Thus, the line of action of F_{12} is along OC. For equilibrium,

$$F_1 + F_2 + F_{12} = 0$$

The above vector equation is solved for two unknowns. The elements of the vectors are listed as follows:

Vector	Magnitude	Direction
F_2	F_2	Along CB
F_1	?	Along CA
F_{12}	?	Along OC

A force polygon is drawn to determine the magnitudes and directions of F_1 and F_{12}. Draw a line representing F_2 to a suitable scale. From one end, draw a line parallel to line CA. From the other end, draw a line parallel to OC. The obtained triangle is the force polygon. The magnitudes are measured. The directions are such that the three vectors are in the same sense (Figure 7.5).

EXAMPLE 7.2

For the engine mechanism shown in Figure 7.6a, the driving force P acting on piston (4) is given. Find the resisting torque on the crank (2), the forces acting on the connecting rod (3), and the transmitted force from the mechanism to the frame (1).

SOLUTION

Since there is no moment applied on link (3), the forces acting on it, F_{43} and F_{23}, are along the centerline of the link. Consequently, F_{34} and F_{32} are along the link. For the equilibrium of link (4) (Figure 7.6b),

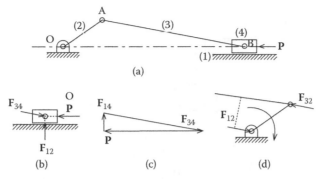

FIGURE 7.6   Force analysis for the engine mechanism. (a) Engine mechanism (b) forces acting on the piston (c) forces acting on the connecting rod (d) forces acting on the crank.

$$\mathbf{P} + \mathbf{F}_{14} + \mathbf{F}_{34} = 0$$

The force polygon is shown in Figure 7.6c. Forces $\mathbf{F}_{14}$ and $\mathbf{F}_{34}$ are determined in magnitude and direction. Thus,

$$\mathbf{F}_{34} = -\mathbf{F}_{43} = -\mathbf{F}_{32} = \mathbf{F}_{23}$$

For the equilibrium of the crank (2) (Figure 7.6d), the driving torque is given by

$$T = h \times F_{32} \quad \text{counterclockwise}$$

$$\mathbf{F}_{12} = -\mathbf{F}_{32} = -\mathbf{F}_{21}$$

The transmitted force $\mathbf{F}_{TR}$ to the frame is the sum of $\mathbf{F}_{21}$ and $\mathbf{F}_4$.

$$\mathbf{F}_{41} = -\mathbf{F}_{14}$$

Referring to Figure 7.6c, we get $\mathbf{F}_{TR} = \mathbf{P}$.

EXAMPLE 7.3

For the shaper mechanism shown in Figure 7.7a, the resisting force P acting on ram (6) is given. Find the driving torque on the crank.

SOLUTION

For the equilibrium of ram (6),

$$\mathbf{P} + \mathbf{F}_{16} + \mathbf{F}_{56} = 0$$

$\mathbf{P}$ is known in magnitude and direction.
The direction of $\mathbf{F}_{56}$ is along BC [along link (5)].

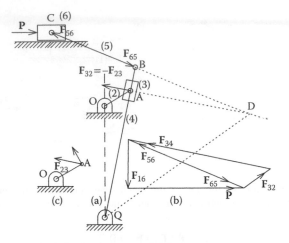

FIGURE 7.7   Force analysis in the shaper mechanism.

The direction of $\mathbf{F}_{16}$ is vertical.

The triangle of forces is drawn (Figure 7.7b). The magnitudes and directions of $\mathbf{F}_{16}$ and $\mathbf{F}_{56}$ are determined. For the equilibrium of link (4),

$$\mathbf{F}_{54} + \mathbf{F}_{34} + \mathbf{F}_{14} = 0$$

$\mathbf{F}_{54}$ is known in magnitude and direction ($\mathbf{F}_{54} = -\mathbf{F}_{56}$).
$\mathbf{F}_{34}$ is normal to link (4), that is, normal to QB.

The direction of $\mathbf{F}_{14}$ is along the line joining Q and the intersection of the two forces $\mathbf{F}_{54}$ and $\mathbf{F}_{34}$. The force polygon is drawn in continuation to the previous polygon. For block (3)

$$\mathbf{F}_{34} = -\mathbf{F}_{43} = \mathbf{F}_{23} = -\mathbf{F}_{32}$$

The driving torque is equal to the component of $\mathbf{F}_{23}$ normal to the crank times the length of the crank.

$$T = F_{23}^{T^*}\, \mathrm{OA}$$

EXAMPLE 7.4

Figure 7.8a shows a double-slider mechanism. A force of 100 N is acting on slider (6), and a force of 50 N is acting on slider (4). Find the driving torque on crank (2).

$$\mathrm{OA} = 30 \text{ cm}, \ \mathrm{AB} = 80 \text{ cm}, \ \mathrm{AC} = 110 \text{ cm}, \ \mathrm{CD} = 80 \text{ cm}$$

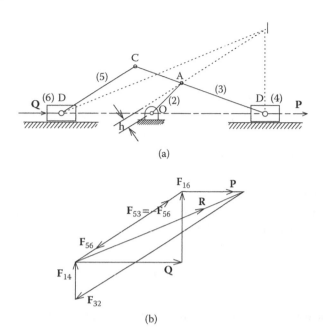

FIGURE 7.8   (a) Force analysis in a double slider mechanism (b) force polygon.

SOLUTION

First, we consider the equilibrium of (6).

$$Q + F_{16} + F_{56} = 0$$

Draw the triangle of forces to determine F_{16} and F_{56} (Figure 7.8b). Consider the equilibrium of links (3) and (4) together.

$$P + F_{14} + F_{53} + F_{23} = 0$$

To simplify the analysis, we obtain the resultant of P and F_{53} ($F_{53} = -F_{56}$).

$$R = P + F_{53}$$

The line of action of R passes by point D. Now, we can consider that links (3) and (4) are in equilibrium under the three forces, R, F_{14}, and F_{23}, which intersect at one point (point I). The force polygon is completed to obtain F_{23}. The torque on the crank is equal to $F_{23} \times h = 2725.44$ N.

7.2.2   STATIC FORCES IN GEARS

Determination of the forces acting on the gear teeth is important to the design of gears and the shafts supporting the gears. The torque transmitted by the gear is usually known. The force F_n transmitted from one tooth to another is normal to the surfaces of contact. The force analysis for each type of gear is explained in the subsequent sections.

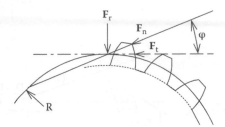

FIGURE 7.9   Forces acting on a tooth of a spur gear.

7.2.2.1   Spur Gears

The normal force F_n acting on the gear is along the line of action, which makes an angle φ (the pressure angle) with the pitch line (Figure 7.9). The tangential force $\mathbf{F}_t$ is along the pitch line and is given by

$$F_t = \frac{T}{R}$$

where T is the transmitted torque and R is the pitch radius. Thus,

$$F_n = \frac{F_t}{\cos\varphi}$$

The radial force F_r is given by,

$$F_r = F_t \tan\varphi$$

7.2.2.2   Helical Gears

The transmitted force $\mathbf{F}_n$ is normal to the tooth surface. It has three components, namely, the tangential component $\mathbf{F}_t$, the radial component F_r, and the axial component $\mathbf{F}_a$. The tangential component $\mathbf{F}_t$ is given by,

$$F_t = \frac{T}{R}$$

According to Figure 7.10, the values of the radial force $\mathbf{F}_r$ and the axial force $\mathbf{F}_a$ are given by

$$F_r = F_t \tan\varphi$$

$$F_a = F_t \tan\psi$$

where φ is the pressure angle and ψ is the helix angle. The normal force F_n is given by

$$F_n = F_t \sqrt{1 + \tan^2\varphi + \tan^2\psi}$$

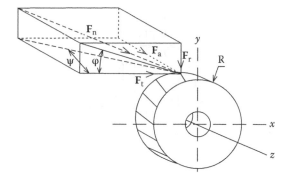

FIGURE 7.10 Forces acting on a tooth of a helical gear.

7.2.2.3 Planetary Gears

The forces are transmitted from one gear to another through the line of action. It is easier to carry on the analysis by using the tangential forces.

EXAMPLE 7.5

Consider the epicycle gear train of Figure 7.11. Gears 1 and 5 are sun gears, while the planets 2 and 3 are attached together and rotate freely on arm A. The arm rotates freely about the central axis, which coincides with the axes of the input and output shafts. The internal gear 4 is fixed in the frame. The input shaft (gear 1) rotates at 700 rpm clockwise. If the unit transmits 20 kW, determine the following:

1. The forces in all the gears
2. The braking torque
3. The speed of the output (gear 5)

All gears have a 4-mm module. The number of teeth is as follows: $N_1 = 25$, $N_2 = 30$, $N_3 = 20$

SOLUTION

The radii of the gears are

$$R_1 = 50 \text{ mm}, R_2 = 60 \text{ mm}, R_3 = 40 \text{ mm}, R_4 = 150 \text{ mm}, R_5 = 70 \text{ mm}$$

The input speed is 700 rpm clockwise. Thus, the input torque is

$$T_{in} = \frac{20 \times 60}{2 \times \pi \times 700} = 0.273 \text{ Nm clockwise}$$

The force F_{12} transmitted by gear 1 is

$$F_{12} = \frac{273}{50} = 5.46 \text{ N}$$

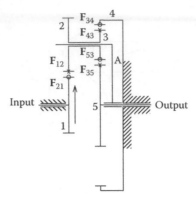

FIGURE 7.11 Epicyclic gear train of EXAMPLE 7.5.

This force is equal to F_{12}, which is the force transmitted from gear 1 to gear 2. It should be noted that the driving torque is in the direction of rotation. In this case, it is clockwise. According to Figure 7.11, the direction of rotation, clockwise, is represented by an upward arrow. The directions of the forces in the gears are normal to the plane of the paper. To differentiate between the directions, a cross (x) is used for the inward direction and a circle (o) is used for the outward direction, as shown in the figure.

$$F_{12} = 5.46\,\text{N}$$

Consider the equilibrium of gears 2 and 3. The moment about the axis gives

$$\left(F_{12} \times R_2\right) - \left(F_{43} \times R_3\right) - \left(F_{53} \times R_3\right) = 0 \tag{a}$$

The sum of the forces gives

$$F_{12} + F_{43} - F_{53} = 0 \tag{b}$$

Solving Equations a and b simultaneously, we get

$$F_{53} = \frac{F_{12}}{2}\left(1 + \frac{R_2}{R_3}\right) = 6.825\,\text{N}$$

$$F_{43} = \frac{F_{12}}{2}\left(\frac{R_2}{R_3} - 1\right) = 1.365\,\text{N}$$

The braking torque T_b is given by

$$T_b = F_{34} \times R_4 = 1.365 \times 0.15 = 0.204\,\text{N}\cdot\text{m}$$

The direction of T_b is opposite to the moment of F_{34}, which is clockwise. To obtain the speed of gear 5, we apply the condition that the toque times the speed is constant. The output torque is given by

$$T_b = F_{34} \times R_4 = 1.365 \times 0.15 = 0.204\,\text{N}\cdot\text{m}$$

Thus,

$$T_{in} \times n_1 = T_o \times n_5$$

$$0.273 \times 700 = 0.478 \times n_5$$

$$n_5 = 400 \, \text{rpm}$$

The direction of n_5 is opposite to the resisting torque. The direction of the resisting torque is opposite to the direction of the moment of F_{35}, which is clockwise. Therefore, the direction of n_5 is clockwise.

7.3 FRICTION FORCE ANALYSIS

Due to the transmitted force and the relative motion between the links in a mechanism, friction forces are generated that oppose the relative motion. Friction forces are harmful to machines. They consume energy and, consequently, decrease the efficiency of the mechanism. Besides, they cause rapid wear in the moving parts. Friction occurs at the sliding and the turning joints. To make a complete friction study for any mechanism, the nature of friction in both joints should be considered.

7.3.1 SLIDING JOINT

Consider the sliding joint shown in Figure 7.12a. Link (2) is sliding relative to link (1). The slider is subjected to an external force **P** and a driving force **F**. Suppose that the slider moves in the direction of the driving force. The reaction between the two links has two components, namely, the normal component **N** and the friction force $F_f = \mu N$. The direction of the friction force is opposite to the relative motion.

The resultant of N and μN is the reaction force of link (1) on link (2) and is given by

$$\mathbf{F}_{12} = \mathbf{N} + \mathbf{F}_f$$

$\mathbf{F}_{12}$ makes an angle λ, called the friction angle, with **N** such that

$$\tan \lambda = \mu$$

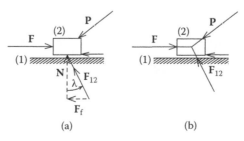

(a)   (b)

FIGURE 7.12 (a) Forces transmitted by a sliding joint including friction force (b) intersection of the forces.

For the equilibrium of link (2), the three forces intersect at one point. Thus, the line of action of $\mathbf{F}_{12}$ is determined as shown in Figure 7.12b. A force polygon is drawn to represent the forces.

Conclusion

For link (i) sliding on link (j), the steps required to determine $\mathbf{F}_{ij}$ are as follows:

*   Determine the direction of $\mathbf{N}_{ij}$ without friction.
*   Determine the direction of the velocity of j relative to i, V_{ji}.
*   Move the tail of the vector $\mathbf{N}_{ij}$ in the direction of V_{ji} until it makes an angle λ with the original vector.

7.3.2 TURNING JOINTS

The turning joint is represented by a journal (2) rotating in a bearing (1) (Figure 7.13a). The journal is subjected to an external force $\mathbf{F}$. Suppose that the angular velocity of the journal relative to bearing ω_{21} is known, say clockwise.

The resultant of the normal reaction $\mathbf{N}$ and the friction force $\mathbf{F}_f$ is $\mathbf{F}_{12}$. For the equilibrium of the joint,

$$\mathbf{F} + \mathbf{F}_{12} = 0$$

This means that $\mathbf{F}_{12}$ is equal to $\mathbf{F}$ in magnitude and opposite in direction. For the moment balance, link (2) imposes a torque to overcome this torque, which is known as the friction torque and is given by

$$T_f = \mu N \times R$$

where R is the radius of the pin of link (2). $\mathbf{F}_{12}$ must be at a distance r from $\mathbf{F}$ such that

$$T_f = F_{12} \times r = \mu N \times R$$

$$r = \frac{\mu R}{\cos \lambda}$$

Usually, the coefficient of friction is small. Thus, λ is small and $\cos \lambda$ is approximately equal to one. Therefore,

$$r = \mu R$$

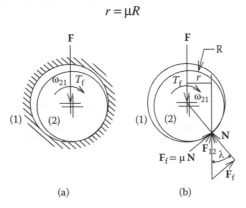

(a)   (b)

FIGURE 7.13 Forces and moments acting on (a) turning joint including friction force (b) equivalent friction force.

Conclusion

The force in a turning joint does not pass through the center of the joint. It is at a distance $r = \mu R$ and gives a moment opposite to the relative angular velocity.

In general, for links (i) and (j) connected by a turning joint and having a relative angular velocity ω_{ij}, $\mathbf{F}_{ji}$ is tangent to a circle, called the friction circle, such that it gives a moment opposite to ω_{ij} or in the direction of ω_{ji}.

To demonstrate the above analysis, consider link (i), which transmits a force from (j) to link (k). The links are connected by turning joints. The steps are as follows:

- Determine the directions of $\mathbf{F}_{ij}$ and $\mathbf{F}_{ik}$ without friction as in the preceding section.
- Determine the relative angular velocities between the links, ω_{ji} and ω_{ki}.
- Draw the friction circles at the ends of link (i).
- Shift the line of action $\mathbf{F}_{ij}$ and $\mathbf{F}_{ik}$ to be tangent to the friction circles according to the directions of ω_{ji} and ω_{ki} as shown in Figure 7.14a through d.

The directions of the relative angular velocities can be obtained by drawing the velocity polygon, determining the angular velocity of each link, and then determining

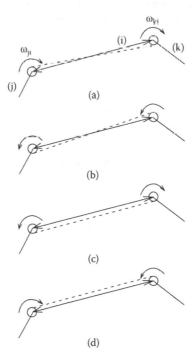

FIGURE 7.14 Configuring the direction of transmitted force in an intermediate considering friction. (a) Both angular velocities are clockwise (b) both angular velocities are counter clockwise (c) angular velocities are counter clockwise–clockwise (d) angular velocities are clockwise–counter clockwise.

the relative angular velocities. However, it is possible to obtain these values directly by using Mostafa's theorem.

Mostafa's Theorem

The relative angular velocity of two links is equal to the component of the relative velocity of any two points, one on each link, along the line joining them divided by the normal distance from the common joint of the links to the line. Its direction is the same as the moment of this component about the joint center.

Proof

Figure 7.15 shows links (1) and (2), which are connected by a turning joint at point C. Point A lies on link (1) and point B lies on link (2). The angular velocities of the links are ω_1 and ω_2 positive in the counterclockwise directions. Let $\mathbf{V}_A$ and $\mathbf{V}_B$ be the absolute velocities of points A and B, respectively. Thus,

$$\mathbf{V}_{BA} = \mathbf{V}_B - \mathbf{V}_A$$

There is no loss of generality if we let the x-axis to be along AB. $\mathbf{V}_{BA}$ has two components along the x- and y-axis. Using complex numbers gives

$$V_{BA} = V_{BA}^x + i V_{BA}^y \tag{a}$$

Also,

$$\mathbf{V}_A = \mathbf{V}_C - i\omega_1 \mathbf{r}_1$$

$$\mathbf{V}_B = \mathbf{V}_C - i\omega_2 \mathbf{r}_2$$

Subtracting, then

$$\mathbf{V}_{BA} = -i\omega_2 \mathbf{r}_2 + i\omega_1 \mathbf{r}_1$$

$$= -i\,\omega_2 \left(r_2^x + i h \right) + i\omega_1 \left(r_1^x + i h \right)$$

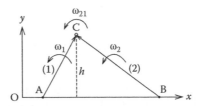

FIGURE 7.15   Relative angular velocity of two links.

where r_1^x and r_2^x are the projections of $\mathbf{r}_1$ and $\mathbf{r}_2$ on line AB, and h is the normal distance from C to AB. Thus,

$$\mathbf{V}_{BA} = h\left(\omega_2 - \omega_1\right) + i\left(\omega_1\, r_1^x - \omega_2\, r_2^x\right) \qquad\qquad \text{(b)}$$

Comparing Equations a and b and equating the real parts, we arrive at

$$V_{BA}^x = h\left(\omega_2 - \omega_1\right) = h\omega_{21}$$

It is clear that the moment of V_{BA}^x about C is in the direction of ω_{21}.

EXAMPLE 7.6

For the engine mechanism shown in Figure 7.16a, the driving force P acting on piston (4) is given. Also, the sliding coefficient of friction, the coefficient of friction in the turning joints, and the radii of the pins in the turning joint are given. The crank rotates counterclockwise. Find the resisting torque on the crank.

SOLUTION

- Determine the directions of $\mathbf{F}_{14}$, $\mathbf{F}_{34}$, and $\mathbf{F}_{42}$ without friction and place them on the links.
- Determine the directions of ω_{32}. To apply the theorem of Mostafa, choose O as a point on link (2) and point B as a point on link (3). The velocity of B relative to O is to the left. Thus, the direction of ω_{32} is clockwise.
- For ω_{34}, choose point A on link (3) and a point at infinity on link (4) [the velocity of a point at infinity on link (4) is zero]. The moment of the relative velocity is clockwise.
- ω_{21} is counterclockwise.
- The sliding velocity V_{41} is to the left.
- Shift the line of action of $\mathbf{F}_{14}$, $\mathbf{F}_{34}$, $\mathbf{F}_{42}$, and $\mathbf{F}_{12}$ as outlined previously and is shown in Figure 7.16a.
- Draw the force polygon (Figure 7.16b).
- Measure the normal distance h, and then determine the torque.

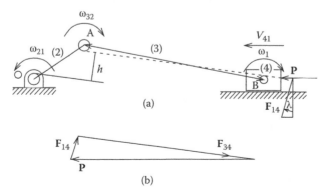

(a)

(b)

FIGURE 7.16 Friction force analysis for an engine mechanism. (a) Orientation of the force in the connecting rod (b) force analysis.

EXAMPLE 7.7

For the shaper mechanism shown in Figure 7.17a, the following data are given:

- The resisting force **P**.
- The direction of rotation of the crank is clockwise.
- The coefficients of friction and the diameter of the pins.

Find the driving torque on the crank considering friction.

SOLUTION

- The direction of the force F_{16}, F_{56}, F_{54}, F_{43}, F_{32}, F_{14}, and F_{12} without friction are obtained and shown in Figure 7.17a.
- The direction of the relative angular velocities ω_{56}, ω_{54}, ω_{32}, and ω_{21} are obtained and are indicated on the corresponding joints.
- The directions of the sliding velocities V_{61} and V_{34} are obtained and are indicated at the corresponding joints.
- The directions of F_{16} and F_{56} are modified for friction as shown in Figure 7.16b.
- The direction of F_{32} is modified for friction as shown in Figure 7.17c.
- The force polygon is shown in Figure 7.17d.

F_{12} is equal to F_{34}. The driving torque is equal to F_{12} times h in the clockwise direction.

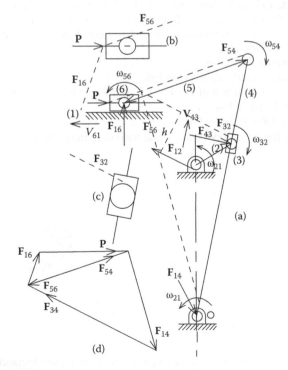

FIGURE 7.17   Friction force analysis for a shaper mechanism.

7.3.3 FORCES IN WORM GEARS

Figure 7.18a shows an outline of the worm teeth. The thread makes an angle λ, called the lead angle, with the circumference of the worm. The face of the tooth is inclined with an angle φ, which is the pressure angle. The force transmitted between the worm and worm gear is the normal force F_n, which develops a friction force F_f due to the sliding action.

$$F_f = \mu F_n$$

where μ is the coefficient of friction. For the worm, F_n has three components, namely, the tangential force F_{tW}, the axial force F_{aW}, and the radial force F_{rW}. The relations between the forces (Figure 7.18b) are

$$F_{tW} = F_{aW} \tan \lambda$$

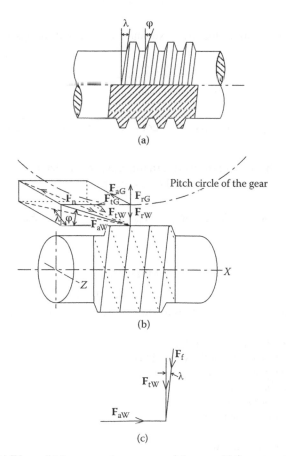

(a)

(b)

(c)

FIGURE 7.18   (a) Worm (b) forces on the worm and the gear (c) forces on the worm.

or

$$F_{aW} = F_{tW} \cot \lambda$$

$$F_{rW} = F_{tW} \cot \lambda \tan \varphi$$

$$F_n = F_{tW} \times \cot \lambda \times \sqrt{1 + \tan^2 \lambda + \tan^2 \varphi}$$

The friction force F_f is shown in the horizontal projection (Figure 7.18c). Its direction is along the face of the tooth. The total tangential and axial forces on the worm are given by

$$R_{tW} = F_{tW} + F_f \cos \lambda$$

$$R_{aW} = F_{aW} - F_f \sin \lambda$$

When the worm is subjected to an external torque T, it develops a tangential force such that

$$\frac{T}{R} = R_{tw} = F_{tW} + \mu F_n \cos \lambda$$

$$= F_{tW} \left(1 + \mu \cos \lambda \times \cot \lambda \times \sqrt{1 + \tan^2 \lambda + \tan^2 \varphi} \right)$$

where R is the pitch radius of the worm. In this case, F_{tW} is determined. Consequently, the rest of the forces are determined. For the gear,

$$R_{tG} = -R_{aW}, \quad R_{aG} = -F_{tW}, \quad \text{and} \quad F_{rG} = -F_{rW}$$

EXAMPLE 7.8

A worm–worm gear drive is used in an elevator. The maximum power transmitted is 5 kW; the input speed is 1750 rpm. The following data are provided:

- The speed reduction is 1:20.
- The pitch diameter of the worm D is 150 mm.
- The pitch p is 45 mm.
- The number of threads n is 2.
- The pressure angle φ is 20°.
- The coefficient of friction μ is 0.05.

Find the forces in the worm and the worm gear, the torque delivered by the worm gear, and the efficiency of the reduction unit.

SOLUTION

- The input torque

$$T = \frac{5 \times 1000}{\omega} = \frac{5000 \times 60}{2 \times \pi \times 1750} = 27.284 \text{ N} \cdot \text{m}$$

- The resultant tangential force of the worm

$$R_{tW} = \frac{27.284}{0.075} = 363.78 \text{ N}$$

- The lead angle

$$\tan \lambda = \frac{2 \times p}{\pi \times D} = 0.191$$

$$\lambda = 10.81°$$

- The tangential force of the worm

$$R_{tw} = F_{tW}\left(1 + \mu \cos \lambda \times \cot \lambda \times \sqrt{1 + \tan^2 \lambda + \tan^2 \varphi}\right)$$

$$F_{tW} = 284.644 \text{ N}$$

- The axial force of the worm

$$F_{aW} = F_{tW} \cot \lambda = 1490 \text{ N}$$

- The normal force is

$$F_n = F_{tW} \times \cot \lambda \times \sqrt{1 + \tan^2 \lambda + \tan^2 \varphi} = 1611 \text{N}$$

- The friction force is

$$F_f = \mu F_n = 80.569 \text{ N}$$

- The total axial force of the worm

$$R_{aW} = F_{aW} - F_f \sin \lambda = 1475 \text{ N}$$

For the gear, $R_{tG} = -R_{aW} = 1475$ N. The radius of the gear is given by

$$2 \pi R_G = \text{no. of teeth of worm} \times \text{speed ratio} \times p$$

$$R_G = 286.5 \text{ mm}$$

The torque delivered by the worm gear

$$T_G = R_{tG} \times R_G = 422.636 \, \text{N} \cdot \text{m}$$

To calculate the efficiency of the reduction unit, we should estimate the torque that should be delivered without friction T'_G, which us given by

$$T'_G = T \times \text{speed ratio} = 545.68 \, \text{N} \cdot \text{m}$$

Therefore, the efficiency η,

$$\eta = \frac{T_G}{T'_G} = \frac{422.636}{545.68} = 0.775$$

7.4   DYNAMIC FORCE ANALYSIS

7.4.1   INERTIA EFFECTS

According to Newton's first law, a body remains in a state of rest or uniform motion unless there is an external effect is applied on it. Newton's second law states that when a force **F** acts on a body of mass m, it causes it to accelerate in the same direction of the force such that

$$\mathbf{F} = m\mathbf{A} \tag{7.3}$$

where **A** is the acceleration. It is in the direction of the force (Figure 7.19).

When a group of forces $\mathbf{F}_1, \mathbf{F}_2, \mathbf{F}_3 \ldots$ acts on the body (Figure 7.20), the body has acceleration in the direction of the resultant **R** of the forces.

$$R = \sum F_i = mA \tag{7.4}$$

FIGURE 7.19   Dynamic force on a mass.

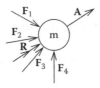

FIGURE 7.20   Effect of a group of forces on a mass.

Equation 7.2 can be put in the form

$$\mathbf{R} - m\mathbf{A} = 0$$

$$\mathbf{R} + (-m\mathbf{A}) = 0 \tag{7.5}$$

Although Equations 7.2 and 7.3 are the same, they have completely different meanings. The first equation deals with acceleration. The second equation can be understood as the body is to be in equilibrium under the effect of the external force $\mathbf{F}_i$ and a fictitious force $(-m\mathbf{A})$. This fictitious force, literally, is called the inertia force. In this case, we can consider that the body is at rest and in equilibrium under the effect of all the forces, including the inertia force $\mathbf{IF}$ (Figure 7.21).

$$\mathbf{IF} = -m\mathbf{A} \tag{7.6}$$

This is called D'Alambert principle.

The above analysis holds in the case of rotational motion. When a body with the mass moment of inertia I_G about the mass center is acted on by an external torque T, it gets an angular acceleration α (Figure 7.22).

$$T = I_G \alpha$$

The body is considered to be in equilibrium under the effect of the external torque T and the inertia torque IT.

$$IT = -I_G \alpha$$

The mass moment of inertia I is sometimes expressed in terms of the radius of gyration k such that

$$I = mk^2$$

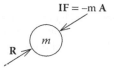

FIGURE 7.21   D'Alambert principal.

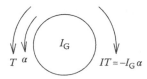

FIGURE 7.22   Moments acting on a rotating body.

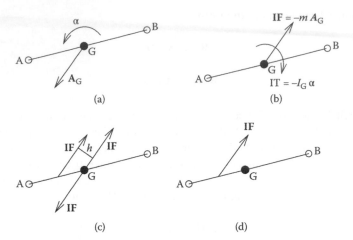

FIGURE 7.23 Inertia effects on a link. (a) Accelerations (b) inertia force and moment (c) forces equivalent to the inertia moment (d) the equivalent inertia force.

7.4.2 INERTIA EFFECTS OF A LINK

Figure 7.23 shows link AB with a mass m at the center of gravity G and a mass moment of inertia I_G about the center of gravity G. Suppose that the center of gravity has a linear acceleration $\mathbf{A_G}$ and the link has an angular acceleration α (Figure 7.23a). The acceleration and the angular acceleration are replaced by the inertia force **IF** and inertia torque IT as shown in Figure 7.23b. The inertia torque is replaced by two forces, **IF** and **−IF**, with a normal distance h apart (Figure 7.23c).

$$h = \frac{I_G \alpha}{m A_G}$$

Therefore, the net resultant inertia is a single force equal to $m A_G$ opposite to **A** and at a distance $h = \dfrac{I_G \alpha}{m A_G}$ from G such that it gives a moment opposite to α (Figure 7.23c).

EXAMPLE 7.9

For the four-bar mechanism shown in Figure 7.24a, crank OA rotates with an angular velocity of 30 rad/s clockwise and with an angular acceleration of 200 rad/s² counterclockwise. The available data of the mechanism are as follows:

$$OA = 5 \text{ cm}, \ AB = 7.5 \text{ cm}, \ QB = 9 \text{ cm}, \ OQ = 10 \text{ cm}$$

$$m_2 = 5 \text{ kg}, \ m_3 = 7.5 \text{ kg}, \ m_4 = 9 \text{ kg}$$

$$I_{G2} = 0.004 \text{ kg} \cdot \text{m}^2, \ I_{G3} = 0.005 \text{ kg} \cdot \text{m}^2, \ I_{G4} = 0.013 \text{ kg} \cdot \text{m}^2$$

$$OG_2 = 2.5 \text{ cm}, \ AG_3 = 3.75 \text{ cm}, \ QG_4 = 4.5 \text{ cm}$$

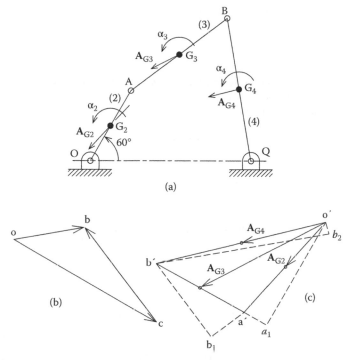

FIGURE 7.24 (a) Accelerations of a four-bar mechanism (b) velocity polygon (c) acceleration polygon.

Determine the following:

1. The reaction forces on the links
2. The driving torque on the crank
3. The shaking force on the frame

SOLUTION

The solution is set up using the following steps:

1. Draw the velocity and acceleration polygons (Figure 7.24b and c).
2. Find the accelerations of the centers of gravity of the links.

$$A_{G2} = 23 \text{ m/s}^2$$

$$A_{G3} = 54 \text{ m/s}^2$$

$$A_{G4} = 33.25 \text{ m/s}^2$$

3. Determine the angular accelerations of the links.

$$\alpha_2 = 200 \text{ rad/s}^2, \text{ counterclockwise}$$

$$\alpha_3 = 460.8 \text{ rad/s}^2, \text{ counterclockwise}$$

$$\alpha_4 = 736 \text{ rad/s}^2, \text{ counterclockwise}$$

4. Locate the accelerations and the angular accelerations on the links in the mechanism (Figure 7.24a).
5. Calculate the inertia forces and inertia torques.

$$IF_2 = 5.0 \times 23 = 115.0 \text{ N}$$

$$IF_3 = 7.5 \times 54 = 405.0 \text{ N}$$

$$IF_4 = 9 \times 33.25 = 299.25 \text{ N}$$

$$IT_2 = 0.004 \times 200 = 0.800 \text{ m} \cdot \text{N}$$

$$IT_3 = 0.005 \times 460.8 = 2.304 \text{ m} \cdot \text{N}$$

$$IT_4 = 0.013 \times 736 = 9.568 \text{ m} \cdot \text{N}$$

Find the normal distances $h = \dfrac{I_G \alpha}{m A_G}$.

$$h_2 = \frac{0.8}{115.0} = 0.696 \text{ cm}$$
$$h_3 = \frac{2.304}{405.0} = 0.55 \text{ cm}$$
$$h_4 = \frac{9.568}{299.25} = 3.2 \text{ cm}$$

6. Draw the inertia force at points E_2, E_3, and E_4 on each link after shifting them with a suitable scale (Figure 7.25).
7. The force analysis in each link is performed. It is clear that the number of unknowns in each link is four, which are the magnitudes and direction at each joint. However, it is possible to find the reaction between links (3) and (4) by using the concept of transverse and radial components, which is summarized as follows.
   a. Resolve the forces on each link to transverse and radial components normal and along the link, respectively. For link (4) (Figure 7.26a),

$$\mathbf{IF_4} = \mathbf{IF_4^{TR}} + \mathbf{IF_4^R}$$
$$\mathbf{F_{34}} = \mathbf{F_{34}^{TR}} + \mathbf{F_{34}^R}$$

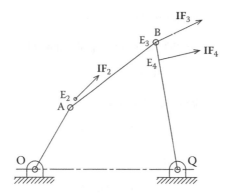

FIGURE 7.25   Inertia forces in a four-bar mechanism.

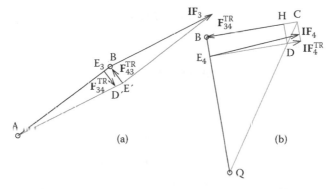

FIGURE 7.26   (a) Transverse and radial components on the coupler (b) transverse and radial components on the rocker.

b.   Taking moment about point Q, we can obtain F_{34}^{TR}

$$F_{34}^{TR} = IF_4^{TR} \frac{QE_4}{QB}$$

The value of $\mathbf{F}_{34}^{TR}$ can be obtained graphically by projecting $\mathbf{IF}_4^{TR}$ on a line normal to the link at point B, line BC, joining QC to intersect $\mathbf{IF}_4^{TR}$ at point D, and then projecting point H to point E on line BC. HB represents $\mathbf{F}_{34}^{TR}$ to the scale (Figure 7.26a). Similarly, we can perform the same steps for link (3) (Figure 7.26b). Line E′B is $\mathbf{F}_{43}^{TR}$ to the scale. Having obtained $\mathbf{F}_{34}^{TR}$ and $\mathbf{F}_{43}^{TR}$, the transverse reactions $\mathbf{F}_{34}$ and $\mathbf{F}_{43}$ can be obtained.

$$\mathbf{F}_{34} + \mathbf{F}_{43} = 0$$
$$\mathbf{F}_{34}^{TR} + \mathbf{F}_{34}^{R} + \mathbf{F}_{43}^{TR} + \mathbf{F}_{43}^{R} = 0$$

$$\mathbf{F}_{34}^{TR} = 256\,N$$

$$\mathbf{F}_{43}^{TR} = 98\,N$$

c. A force polygon representing the above equation is shown in Figure 7.27.

From the polygon, $F_{34} = 353$ N. For the equilibrium of link (4),

$$IF_4 + F_{34} + F_{14} = 0$$

Thus, F_{14} and F_{41} are determined (Figure 7.28).

$$F_{14} = F_{41} = 188 \text{ N}$$

For the equilibrium of link (3),

$$IF_3 + F_{43} + F_{23} = 0$$

Thus, F_{23} and F_{32} are determined (Figure 7.28)

$$F_{23} = F_{32} = 720 \text{ N}$$

For the equilibrium of link (2),

$$IF_2 + F_{32} + F_{12} = 0$$

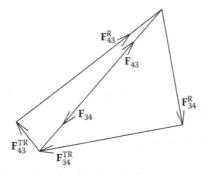

FIGURE 7.27   Force polygon for a four-bar mechanism.

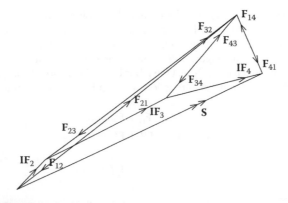

FIGURE 7.28   The shaking force on the four-bar mechanism.

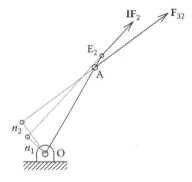

FIGURE 7.29 The driving torque on the crank.

Thus, $\mathbf{F}_{12}$ and $\mathbf{F}_{21}$ are determined (Figure 7.28).

$$F_{12} = F_{21} = 834\,\text{N}$$

The shaking force **S** is the sum of all forces connected to link (1). Thus,

$$\mathbf{S} = \mathbf{F}_{41} + \mathbf{F}_{21}$$

$$S = 805\,\text{N}$$

It is clear that

$$\mathbf{S} = \mathbf{IF}_2 + \mathbf{IF}_3 + \mathbf{IF}_4$$

That is, the shaking force is the vector sum of all inertia forces. $S = 599$ N. The driving torque T_{dr} is obtained by taking moment about point O (Figure 7.29).

$$n_1 = 12.4\ \text{cm}, \qquad n_2 = 19.6\ \text{cm}$$

$$T_{dr} = IF_2\,n_1 + F_{32}\,n_2 = 15.538\,\text{mN} \quad \text{counterclockwise}$$

Note: Usually, the friction forces are neglected in the dynamic force analysis.

EXAMPLE 7.10

For the shaper mechanism shown in Figure 7.30a, the forces acting on each link are indicated. These forces may be a combination of static and inertia forces. It is required to determine the reaction forces on the links and the driving torque.

SOLUTION

It is worth to realize that force analysis of mechanisms always starts from the last link. For this mechanism, the last chain is an engine chain. For the engine chains, it is easier to consider the equilibrium of the piston and the connecting rod together.

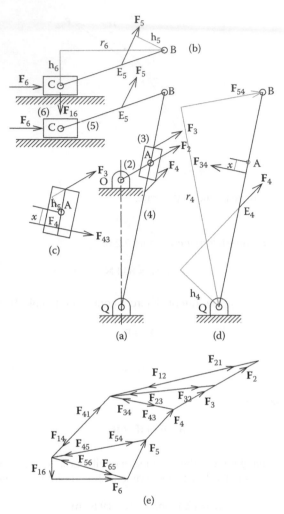

FIGURE 7.30 (a) Inertia forces in the shaper mechanism (b) equilibrium of links 5 and 6 together (c) equilibrium of link 3 (d) equilibrium of link 4 (e) force polygon.

For links (5) and (6), we take moment at point B to determine F_{16} (Figure 7.30b).

$$F_{16} = \frac{F_6 h_6 - F_5 h_5}{r_6}$$

For the equilibrium of link (6),

$$F_6 + F_{16} + F_{56} = 0$$

A force polygon is drawn to scale (Figure 7.30e) to determine F_{56} and F_{54}. For link (4),

$$F_4 + F_{54} + F_{34} + F_{14} = 0$$

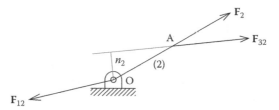

FIGURE 7.31 The driving torque on the crank of the shaper mechanism.

The magnitude and the position of F_{34} are unknown. In this case, we can consider the equilibrium of link (3). We assume that the position of F_{43} is at a distance x from point A. Taking moment about point A (Figure 7.30c) gives

$$F_3 h_3 = F_{43} x \tag{a}$$

For link (4) (Figure 7.30d), taking moment about point Q gives

$$F_4 h_4 + F_{54} r_4 = F_{34} \left(QA - x \right) \tag{b}$$

From Equations a and b, the value and position of F_{34} and F_{43} are determined. By completing the force polygon (Figure 7.30c) for link (4), F_{14} is determined. For link (3), F_{23}, and hence F_{32}, are determined. For link (2), F_{12} is determined.

The driving torque is determined by taking the moment at point O for link (2) (Figure 7.31).

7.5 ANALYTICAL FORCE ANALYSIS

In the previous sections, force analysis for mechanisms was performed graphically. It is clear that it takes a lot of effort and time to perform the analysis only for one position as in the case for position, velocity, and acceleration analysis. Analytical force analysis is a good solution for accuracy and is time-saving.

7.5.1 FORCES ON A LINK

Consider link (i) (Figure 7.32), which is connected to link (k) at point A, to link (j) at point B, and to link (n) at point C. These links exert forces F_{ki}, F_{ji}, and F_{ni}. These forces can be expressed in terms of the X and Y components along the x- and y-axis. They are considered positive in the positive directions of the axes. Thus,

$$\mathbf{F}_{ki} = \mathbf{X}_{ki} + \mathbf{Y}_{ki}$$

$$\mathbf{F}_{ji} = \mathbf{X}_{ji} + \mathbf{Y}_{ji}$$

$$\mathbf{F}_{ni} = \mathbf{X}_{ni} + \mathbf{Y}_{ni}$$

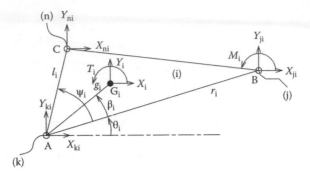

FIGURE 7.32   Forces acting on a general link.

Usually X_{ni} and Y_{ni} are known from the analysis of link (n).
The data provided for link (i) are as follows:

- The mass of the link, m_i
- The position of the center of gravity G_i, which is determined by the distance g_i from point A and the angle β_i from line AB.
- The mass moment of inertia I_{Gi} about G_i.
- The external moment M_{1i} (for links connected to the frame, and is equal to zero for the others).

The lengths r_i and l_i, the angles θ_i and ψ_i, the angular velocity ω_i, the angular acceleration α_i, and the acceleration components of point A (A_A^x, A_A^y) are already known from the kinematics analysis.

It is required to determine the reaction forces X_{ji}, Y_{ji}, X_{ki}, and Y_{ki}.

Analysis:

1. Inertia forces

$$\mathbf{F}_i = -m_i \mathbf{A}_{Gi}$$

The value of $\mathbf{A}_{Gi}$ can be obtained by using Equation 2.72 (Section 2.6.6). Thus,

$$\mathbf{F}_i = -m_i \left[(A_A^x + iA_A^y) + g_i(-\omega_i^2 + i\alpha_i) e^{i(\theta 3 + \beta 3)} \right]$$

$$= X_i + iY_i$$

Hence,

$$X_i = -m_i \left[A_A^x - g_i\omega_i^2 \cos(\theta_i + \beta_i) - g_i\alpha_i \sin(\theta_i + \beta_i) \right]$$

$$X_i = -m_i \left[A_A^x - g_i\omega_i^2 \cos(\theta_i + \beta_i) - g_i\alpha_i \sin(\theta_i + \beta_i) \right]$$

The position vectors are defined in terms of the horizontal and vertical components. Thus,

$$\mathbf{g}_i = g_{ix}\mathbf{i} + g_{iy}\mathbf{j}$$

where $\mathbf{i}$ and $\mathbf{j}$ are unit vectors in the directions of the x- and y-axis.

$$g_{ix} = A_A^x g_i \cos(\theta_i + \beta_i) \tag{7.7}$$

$$g_{iy} = A_A^x g_i \sin(\theta_i + \beta_i) \tag{7.8}$$

Thus,

$$X_i = -m_i\left(A_A^x - g_{ix}\omega_i^2 - g_{iy}\alpha_i\right) \tag{7.9}$$

$$Y_i = -m_i\left(A_A^x - g_{iy}\omega_i^2 + g_{ix}\alpha_i\right) \tag{7.10}$$

2. Inertia torque

$$T_i = -I_{Gi}\alpha_i \tag{7.11}$$

where I_{Gi} is the mass moment of inertia of link (i) about its center of gravity.

3. Equilibrium

It is important to consider that the forces and inertia torque are positive in the indicated direction. Also, we put the vectors $\mathbf{r}_i$ and $\mathbf{l}_i$ in terms of their components. Thus,

$$r_{ix} = r_i \cos\theta_i \tag{7.12}$$

$$r_{iy} = r_i \sin\theta_i \tag{7.13}$$

$$l_{ix} = l_i \cos(\theta_i + \beta_i) \tag{7.14}$$

$$l_{iy} = l_i \sin(\theta_i + \beta_i) \tag{7.15}$$

Taking moment about point A, we get

$$M_{1i} + T_i - X_i g_{iy} + Y_i g_{ix} - X_{ni} l_{iy} + Y_{ni} l_{ix} - X_{ji} r_{iy} + Y_{ji} g_{ix} = 0$$

Let

$$K_i = M_{1i} + T_i - X_i g_{iy} + Y_i g_{ix} - X_{ni} l_{iy} + Y_{ni} l_{ix} \qquad (7.16)$$

It is clear that all the elements on the right-hand side of Equation 7.16 are known. This means that K_i can be calculated for each link.

7.5.2 FOUR-BAR CHAIN

The four-bar chain is represented by links (3) and (4) (Figure 7.33). The position analysis for the four-bar chain is presented in Section 1.9.3.2. The velocity and acceleration analysis is presented in Section 2.6.2. Suppose that links (i) and (n) are attached to link (3) at points A and C, respectively. Link (j) is attached to link (4) at point D as shown in the figure.

The forces X_{n3}, Y_{n3}, X_{j4}, and Y_{j4} usually are determined from the chains connected to them and are known. The data given for the four-bar chain are as follows:

- X_{n3}, Y_{n3}, m_3, I_{G3}, r_3, I_3, g_3, θ_3, ψ_3, β_3, ω_3, α_3
- X_{j4}, Y_{j4}, m_4, I_{G4}, r_4, I_4, g_4, θ_4, β_4 (negative), ψ_4 (negative), ω_4, α_4, and the external moment M_{14}

It is required to determine

$$X_{34},\ Y_{34},\ X_{i3},\ Y_{i3},\ X_{14},\ Y_{14},\ X_{43},\ Y_{43},\ X_{41},\ Y_{41}$$

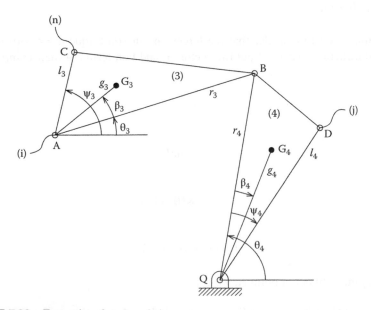

FIGURE 7.33   Forces in a four-bar chain.

Analysis:
Use Equations 7.7 through 7.16 for links (3) and (4) to determine K_3 and K_4. They are negative if clockwise. Note that $M_3 = 0$ and $M_4 = M_{14}$. For link (3), take moment about point A. For link (4), take moment about point Q; then,

$$-X_{43}r_{3y} + Y_{43}r_{3x} + K_3 = 0 \tag{7.17}$$

$$-X_{34}r_{4y} + Y_{34}r_{4x} + K_4 + M_{14} = 0 \tag{7.18}$$

Using

$$X_{34} = -X_{43}$$

$$Y_{34} = -Y_{43}$$

Equations 7.17 and 7.18 lead to

$$X_{34} = \frac{K_3\, r_{4x} + \left(K_4 + M_{14}\right) r_{3x}}{r_{3x}r_{4y} - r_{3y}r_{4x}} \tag{7.19}$$

$$Y_{34} = \frac{K_3\, r_{4y} + \left(K_4 + M_{14}\right) r_{3y}}{r_{3x}r_{4y} - r_{3y}r_{4x}} \tag{7.20}$$

Considering the equilibrium of link (4), we arrive at

$$X_{14} = -X_{34} - X_4 - X_{j4} \tag{7.21}$$

$$Y_{14} = -Y_{34} - Y_4 - Y_{j4} \tag{7.22}$$

For the equilibrium of link (3),

$$X_{i3} = -X_{43} - X_{n3} - X_3 \tag{7.23}$$

$$Y_{i3} = -Y_{43} - Y_{n3} - Y_3 \tag{7.24}$$

Remember that the direction of the forces is positive along the x- and y-axis. The forces on link (i) are

$$X_{3i} = -X_{i3} \tag{7.25}$$

$$Y_{3i} = -Y_{i3} \tag{7.26}$$

7.5.3 ENGINE CHAIN

The engine chain is represented by links (3) and (4) (Figure 7.34). The position analysis for the engine chain is presented in Sections 1.9–3.3. The velocity and acceleration analysis is presented in Section 2.6.3.

Suppose that link (i) and (n) are attached to link (3) at points A and C as shown in the figure.

Given:

- X_{n3}, Y_{n3}, m_3, I_{G3}, r_3, I_3, g_3, θ_3, ψ_3, β_3, ω_3, α_3, and the external force P_4, positive along $e^{i\alpha}$
- m_4, A_4

Find:

$$X_{i3},\ Y_{i3},\ X_{14},\ Y_{14}$$

Analysis:

Use Equations 7.7 through 7.16 for link (3) to determine K_3. $M_3 = 0$. For link (4), the inertia force IF_4 is given by

$$\text{IF}_4 = -m_4 A_4 \tag{7.27}$$

The resultant force on link (4) is F_4 and is given by

$$F_4 = \text{IF}_4 + P_4 \tag{7.28}$$

F_{14} is normal to the sliding surfaces. It is considered positive in the direction $ie^{i\alpha}$. If we assume that forces F_4 and P_4 pass through point B, then F_{14} also passes through point B. For the equilibrium of links (3) and (4) together, thus,

$$K_3 + F_4 r_3 \sin(\alpha - \theta_3) + F_{14} r_3 \cos(\alpha - \theta_3) = 0$$

Therefore, for link (4)

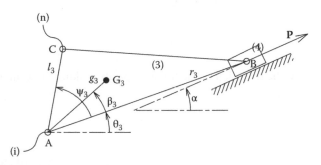

FIGURE 7.34 Orces in an engine chain.

$$F_{14} = -\frac{K_3 + F_4 r_3 \sin(\alpha - \theta_3)}{r_3 \cos(\alpha - \theta_3)} \tag{7.29}$$

$$X_{14} = -F_{14} \sin\alpha \tag{7.30}$$

$$Y_{14} = F_{14} \cos\alpha \tag{7.31}$$

$$X_{41} = -X_{14} \tag{7.32}$$

$$Y_{41} = -Y_{14} \tag{7.33}$$

$$X_{34} = -F_4 \cos\alpha - X_{14} \tag{7.34}$$

$$Y_{34} = -F_4 \sin\alpha - Y_{14} \tag{7.35}$$

For the equilibrium of link (3)

$$X_{43} = -X_{34} \tag{7.36}$$

$$Y_{43} = -Y_{34} \tag{7.37}$$

$$X_{i3} = -X_3 - X_{n3} - X_{43} \tag{7.38}$$

$$Y_{i3} = -Y_3 - Y_{n3} - Y_{43} \tag{7.39}$$

The forces on link (i) are

$$X_{3i} = -X_{i3} \tag{7.40}$$

$$Y_{3i} = -Y_{i3} \tag{7.41}$$

7.5.4 SHAPER CHAIN

The shaper chain is represented by links (3) and (4) (Figure 7.35). The position analysis for the shaper chain is presented in Sections 1.9–3.4. The velocity and acceleration analysis is presented in Section 2.6.4.

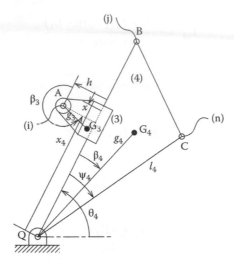

FIGURE 7.35 Forces in a shaper chain.

Given:

- m_3, I_{G3}, x_3, I_3, g_3, θ_3, β_3, $\omega_3 = \omega_4$, $\alpha_3 = \alpha_4$, h
- X_{j4}, Y_{j4}, X_{n4}, Y_{n4}, m_4, I_{G4}, x_4, I_4, g_4, θ_4, β_4 (negative), ψ_4 (negative), ω_4, α_4, and the external moment M_{14}.

Find:

$$F_{34},\ F_{43},\ X_{14},\ Y_{14},\ X_{41},\ Y_{41},\ X_{i3},\ Y_{i3},\ X_{3i},\ Y_{3i}$$

Analysis:

Link (3) is connected only to links (i) and (4). Use Equations 7.7 through 7.16 to determine K_3 and K_4 for link (3). F_{34} and F_{43} are considered positive in the direction of $ie^{i\theta^4}$ and are at a distance x from point A. For link (3), take moment about point A, and for link (4) take moment about point Q. Then,

$$K_3 + F_{43}x = 0 \tag{7.42}$$

$$K_4 + F_{34}(x_4 + x) = 0 \tag{7.43}$$

From Equations 7.42 and 7.43

$$F_{34} = -\frac{K_3 + K_4}{x_4} \tag{7.44}$$

$$F_{43} = -F_{34} \tag{7.45}$$

$$x = -\frac{K_3}{F_{43}} \tag{7.46}$$

$$X_{14} = -X_4 - X_{n4} - X_{j4} + F_{34}\sin\theta_4 \tag{7.47}$$

$$Y_{14} = -Y_4 - Y_{n4} - Y_{j4} - F_{34}\cos\theta_4 \tag{7.48}$$

$$X_{i3} = -X_3 + F_{43}\sin\theta_4 \tag{7.49}$$

$$Y_{n3} = -Y_3 - F_{43}\cos\theta_4 \tag{7.50}$$

7.5.5 Tilting Block Chain

The tilting block chain is shown in Figure 7.36. The analysis of this chain is similar to that of the shaper chain. We can deduce the reaction forces by exchanging the subscripts (3) and (4). Therefore,

$$F_{43} = -\frac{K_3 + K_4}{x_4} \tag{7.51}$$

$$F_{34} = -F_{43} \tag{7.52}$$

$$x = -\frac{K_4}{F_{34}} \tag{7.53}$$

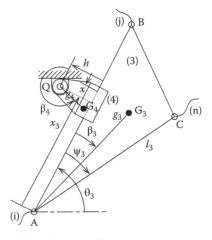

FIGURE 7.36 Forces in a tilting block chain.

$$X_{i3} = -X_3 - X_{j3} - X_{n3} + F_{43} \sin\theta_3 \qquad (7.54)$$

$$Y_{i3} = -Y_3 - Y_{j3} - Y_{n3} - F_{43} \cos\theta_3 \qquad (7.55)$$

$$X_{14} = -X_4 + F_{34} \sin\theta_3 \qquad (7.56)$$

$$Y_{14} = -Y_4 - F_{34} \cos\theta_3 \qquad (7.57)$$

7.5.6 CRANK

The crank (Figure 7.37) is usually the driving link. It is connected, in general, to links (j) and (k), respectively. The forces of these links on the crank are known from previous analysis.

Given:

$$X_{j2}, \ Y_{j2}, \ X_{k2}, \ Y_{k2}, \ m_3, \ I_{G2}, \ r_2, \ l_2, \ g_2, \ \theta_2, \ \psi_2, \ \beta_2, \ \omega_2, \ \alpha_2$$

Find:

$$M_{12}, \ X_{12}, \ Y_{12}$$

Analysis:

Use Equations 7.7 through 7.16 to calculate g_{2x}, g_{2y}, X_2, Y_2, T_2, r_{2x}, r_{2y}, l_{2x}, and l_{2y}. Taking moment about point O, we get

$$M_{12} = -T_2 + X_2 g_{2y} - Y_2 g_{2x} + X_{j2} r_{2y} - Y_{j2} r_{2x} + X_{k2} l_{2y} - Y_{k2} l_{2x}$$

$$T_2 = -I_{G2}\alpha_2$$

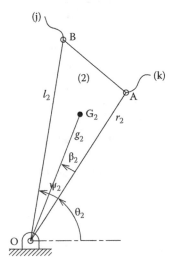

FIGURE 7.37   Forces on the crank.

Considering the equilibrium of forces gives

$$X_{12} = -X_2 - X_{k2} - X_{j2}$$

$$Y_{12} = -Y_2 - Y_{k2} - Y_{j2}$$

7.5.7 Shaking Force and the Shaking Moment

The shaking effects are the sum of the acting forces and moments of all links attached to the frame, link (1). The shaking force **SF** is given by

$$SF_x = \sum_{i=1}^{p} X_{i1}$$

$$SF_y = \sum_{i=1}^{p} Y_{i1}$$

where p is the number links attached to the frame. The shaking force is also equal to the sum of the inertia forces of all the links in the mechanism. If the number of links is n, then

$$SF_x = \sum_{i=1}^{n} X_i$$

$$SF_y = \sum_{i=1}^{n} Y_i$$

The magnitude of the shaking force SF is given by

$$SF = \sqrt{\left(\sum_{i=1}^{n} X_i\right)^2 + \left(\sum_{i=1}^{n} Y_i\right)^2}$$

The shaking moment is given by

$$SM = \sum_{i=1}^{n} M_{i1} + \text{(the moment of the shaking force depending on its location)}$$

7.5.8 Applications

In analyzing mechanisms, it is useful to combine the position analysis (Section 1.9), the velocity and acceleration analysis (Section 2.6), and the force analysis (Section 7.5) in one algorithm. The analysis of the chains is listed in the Appendix. The algorithms are written in the format of MathCAD software. The crank is included in each chain in order to demonstrate the sequence of the analysis.

EXAMPLE 7.11

Use the data of the four-bar mechanism given in Example 7.9 to determine the reaction forces on the links, the driving torque on the crank, and the shaking force on the frame over a complete crank rotation. The data are

$r_1 = 0.10$ m, r_2 (OA) $= 0.05$ m, $r_3 = 0.075$ m, $r_4 = 0.09$ m
$m_2 = 5$ kg, $m_3 = 7.5$ kg, $m_4 = 9$ kg
$I_{G2} = 0.004$ kg·m², $I_{G3} = 0.005$ kg·m², $I_{G4} = 0.013$ kg·m²
$g_2 = 0.025$ m, $g_3 = 0.0375$ m, $g_4 = 0.045$ m

Analysis:

The algorithm of this mechanism is listed in Appendix A-1. It is adapted to use MathCAD software.

Results:

A program was written in MathCAD format and was run. It yielded the following results.

Figure 7.38 shows the history of the reaction forces and the shaking force over one crank revolution: Figure 7.38a for F_{34}, Figure 7.38b for F_{23}, Figure 7.38c for F_{12}, Figure 7.38d for F_{14}, Figure 7.38e for the shaking force, and Figure 7.38f for the driving torque.

Figure 7.39 shows plots between the x- and y-component of the reaction forces to show their directions: Figure 7.39a is for $\mathbf{F}_{34}$, Figure 7.39b is for $\mathbf{F}_{23}$, Figure 7.39c for $\mathbf{F}_{14}$, and Figure 7.39d are for $\mathbf{F}_{12}$.

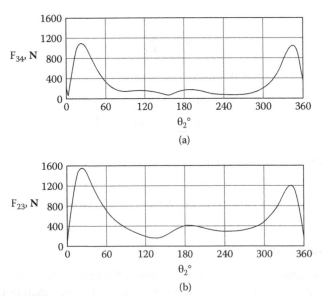

FIGURE 7.38 Reaction forces, shaking force, and the driving torque of the four-bar mechanism. (a), (b), (c), and (d) are the reaction forces in the links (e) shaking force (f) driving moment on the crank.

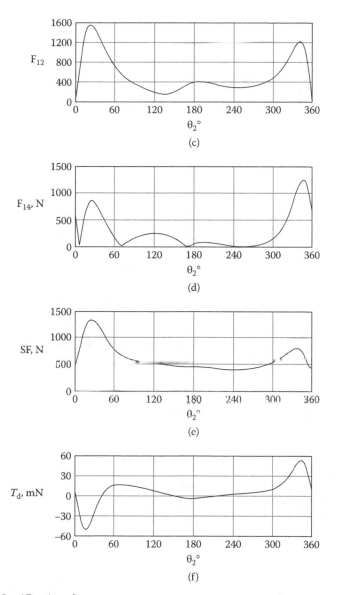

FIGURE 7.38 *(Continued)*

7.5.9 COMBINED STATIC AND DYNAMIC FORCE ANALYSIS

While using the graphical method for the force analysis, usually, each of the static forces and the dynamic forces is treated separately. The total forces acting on each link are obtained by the vector sum of the effect of the static and dynamic forces. In the analytical analysis, it is possible to include the effect of the static forces with the dynamic forces as demonstrated in the following example.

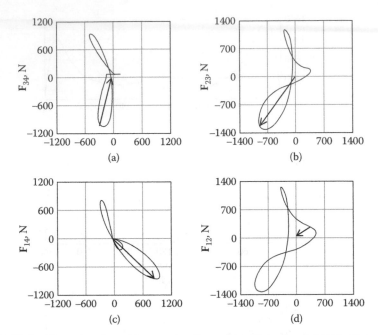

FIGURE 7.39 The polar plot of the reaction forces (a) in links 3 and 4 (b) in links 2 and 3 (c) in links 1 and 4 (d) in links 4 and 1.

EXAMPLE 7.12

For the single-cylinder engine shown in Figure 7.40, the following data are provided:

Speed of crank,   3000 rpm counterclockwise
Length of crank, $r = 10$ cm
Length of connecting rod, $r_3 = 40$ cm
Mass of connecting rod, $m_3 = 1.0$ kg
Mass moment of inertia of the connecting rod, $I_{G3} = 0.00137$ kg·m²
The mass of the piston, $m_4 = 0.9$ kg
Area of the piston, $A_p = 45.5$ cm²

The mass of the crank is counterbalanced by a balancing mass m_b; thus, it has no effect. The gas pressure on the piston during one cycle is shown in Figure 7.41.

The gas pressure is approximated by the following equations for different regions:

For $0 \leq \theta \leq 8°$

$$p_1 = 140 + (1504 \times \theta) \text{ N/cm}^2$$

For $8° \leq \theta \leq 270°$

$$p_2 = -3.653 + (406.65 \times e^\theta) \text{ N/cm}^2$$

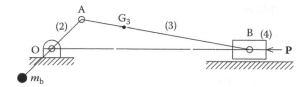

FIGURE 7.40 Static and dynamic force analysis of engine mechanism.

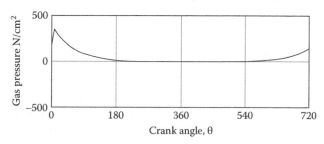

FIGURE 7.41 The gas pressure in the engine mechanism.

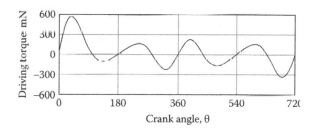

FIGURE 7.42 The driving torque in the engine mechanism.

For $270° \leq \theta \leq 540°$

$$p_3 = 0 \, N/cm^2$$

For $540° \leq \theta \leq 720°$

$$p_4 = -6.323 + \left(0.00051 \times e^\theta\right) N/cm^2$$

Find the driving torque.

Results:

The algorithm of the engine chain is written using the MathCAD program and is listed in Appendix A-2. The driving torque is shown in Figure 7.42.

MathCAD programs for the shaper and tilting block chains are listed in Appendices A-3 and A-4.

7.6 TORQUE DIAGRAM AND FLYWHEELS

As shown in Figure 7.42, the torque delivered by the single-cylinder engine varies considerably over one whole cycle. In fact, parts of the diagram are negative. Usually, the resisting torque is constant. This causes a speed variation in the system as will be described later. Motor cars use engines with four or more cylinders. For a four-cylinder engine with the same data of Example 7.12, the torque diagram is shown in Figure 7.43.

Even with using multiple cylinders, there is still a variation in the torque diagram. To overcome the problem of speed variation, we use a flywheel.

7.6.1 TORQUE DIAGRAM

In any mechanical system, there is a driving member, which may be an electric motor, engine, turbine, and so forth, producing a diving torque T_d, and a driven member, which may be a machine producing a resisting torque T_r. In general, the driving and resisting torques have any form as shown in Figure 7.44. The cycle is defined as the interval, time, or rotational angles, after which the driving and resisting torques are exactly repeated. At any instant, the driving torque is different from the driven torque. This causes acceleration in the system and, consequently, speed variation. It is not possible to eliminate the speed variation completely. However, it is possible to control the amount of speed variation by adding a flywheel of mass moment of inertia I. The magnitude of I depends on the permissible amount of speed variation.

Consider a flywheel of mass moment of inertia I that is subjected to a driving torque T_d and a resisting torque T_r (Figure 7.45).

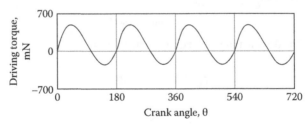

FIGURE 7.43 Torque diagram.

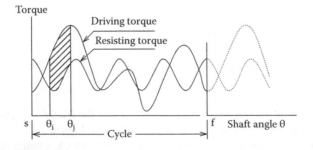

FIGURE 7.44 Driving and resisting torques.

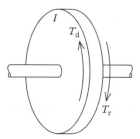

FIGURE 7.45 A flywheel.

The angular acceleration of the flywheel is given by

$$T_d - T_r = I\ddot{\theta}$$

$$= I\dot{\omega}$$

$$= I \frac{d\omega}{d\omega} \frac{d\theta}{dt}$$

Thus,

$$(T_d - T_r) d\theta = I\omega \, d\omega$$

Integrating both sides from position i to position j (Figure 7.44), we arrive at

$$\int_{\theta_i}^{\theta_j} (T_d - T_r) \, d\theta = \int_{\omega_i}^{\omega_j} I\omega \, d\omega$$

$$= \frac{1}{2} I \left(\omega_j^2 - \omega_i^2 \right) \qquad (7.58)$$

where ω_i and ω_j are the angular velocities of the system at θ_i and θ_j. The integration on the left-hand side represents the shaded area between the driving torque and the driven torque bounded by θ_i and θ_j. It represents the energy delivered to the system. It is equal to the change in the kinetic energy, which increases the speed from ω_i and ω_j. Equation 7.58 shows that the change in the delivered energy is equal to the change in the kinetic energy. It is clear that in the regions where the resisting torque is larger than the driving torque, the energy is negative. This means that the system consumes energy, and the speed decreases. Consider the integration to be over the complete cycle. In this case, Equation 7.58 is written as

$$\int_{0}^{f} (T_d - T_r) \, d\theta = \frac{1}{2} I \left(\omega_f^2 - \omega_s^2 \right) \qquad (7.59)$$

It is essential that ω_s to be equal to ω_f; otherwise, the seed changes indefinitely. Thus,

$$\int_o^f (T_d - T_r)\,d\theta = 0$$

This means that the area under the driving torque must be equal to the area under the driven torque over a whole cycle.

7.6.2 SPEED VARIATION DIAGRAM

The speed variation during the cycle can be determined from Equation 7.58.

$$\int_{\theta_i}^{\theta_j} (T_d - T_r)\,d\theta = \frac{1}{2}I\left(\omega_j^2 - \omega_i^2\right)$$

or

$$a_{ij} = \frac{1}{2}I\left(\omega_j^2 - \omega_i^2\right)$$

where

$$a_{ij} = \int_{\theta_i}^{\theta_j} (T_d - T_r)\,d\theta$$

$$= \text{area between } T_d \text{ and } T_r \text{ from position i to position j}$$

Thus,

$$\omega_j^2 = \omega_i^2 + c_{ij}$$

where c_{ij} is proportional to a_{ij}.

However, it is not necessary to obtain the exact values of the angular velocities. The main objective is to locate the positions where the maximum and minimum angular velocities occur. This is obtained by drawing a simplified speed diagram. At the points of intersection of the driving and driven torque diagrams, the angular acceleration is zero. This is where the angular velocity starts to increase or decrease. The determination of the simplified speed diagram is outlined as follows:

- Locate the points of intersection of the torque diagrams: points 1, 2 ... 6 (Figure 7.46a).
- Calculate the areas between the driving and resisting torques between the intersection points O: 1, 2 ..., f. Call them $a_{O1}, a_{12} \ldots a_{6c}$. Notice that in the regions where the resisting torque is larger than the diving torque, the areas are negative. That is, the angular velocities decrease. It increases in the other regions. It should be noted that the sum of the areas is zero.
- Assume that the angular velocity at point O is ω_s.
- Draw line $\omega_s - \omega_1$ with a drop equal to a_{O1} (a_{O1} is negative) (Figure 7.46b).

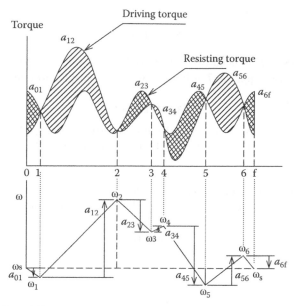

FIGURE 7.46   Speed diagram.

- From ω_1 draw line $\omega_7 \cdot \omega_7$ with a rise equal to a_{12} (a_{12} is positive)
- Repeat until reaching point ω_f, which is equal to ω_s.
- From the diagram, we locate the positions where the maximum and minimum angular velocities (ω_{max} and ω_{min}) occur. From Figure 7.46b, ω_{max} is at point 2 and ω_{min} is at point 5.

Thus, the energy that causes the maximum variation in the angular speed is between points 2 and 5 with magnitude ΔE. It is represented by the sum of the areas a_{23}, a_{34}, and a_{45}. Thus,

$$\Delta E = \frac{1}{2} I \left(\omega_{max}^2 - \omega_{min}^2 \right) \tag{7.60}$$

The right-hand side of Equation 7.60 is expanded in the form

$$\frac{1}{2} I \left(\omega_{max}^2 - \omega_{min}^2 \right) = \frac{1}{2} I \left(\omega_{max} - \omega_{min} \right) \left(\omega_{max} + \omega_{min} \right)$$

$$= I c \omega_0^2$$

where ω_0 is termed as the mean angular velocity and c is termed by the coefficient of speed variation. They are given by

$$\omega_0 = \frac{1}{2} \left(\omega_{max} + \omega_{min} \right)$$

$$c = \frac{\left(\omega_{max} - \omega_{min} \right)}{\omega_0}$$

Therefore,

$$\Delta E = I c \omega_0^2 \tag{7.61}$$

In some cases, ΔE is negative and the right-hand side of Equation 7.60 is also negative. Therefore, Equation 7.61 holds if we consider the absolute value of ΔE.

EXAMPLE 7.13

A machine is performing a repeated job every 100 seconds. The job requires a torque as indicated in Figure 7.47. The machine operates at a mean speed of 300 rpm. Determine:

- The power of the driving electric motor.
- The moment of inertia of the flywheel if the speed variation is 3%.

SOLUTION

The angular velocity is given by

$$\omega = \frac{2 \times \pi \times 300}{60} = 31.4 \text{ rad/s}$$

The average torque (which is the driving torque) is obtained by estimating the area under the resisting torque and dividing by the cycle.

$$T_d = \frac{800 \times 10 + 600 \times 10 + 800 \times 10 + 400 \times 10}{100} = 260 \text{ N·m}$$

The motor power P is given by

$$P = T_d \times \omega = 8.168 \text{ kW}$$

Now we calculate the areas between the driving and the resisting torques:

$$a_1 = -5400 \text{ N·m·s}$$

$$a_2 = 2600 \text{ N·m·s}$$

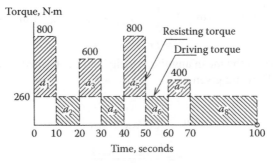

FIGURE 7.47   Torque diagrams for EXAMPLE 7.13.

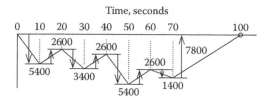

FIGURE 7.48 Speed diagram for EXAMPLE 7.13.

$$a_3 = -3400 \text{ N} \cdot \text{m} \cdot \text{s}$$

$$a_4 = 2600 \text{ N} \cdot \text{m} \cdot \text{s}$$

$$a_5 = -5400 \text{ N} \cdot \text{m} \cdot \text{s}$$

$$a_6 = 2600 \text{ N} \cdot \text{m} \cdot \text{s}$$

$$a_7 = -1400 \text{ N} \cdot \text{m} \cdot \text{s}$$

$$a_8 = 7800 \text{ N} \cdot \text{m} \cdot \text{s}$$

Draw the speed variation diagram (Figure 7.48).

From the diagram, we find that the minimum angular speed occurs at 50 seconds and the maximum occurs at 100 seconds. The energy causing the maximum speed variation is between 50 seconds and 100 seconds, that is, the sum of a_6, a_7, and a_8 adjusted to units of energy by multiplying by ω. Thus,

$$\Delta E = (2600 - 1400 + 7800) \times 10\pi = 2.827 \times 10^5 \text{ mN}$$

The coefficient of speed variation $c = 0.03$. Applying Equation 7.61,

$$2.827 \times 10^5 = I \times 0.03 \times 31.4^2$$

Therefore,

$$I = 9509 \text{ kg} \cdot \text{m}^2$$

In some applications, the driving and resisting torques are represented by equations. The procedure for manipulating such cases is demonstrated by the following example.

EXAMPLE 7.14

In a system, the driving and resisting torques are given by

$$T_d = 200 + 50\sin\theta \ \text{N·m}$$

$$T_r = K + 30\cos 1.5\theta \ \text{N·m}$$

The mean speed is 150 rpm. Find the inertia of the flywheel so that the speed variation is limited to 3%.

SOLUTION

We have concluded in the previous section that the areas under the driving and resisting torques should be equal over one cycle. Also, it is known that the area of a harmonic function is also zero over one cycle. Thus, we conclude that $K = 200$ Nm.

The first step is to determine the common cycle of the driving and resisting torques.

The cycle of the T_d is 2π, while the cycle for T_r is $\dfrac{2\pi}{1.5}$. The common cycle should contain a whole number of cycles for each torque. Let n be the number of cycles for T_d and m be the number of cycles for T_r. The common cycle q is given by

$$q = n \times 2\pi = m \times \frac{2\pi}{1.5}$$

n and m must be integers. So, we find the minimum values of n and m to satisfy the above equation.

$$\frac{n}{m} = \frac{1}{1.5} = \frac{2}{3}$$

Therefore, $n = 2$ and $m = 3$. The common cycle is 4π.

The second step is to find the points of intersection of T_d and T_r. The simplest way is the graphical method. It is to draw T_d and T_r and then determine the points of intersection of both curves (Figure 7.49).

From the graph, the points of intersection are at 0.15π, π, 1.85π, 3π, and 4π.

The third step is to calculate the areas between the driving and resisting torques between the points of intersection starting from 0 ending at 4π.

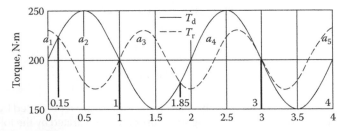

FIGURE 7.49   Torque diagrams for EXAMPLE 7.14.

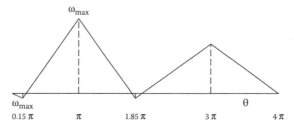

FIGURE 7.50 Speed diagram for EXAMPLE 3.14.

$$a_1 = \int_0^{0.15\pi} (50\sin\theta - 30\cos\theta)\,d\theta = -7.54\ \text{N}\cdot\text{m}$$

$$a_2 = \int_{0.15\pi 0}^{\pi} (50\sin\theta - 30\cos\theta)\,d\theta = 127.54\ \text{N}\cdot\text{m}$$

$$a_3 = \int_{\pi}^{1.85\pi} (50\sin\theta - 30\cos\theta)\,d\theta = -127.54\ \text{N}\cdot\text{m}$$

$$a_4 = \int_{1.85\pi}^{3\pi} (50\sin\theta - 30\cos\theta)\,d\theta = 87.54\ \text{N}\cdot\text{m}$$

$$a_5 = \int_{3\pi}^{4\pi} (50\sin\theta - 30\cos 0)\,d\theta = -80\ \text{N}\cdot\text{m}$$

The fourth step is to draw the speed variation diagram to determine the location of the maximum and minimum speeds (Figure 7.50). The minimum angular velocity occurs at $\dfrac{\theta}{\pi} = 0.15$ and the maximum angular velocity occurs at $\dfrac{\theta}{\pi} = 1$. In this case, ΔE is equal to $a_2 = 127.54$ N·m.

$c = 0.03$ and $\omega = 15.7$ rad/s. Applying Equation 7.61, we get

$$127.54 = I \times 0.03 \times 15.7^2$$

$$I = 17.25\ \text{kg}\cdot\text{m}^2$$

7.7 FORCE ANALYSIS IN CAMS

Figure 7.51a shows a cam drive that consists of

- A cam
- A translating follower
- A retaining spring
- Follower guides at A and B

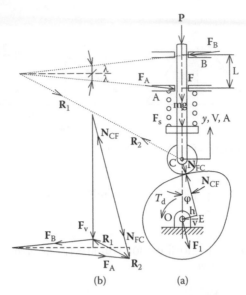

FIGURE 7.51   Force analysis in a cam drive. (a) Cam drive (b) force analysis.

The cam is acted on by the normal reaction $\mathbf{N}_{FC}$, the support reaction $\mathbf{F}_1$ at point O, and the driving torque T_d.

The follower, at any instant, is at a displacement y from its lowest position and is moving upward with a velocity $\dot{y}$ and an acceleration $\ddot{y}$. The forces acting on the follower are

- The external force P.
- The weight of the follower system mg.
- The spring force $F_s = K(y + \delta)$, where K is the spring stiffness and δ is the initial deflection in the spring.
- The inertia force $\mathrm{IF} = m\ddot{y}$, positive downward since the acceleration is positive upward.
- The normal reaction N_{CF} between the cam and the follower.
- The reaction forces at the guides $\mathbf{F}_A$ and $\mathbf{F}_B$; the distance between the guides is L. If the friction is neglected, these forces are normal to the follower. If the friction is considered, these forces are inclined an angle λ as is outlined in Section 7.3.1; $\tan \lambda = \mu$, where μ is the coefficient of friction.

7.7.1   Force Analysis

Consider the forces acting on the follower. Let F_v be the sum of the vertical forces.

$$F_v = P + F_s + mg + \mathrm{IF} \tag{7.62}$$

Let $\mathbf{R}_1$ be the resultant of N_{CF} and F_v. Its line of action passes through point C, which is at the intersection of $\mathbf{N}_{CF}$ and $\mathbf{F}_v$.

$$\mathbf{R}_1 = \mathbf{N}_{CF} + \mathbf{F}_v$$

Let $\mathbf{R}_2$ be the resultant of $\mathbf{F}_A$ and $\mathbf{F}_B$. Its line of action passes through point D, which is at the intersection of $\mathbf{F}_A$ and $\mathbf{F}_B$.

$$\mathbf{R}_2 = \mathbf{F}_A + \mathbf{F}_B$$

Now, we can consider the forces acting on the follower are $\mathbf{R}_1$ and $\mathbf{R}_2$. Since there is no external moment acting on the follower, $\mathbf{R}_1$ and $\mathbf{R}_2$ must have the same line of action, which is line CD. For the equilibrium of the follower,

$$\mathbf{F}_v + \mathbf{N}_{CF} + \mathbf{R}_2 = 0$$

Usually, $\mathbf{F}_v$ is known. A force polygon is drawn (Figure 7.50b) to determine the magnitudes of $\mathbf{N}_{CF}$ and $\mathbf{R}_2$. Consequently, the magnitudes of $\mathbf{F}_A$ and $\mathbf{F}_B$ can be determined. The normal and friction forces at the guides can be determined by resolving $\mathbf{F}_A$ and $\mathbf{F}_B$ to their vertical and horizontal components as shown in the figure. The driving torque is obtained by multiplying N_{CF} by the normal distance h.

$$T_d = N_{CF} \times h$$

If friction is neglected, the resultant of $\mathbf{F}_A$ and $\mathbf{F}_B$ is horizontal. In this case,

$$F_v = N_{CF} \cos \varphi \tag{7.63}$$

The values of $\mathbf{F}_A$ and $\mathbf{F}_B$ can be obtained by taking moments about points B and A, respectively.

The driving torque is given by

$$T_d = F_{CF} \times h = \frac{F_v \times h}{\cos \varphi}$$

$$\frac{h}{\cos \varphi} = OE = v$$

According to Section 3.5.1.5.1, is equal to the reduced velocity. Therefore,

$$T_d = F_v \times v$$

or

$$T_d = \left[P + K(y+\delta) + mg + m\ddot{y} \right] v \tag{7.64}$$

Another method to obtain the driving torque is to consider the power in the system.

$$T_d \omega = F_v \dot{y}$$

but $v = \dfrac{y}{\omega}$

Therefore,

$$T_d = F_v v = \left[P + k(y+\delta) + mg + m\ddot{y} \right] v$$

7.7.2 PREVENTION OF SEPARATION

One of the major problems in cams is the separation of the follower from the cam. This is more likely to occur during the retardation period of the follower when the inertia force is upward and overcomes the external load, the weight of the follower, and the spring force. In this case, there is no normal reaction between the cam and the follower. This problem needs special attention especially for high-speed rotating cams. The proper spring stiffness must be selected to insure continuous contact. That is,

$$N_{CF} \geq 0$$

or

$$P + k(y + \delta) + mg + m\ddot{y} \geq 0$$

The minimum value for the spring stiffness then is obtained from

$$P + K(y + \delta) + mg + m\ddot{y} = 0$$

In this case,

$$K = -\frac{P + mg + m\ddot{y}}{y + \delta} \tag{7.65}$$

The value of the stiffness is obtained by maximizing Equation 7.65. Equation 7.65 may be represented graphically by plotting $-my$ versus y (Figure 7.52). Point A is located on the graph with coordinates $(-\delta, P)$. The slope of the line from point A to any point on the graph represents the right-hand side of Equation 7.64. The proper value of K is when the line from A is tangent to the graph. The tangent point is at point M with coordinates $(y*, -my*)$.

The value of K is given by

$$K_{max} = -\frac{P + mg + my*}{y* + \delta}$$

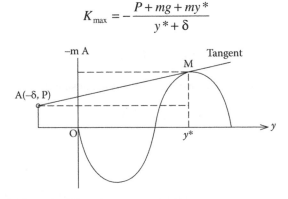

FIGURE 7.52   Maximum pressure angle.

EXAMPLE 7.15

A disk cam actuating a translating follower with a simple harmonic motion has the following data:

- The lift $s = 20$ mm, and the base circle radius is 5 cm.
- The lift angle $\beta_r = 90°$.
- The return angle $\beta_t = 90°$, no upper dwell.
- The speed of the cam is 1500 rpm.
- The mass of the follower system $m = 200$ g.
- The initial deflection of the spring is 5 mm.

Determine the minimum stiffness of the spring in order to prevent separation. Also, plot the normal force and the driving torque during the angle of action.

SOLUTION

We can design the spring according to the rise or the return strokes. Since both strokes have the same motion and the same angle, the spring is designed according to either stroke. The motion of the follower during the rise is described by

$$y = \frac{s}{2}\left(1 - \cos\frac{\pi\theta}{\beta_r}\right) \tag{a}$$

The velocity and acceleration of the follower are given by

$$v = \frac{s\,\omega}{2\beta_t}\sin\frac{\pi\theta}{\beta_r} \tag{b}$$

$$\ddot{y} = \frac{s\,\omega^2}{2\beta_t^2}\cos\frac{\pi\theta}{\beta_r} \tag{c}$$

The relation between $\ddot{y}$ and y can be obtained by substituting Equation a in Equation b. Hence,

$$\ddot{y} = \frac{s\omega^2}{2\beta_t^2}\left(1 - \frac{2y}{s}\right)$$

The relation between $-m\ddot{y}$ and y is a straight line as shown in Figure 7.53. Point A is located at $(-5, 0)$. The maximum value of the stiffness occurs at point M. The values of y^* and $\ddot{y}$ are given by

$$y^* = s = 2 \text{ cm}$$

$$-my^* = m\frac{s\,\pi^2\,\omega^2}{2\beta_t^2} = 197.4 \text{ N}$$

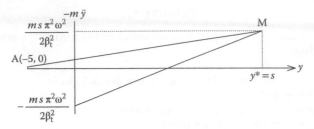

FIGURE 7.53 Maximum pressure angle for EXAMPLE 7.15.

Using the given data and substituting in Equation 7.64, we get

$$K = 78.96 \, \text{N/cm}$$

For safety, we consider that $K = 80$ N/cm. The motion of the follower during the return stroke is represented by

$$y = \frac{s}{2}\left(1 - \cos\frac{\pi(\beta_t - \theta)}{\beta_t}\right)$$

$$v = -\frac{s\omega}{2\beta_t}\sin\frac{\pi(\beta_t - \theta)}{\beta_t}$$

$$\ddot{y} = \frac{s\omega^2}{2\beta_t^2}\cos\frac{\pi(\beta_t - \theta)}{\beta_t}$$

The normal force is given by Equation 7.63:

$$N = \frac{80(y + 0.5) + (0.2 \times 9.8) + (0.2 \times \ddot{y})}{\cos\varphi}$$

According to Equation 3.53

$$\tan\varphi = \frac{v - h}{y_0 + y}$$

Then,

$$\cos\varphi = \frac{y_0 + y}{\sqrt{(y_0 + y)^2 + (v - h)^2}}$$

The driving torque is given by Equation 7.64:

$$T_d = \left[80(y + 0.5) + 0.98 + (0.2 \times \ddot{y})\right]v$$

The plots of the driving torque and the normal force over the angle of action are shown in Figures 7.54 and 7.55.

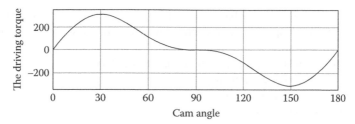

FIGURE 7.54 The driving torque in the cam set up.

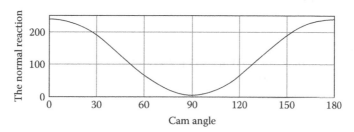

FIGURE 7.55 The normal reaction between the cam and the follower.

APPENDIX

The algorithms for the complete analysis of the chains are written in the MathCAD format.

Remarks

- t refers to the crank angle.
- r2 refers to the length of the crank.
- g refers to the position of the center of gravity.
- bt refers to the angle β of the center of gravity.
- m refers to the mass.
- Ig refers to the mass moment of inertia.
- It is assumed, in general, that link (3) is connected to another chain at point C3.
- It is assumed, in general, that link (4) is connected to another chain at point C4.
- l refers to the position of point C.
- si refers to the angle ψ for point C.
- sgn refers to the orientation sign.
- h refers to a normal distance.
- al refers to angle α in the engine chain.
- The letter x stands for lengths along the x-axis.
- The letter y stands for lengths along the y-axis.
- The letter v stands for the velocity.
- The letter a stands for the acceleration
- X stands for the horizontal component forces.
- Y stands for the vertical component forces.
- T stands for the torque.

- Letter f refers to the four-bar chain, letter e refers to the engine chain, letter s refers to the shaper chain, and letter t refers to the tilting block chain.
- p refers to pressure.
- ap refers to the area of the piston in the engine chain.
- ct refers to the cosine of an angle.
- st refers to the sine of an angle.
- = in MathCAD is written : =

A-1 FOUR-BAR CHAIN

Crank

$$t = 0, 0.01 \ldots 2*\pi$$
$$r2 = 0.05 \; g2 = 0.025 \; bt2 = 0 \; m2 = 5 \; Ig2 = 0.004$$
$$om2 = -30 \; alfa2 = 200$$
$$xa(t) = r2 * \cos(t)$$
$$ya(t) = r2 * \sin(t)$$
$$xg2(t) = g2 * \cos(t+bt2)$$
$$yg2(t) = g2 * \sin(t+bt2)$$
$$vax(t) = -ya(t) * om2$$
$$vay(t) = xa(t) * om2$$
$$aax(t) = -xa(t) * om2^2 - ya(t) * alfa2$$
$$aay(t) = -xa(t) * om2^2 + xa(t) * alfa2$$
$$ag2x(t) = -xg2(t) * om2^2 \cdot yg2(t) * alfa2$$
$$ag2y(t) = -xa(t) * om2^2 + xg2(t) * alfa2$$
$$X2(t) = -m2 * ag2x(t)$$
$$Y2(t) = -m2 * ag2y(t)$$
$$T2 = -Ig2 * alfa2$$

Four-bar chain

$$r3f = 0.075 \quad l3f = 0 \; sgn3f = 1 \quad si3f = 0 \quad g3f = 0.0375 \quad bt2 = 0 \quad m3f = 7.5 \quad Ig3f = 0.005$$
$$xqf = 0.1 \quad yqf = 0 \quad r4f = 0.09 \quad l3f = 0 \quad si3f = 0$$
$$g4f = 0.045 \quad bt2 = 0 \quad m4f = 9.0 \quad Ig4f = 0.005.$$
$$Xc3f(t) = 0 \quad Xc3f(t) = 0 \quad Xc4f(t) = 0 \quad Xc4f(t) = 0$$
$$xaf(t) = xa(t)$$
$$yaf(t) = ya(t)$$
$$vaxf(t) = vax(t)$$
$$vayf(t) = vay(t)$$
$$aaxf(t) = aax(t)$$
$$aayf(t) = aay(t)$$
$$Xc3f(t) = 0$$
$$Yc3f(t) = 0$$
$$Xc4f(t) = 0$$
$$Yc4f(t) = 0$$

$$d(t) = \sqrt{(xaf(t) - xqf)^2 + (yaf(t) - yqf)^2}$$

$$std(t) = \frac{yqf - yaf(t)}{d(t)}$$

$$ctd(t) = \frac{xqf - xaf(t)}{d(t)}$$

$td(t) = angle(ctd(t), std(t))$

$$cbt(t) = \frac{xr3f^2 + d(t)^2 - r4f^2}{2 * r3f * d(t)}$$

$bt(t) = acos(cbt(t))$
$t3f(t) = td(t) + sgn3f * bt(t)$
$xc3f(t) = l3f * cos(t3f(t) + si3f)$
$yc3f(t) = l3f * sin(t3f(t) + si3f)$
$rc3xf(t) = xaf(t) + xc3f(t)$
$rc3yf(t) = yaf(t) + yc3f(t)$
$xg3f(t) = g3f * cos(t3f(t) + bt3f)$
$yg3f(t) = g3f * sin(t3f(t) + bt3f)$

$$st4f(t) = \frac{r3f * \sin(t3f(t)) - d(t) * std(t)}{r4f}$$

$$ct4f(t) = \frac{r3f * \cos(t3f(t)) - d(t) * ctd(t)}{r4f}$$

$t4f(t) = angle(ct4f(t), st4f(t))$
$xc4f(t) = l4f * cos(t4f(t) + si4f)$
$yc4f(t) = l4f * sin(t4f(t) + si4f)$
$rc4xf(t) = xqf + xc4f(t)$
$rc4yf(t) = yqf + yc4f(t)$
$xg4f(t) = g4f * cos(t4f(t) + bt4f)$
$yg4f(t) = g4f * sin(t4f(t) + bt4f)$

$$om3f(t) = \frac{vaxf(t) * \cos(t4f(t)) + vayf(t) * \sin(t4f(t))}{r3f * \sin(t3f(t) - t4f(t))}$$

$$om4f(t) = \frac{vaxf(t) * \cos(t3f(t)) + vayf(t) * \sin(t3f(t))}{r4f * \sin(t3f(t) - t4f(t))}$$

$$alf3f(t) = \frac{\left(\begin{array}{l} aaxf(t) * \cos(t4f(t)) + aayf(t) * \sin(t4f(t)) + r4f * om4f(t)^2 \\ -r3f * om3f(t)^2 * \cos(t3f(t) - t4f(t)) \end{array} \right)}{r3f * \sin(t3f(t) - t4f(t))}$$

$$\text{alf4f(t)} = \frac{\begin{pmatrix} \text{aaxf(t)} * \cos{(\text{t3f(t)})} + \text{aayf(t)} * \sin{(\text{t3f(t)})} - \text{r3f} * \text{om3f(t)}^2 \\ + \text{r4f} * \text{om4f(t)}^2 * \cos(\text{t4f(t)} - \text{t3f(t)}) \end{pmatrix}}{\text{r4f} * \sin{(\text{t3f(t)} - \text{t4f(t)})}}$$

$\text{vc3xf(t)} = \text{vaxf(t)} - \text{yc3f(t)} * \text{om3f(t)}$
$\text{vc3yf(t)} = \text{vayf(t)} + \text{xc3f(t)} * \text{om3f(t)}$
$\text{vc4xf(t)} = -\text{yc4f(t)} * \text{om4f(t)}$
$\text{vc4yf(t)} = \text{xc4f(t)} * \text{om4f(t)}$
$\text{ac3xf(t)} = \text{aaxf(t)} - \text{xc3f(t)} * \text{om3f(t)}^2 - \text{yc3f(t)} * \text{alf3f(t)}$
$\text{ac3yf(t)} = \text{aayf(t)} - \text{yc3f(t)} * \text{om3f(t)}^2 + \text{xc3f(t)} * \text{alf3f(t)}$
$\text{ac4xf(t)} = -\text{xc4f(t)} * \text{om4f(t)}^2 - \text{yc4f(t)} * \text{alf4f(t)}$
$\text{ac4yf(t)} = -\text{yc4f(t)} * \text{om4f(t)}^2 + \text{xc4f(t)} * \text{alf4f(t)}$
$\text{ag3xf(t)} = \text{aaxf(t)} - \text{xg3f(t)} * \text{om3f(t)}^2 - \text{yg3f(t)} * \text{alf3f(t)}$
$\text{ag3yf(t)} = \text{aayf(t)} - \text{yg3f(t)} * \text{om3f(t)}^2 + \text{xg3f(t)} * \text{al3f(t)}$
$\text{ag4xf(t)} = -\text{xg4f(t)} * \text{om4f(t)}^2 - \text{yg4f(t)} * \text{alf4f(t)}$
$\text{ag4yf(t)} = -\text{yg4f(t)} * \text{om4f(t)}^2 + \text{xg4f(t)} * \text{alf4f(t)}$
$\text{X3f(t)} = -\text{m3f} * \text{ag3xf(t)}$
$\text{Y3f(t)} = -\text{m3f} * \text{ag3yf(t)}$
$\text{T3f(t)} = -\text{Ig3f(t)} * \text{alf3f(t)}$
$\text{X4f(t)} = -\text{m4f} * \text{ag4xf(t)}$
$\text{Y4f(t)} = -\text{m4f} * \text{ag4yf(t)}$
$\text{T4f(t)} = -\text{Ig4f(t)} * \text{alf4f(t)}$
$\text{k3f(t)} = -\text{T3f(t)} - \text{X3f(t)} * \text{yg3f(t)} + \text{Y3f(t)} * \text{xg3f(t)} - \text{Xc3f(t)} * \text{yc3f(t)}$
$+ \text{Yc3f(t)} * \text{xc3f(t)}$
$\text{k4f(t)} = -\text{T4f(t)} - \text{X4f(t)} * \text{yg4f(t)} + \text{Y4f(t)} * \text{xg4f(t)} - \text{Xc4f(t)} * \text{yc4f(t)}$
$+ \text{Yc4f(t)} * \text{xc4f(t)}$
$\text{x3f(t)} = \text{r3f} * \cos{(\text{t3f(t)})}$
$\text{y3f(t)} = \text{r3f} * \sin{(\text{t3f(t)})}$
$\text{x4f(t)} = \text{r4f} * \cos{(\text{t4f(t)})}$
$\text{y4f(t)} = \text{r4f} * \sin{(\text{t4f(t)})}$

$$\text{X34f(t)} = \frac{\text{k3f(t)} * \text{x4f(t)} + \text{x3f(t)} * \text{k4f(t)}}{\text{x3f(t)} * \text{y4f(t)} - \text{y3f(t)} * \text{x4f(t)}}$$

$$\text{Y34f(t)} = \frac{\text{k3f(t)} * \text{y4f(t)} + \text{y3f(t)} * \text{k4f(t)}}{\text{x3f(t)} * \text{y4f(t)} - \text{y3f(t)} * \text{x4f(t)}}$$

$$\text{F34f(t)} = \sqrt{\text{X34f(t)}^2 + \text{Y34f(t)}^2}$$

$\text{X14f(t)} = -\text{X34f(t)} - \text{X4f(t)} - \text{Xc4f(t)}$
$\text{Y14f(t)} = -\text{Y34f(t)} - \text{Y4f(t)} - \text{Yc4f(t)}$

$$\text{F14f(t)} = \sqrt{\text{X14f(t)}^2 + \text{Y14f(t)}^2}$$

$$X23f(t) = X34f(t) - X3f(t)$$
$$Y23f(t) = Y34f(t) - Y3f(t)$$

$$F23f(t) = \sqrt{X23f(t)^2 + Y23f(t)^2}$$

$$X12f(t) = X23f(t) - X2f(t)$$
$$Y12f(t) = Y23f(t) - Y2f(t)$$

$$F12f(t) = \sqrt{X12f(t)^2 + Y12f(t)^2}$$

$$Td(t) = X2(t) * yg2(t) - Y2(t) * xg2(t) - X23f(t) * ya(t) + Y23f(t) * xa(t) - T2$$
$$SFxf(t) = - X14f(t) - X12f(t)$$
$$SFyf(t) = - Y14f(t) - Y12f(t)$$

$$SFf(t) = \sqrt{SFxf(t)^2 + SFyf(t)^2}$$

A-2 ENGINE CHAIN

Crank

$$t = 0, 0.01 \ldots 2*\pi$$
$$r2 = 0.05 \ g2 = 0.0 \ bt2 = 0 \ m2 = Ig2 = 0.0$$
$$om2 = 314.159 \ alfa2 = 00$$
$$xa(t) = r2 * \cos(t)$$
$$ya(t) = r2 * \sin(t)$$
$$xg2(t) = g2 * \cos(t+bt2)$$
$$yg2(t) = g2 * \sin(t+bt2)$$
$$vax(t) = -ya(t) * om2$$
$$vay(t) = xa(t) * om2$$
$$aax(t) = -xa(t) * om2^2 - ya(t) * alfa2$$
$$aay(t) = -xa(t) * om2^2 + xa(t) * alfa2$$
$$ag2x(t) = -xg2(t) * om2^{2\cdot} yg2(t) * alfa2$$
$$ag2y(t) = -xa(t) * om2^2 + xg2(t) * alfa2$$
$$X2(t) = -m2 * ag2x(t)$$
$$Y2(t) = -m2 * ag2y(t)$$
$$T2 = -Ig2 * alfa2$$

Engine chain

$$he = 0.0 \ al = 0.0 \quad r3e = 0.2 \quad l3e = 0 \ si3e = 0 \quad sgn4e = 1 \quad g3e = 0.05$$
$$bt3e = 0 \quad m3e = 1.35 \ Ig3e = 0.005$$
$$m4e = 0.9 \ ap = 45.5$$

$Xc3e(t) = 0$ $Xc3e(t) = 0$ $Xc4e(t) = 0$ $Xc4e(t) = 0$
$xae(t) = xa(t)$
$yae(t) = ya(t)$
$vaxe(t) = vax(t)$
$vaye(t) = vay(t)$
$aaxe(t) = aax(t)$
$aaye(t) = aay(t)$
$Xc3e(t) = 0$
$Yc3e(t) = 0$
$p1(t) = 140 + 1504 * t$
$p2(t) = -3.653 + e^{-t}$
$p3(t) = 0$
$p4(t) = -6.323 + 0.00051 * e^{-t}$
$f1(t) = p1(t) * ap$
$f2(t) = p2(t) * ap$
$f3(t) = p3(t) * ap$
$f4(t) = p4(t) * ap$

$$P4e(t) = \begin{vmatrix} f1(t) & \text{if} & 0 \leq t \leq \dfrac{8*\pi}{180} \\[2mm] f2(t) & \text{if} & \dfrac{8*\pi}{180} \leq t \leq \dfrac{270*\pi}{180} \\[2mm] f3(t) & \text{if} & \dfrac{270*\pi}{180} \leq t \leq \dfrac{540*\pi}{180} \\[2mm] f14(t) & \text{if} & \dfrac{540*\pi}{180} \leq t \leq \dfrac{720*\pi}{180} \end{vmatrix}$$

$$s4e(t) = xae(t) * \cos(al) + yae(t) * \sin(al)$$
$$+ sgn4e * \sqrt{r3e^2 - (he + xae(t) * \sin(al) - yae(t) * \cos(al))^2}$$

$$ct3e(t) = \frac{x4e(t) * \cos(al) - he * \sin(al) - xae(t)}{r3e}$$

$$st3e(t) = \frac{x4e(t) * \sin(al) + he * \cos(al) - yae(t)}{r3e}$$

$t3e(t) = angle(ct3e(t), st3e(t))$
$xc3e(t) = l3e * \cos(t3e(t) + si3e)$
$yc3e(t) = l3e * \sin(t3e(t) + si3e)$
$rc3ex(t) = xae(t) + xc3e(t)$
$rc3ey(t) = yae(t) + yc3e(t)$

xg3e(t) = l3e * cos(t3e(t) + bt3e)
yg3e(t) = l3e * sin(t3e(t) + bt3e)

$$om3e\,(t) = \frac{vaxe(t)* \sin{(al)} - vaye(t)*\cos{(al)}}{r3e*\cos{(t3e(t)-al)}}$$

v4e(t) = vaxe(t)*cos(al) + vaye(t)*sin(al) − om3e(t)*r3e*sin(t3e(t) −al)

$$alf3e\,(t) = \frac{aaxe(t)* \sin{(al)} - aaye(t)*\cos{(al)} + om3e(t)^2*\sin{(t3e(t)-al)}}{r3e*\cos{(t3e(t)-al)}}$$

a4e(t) = aaxe(t)*cos(al) + aaye(t)*sin(al) − om3e(t)² * r3e*cos(t3e(t) − al)
−alf3e(t) * r3e * sin(t3e(t) − al)
vc3xe(t) = vaxe(t) − yc3e(t) * om3e(t)
vc3ye(t) = vaye(t) + xc3e(t) * om3e(t)
ac3xe(t) = aaxe(t) − xc3e(t) * om3e(t)² − yc3ye(t) * alf3e(t)
ac3ye(t) = aaye(t) − yc3e(t) * om3e(t)² + xc3ye(t) * alf3e(t)
ag3xe(t) = aaxe(t) − xg3e(t) * om3e(t)² − yg3ye(t) * alf3e(t)
ag3ye(t) = aaye(t) − yg3e(t) * om3e(t)² + xg3ye(t) * alf3e(t)
X3e(t) = − m3e * ag3xe(t)
Y3c(t) = − m3e * ag3ye(t)
T3e(t) = − Ig3c * alf3e(t)
IF4e(t) = − m4c * a4c(t)
F4c(t) = IF4e(t) + P4e(t)
K3e(t) = T3e(t) − X3e(t) * yg3e(t) + Y3e(t) * xg3e(t)
−Xc3e(t) * yc3e(t) + Yc3e(t) * xc3e(t)

$$F14e\,(t) = \frac{-k3e(t) - F4e(t)*r3e* \sin{(al-t3e\,(t))}}{r3e*\cos{(al-t3e(t))}}$$

X14e(t) = −F14e(t) * sin(al)
Y14e(t) = F14e(t) * cos(al)
X34e(t) = −F14e(t) * cos(al) − X14e(t)
Y34e(t) = −F14e(t) * sin(al) − Y14e(t)

$$F34e\,(t) = \sqrt{X34e\,(t)^2 + Y34e\,(t)^2}$$

X23e(t) = X34e(t) − X3e(t)
Y23e(t) = Y34e(t) − Y3e(t)

$$F23e\,(t) = \sqrt{X23e\,(t)^2 + Y23e\,(t)^2}$$

X12e(t) = X23e(t) − X2e(t)
Y12e(t) = Y23e(t) − Y2e(t)

$$F12e\,(t) = \sqrt{X12e\,(t)^2 + Y12e\,(t)^2}$$

Td(t) = X2(t)*yg2(t) − Y2(t)*xg2(t) − X23e(t) * ya(t) + Y23e(t) * xa(t) − T2
SFxe(t) = −X14e(t) − X12e(t) + P4e(t) * cos(al)
SFye(t) = −Y14e(t) − Y12e(t) + P4e(t) * sin(al)

$$SFe(t) = \sqrt{SFxe(t)^2 + SFye(t)^2}$$

A-3 SHAPER CHAIN

Crank

t = 0, 0.01.. 2*π
r2 = 0.05 g2 = 0.025 bt2 = 0 m2 = 5 Ig2 = 0.004
om2 = −30 alfa2 = 200
xa(t) = r2 * cos(t)
ya(t) = r2 * sin(t)
xg2(t) = g2 * cos(t+bt2)
yg2(t) = g2 * sin(t+bt2)
vax(t) = −ya(t) * om2
vay(t) = xa(t) * om2
aax(t) = −xa(t) * om2² − ya(t) * alfa2
aay(t) = −xa(t) * om2² + xa(t) * alfa2
ag2x(t) = −xg2(t) * om2²· yg2(t) * alfa2
ag2y(t) = −xa(t) * om2² + xg2(t) * alfa2
X2(t) = −m2 * ag2x(t)
Y2(t) = −m2 * ag2y(t)
T2 = −Ig2 * alfa2

Shaper chain

hs = 0   m3s = 7.5   Ig3s = 0.005   g3s = 0.0   bt3s = 0
xqs = 0   yqs = 0.11   14s = 0.15   si4s = 0   m4s = 9   Ig4s =.013   g4s =.08
   bt4s = 0
Xc4s(t) = 0   Yc4s(t) = 0
xas(t) = xa(t)
yas(t) = ya(t)
vaxs(t) = vax(t)
vays(t) = vay(t)
aaxs(t) = aax(t)
aays(t) = aay(t)

$$x4s(t) = \sqrt{(xas(t) - xqs)^2 + (yas(t) - yqs)^2 - hs^2}$$

$$c4s(t) = \frac{[x4s(t)*(xas(t) - xqs) + hs*(yas(t) - yqs)]}{(x4s(t)^2 + hs^2)}$$

$$s4s(t) = \frac{[x4s(t)*(yas(t) - yqs) - hs*(xas(t) - xqs)]}{(x4s(t)^2 + hs^2)}$$

t4s(t) = angle(ct4s(t),st4s(t))
xc4s(t) = l4s * cos (t4s(t) + si4s)
yc4s(t) = l4s * sin (t4s(t) + si4s)
rc4xs(t) = xqs + xc4s(t)
rc4ys(t) = yqs + yc4s(t)
xg3s(t) = g3s * cos (t3s(t) + bt3s)
yg3s(t) = g3s * sin (t3s(t) + b3s)
xg4s(t) = g4s * cos (t4s(t) + bt4s)
yg4s(t) = g4s * sin (t4s(t) + bt4s)

$$om4s(t) = \frac{-vaxs(t)*s4s(t)+vays(t)*c4s(t)}{x4s(t)}$$

vsls(t) = −om4s(t) * hs − vaxs(t) * c4s(t) − vays(t) * s4s(t)

$$alf4s = \frac{-aaxs(t)*s4s(t)+aays(t)*c4s(t)+hs*om4s(t)^2+2*om4s(t)*vsls(t)}{x4s(t)}$$

asls(t) = −alf4s(t) * hs − aaxs(t) * c4s(t) − aays(t)*s4s(t) − x4s(t)*om4s(t)²
vc4xs(t) = −yc4s(t) *om4s(t)
vc4ys(t) = xc4s(t) *om4s(t)
ac4xs(t) = −xc4s(t) *om4s(t)² − yc4s(t) * alf4s(t)
ac4ys(t) = −yc4s(t) *om4s(t)² + xc4s(t) * alf4s(t)
ag3xs(t) = aaxs(t) − xg3s(t) *om4s(t)² − yg3s(t) * alf4s(t)
ag3ys(t) = aays(t) − yg3s(t) *om4s(t)² + xg3s(t) * al4s(t)
ag4xs(t) = −xg4s(t) *om4s(t)² − yg4s(t) * alf4s(t)
ag4ys(t) = −yg4s(t) *om4s(t)² + xg4s(t) * alf4s(t)
X3s(t) = −m3s * ag3xs(t)
Y3s(t) = −msf * ag3ys(t)
T3s(t) = −Ig3s(t) * alf3s(t)
X4s(t) = −m4s * ag4xs(t)
Y4s(t) = −m4s * ag4ys(t)
T4s(t) = −Ig4s(t) * alf4s(t)
k3s(t) = −T3s(t) − X3s(t) * yg3s(t) + Y3s(t) * xg3s(t)
k4s(t) = −T4s(t) − X4s(t) * yg4s(t) + Y4s(t) * xg4s(t) − Xc4s(t) * yc4s(t)
+ Yc4s(t)* xc4s(t)

$$F34s(t) = \frac{-k3s(t)-k4s(t)}{x4s(t)}$$

X34s(t) = − F34s(t) * s4s(t)
Y34s(t) = F34s(t) * c4s(t)
X14s(t) = − X34s(t) − X4s(t) − Xc4s(t)
Y14s(t) = − Y34s(t) − Y4s(t) − Yc4s(t)

$$F14s(t) = \sqrt{X14s(t)^2 + Y14s(t)^2}$$

$$X23s(t) = X34s(t) - X3s(t)$$
$$Y23s(t) = Y34s(t) - Y3s(t)$$

$$F23s(t) = \sqrt{X23s(t)^2 + Y23s(t)^2}$$

$$X12s(t) = X23s(t) - X2s(t)$$
$$Y12s(t) = Y23s(t) - Y2s(t)$$

$$F12s(t) = \sqrt{X12s(t)^2 + Y12s(t)^2}$$

$$Td(t) = X2(t)*yg2(t) - Y2(t)*xg2(t) - X23s(t) * ya(t) + Y23s(t) * xa(t) - T2$$
$$SFxs(t) = - X14s(t) - X12s(t)$$
$$SFys(t) = - Y14s(t) - Y12s(t)$$

$$SFs(t) = \sqrt{SFxs(t)^2 + SFys(t)^2}$$

A-4 TILTING BLOCK CHAIN

Crank

$$t = 0, 0.01.. \ 2*\pi$$
$$r2 = 0.05 \ g2 = 0.025 \ bt2 = 0 \ m2 = 5 \ Ig2 = 0.004$$
$$om2 = -30 \ alfa2 = 200$$
$$xa(t) = r2 * \cos(t)$$
$$ya(t) = r2 * \sin(t)$$
$$xg2(t) = g2 * \cos(t+bt2)$$
$$yg2(t) = g2 * \sin(t+bt2)$$
$$vax(t) = -ya(t) * om2$$
$$vay(t) = xa(t) * om2$$
$$aax(t) = -xa(t) * om2^2 - ya(t) * alfa2$$
$$aay(t) = -xa(t) * om2^2 + xa(t) * alfa2$$
$$ag2x(t) = -xg2(t) * om2^{2.} \ yg2(t) * alfa2$$
$$ag2y(t) = -xa(t) * om2^2 + xg2(t) * alfa2$$
$$X2(t) = -m2 * ag2x(t)$$
$$Y2(t) = -m2 * ag2y(t)$$
$$T2 = -Ig2 * alfa2$$

Tilting block chain

$$ht = 0 \quad m4t = 7.5 \quad Ig4t = 0.005 \quad g4t = 0.0 \quad bt4t = 0$$
$$xqt = 0 \quad yqt = 0.11 \quad l3t = 0.15 \quad si3t = 0 \ m3t = 9 \quad Ig3t = .013 \quad g3t$$
$$\qquad =.08 \quad bt3t = 0$$
$$Xc3t(t) = 0 \ Yc3t(t) = 0$$
$$xat(t) = xa(t)$$
$$yat(t) = ya(t)$$
$$vaxt(t) = vax(t)$$

vayt(t) = vay(t)
aaxt(t) = aax(t)
aayt(t) = aay(t)

$$x3t(t) = \sqrt{(xat(t) - xqt)^2 + (yat(t) - yqt)^2 - ht^2}$$

$$c3t(t) = \frac{[x3t(t) * (xqt(t) - xat) + ht * (yqt(t) - yat)]}{(x3t(t)^2 + ht^2)}$$

$$s3t(t) = \frac{[x3t(t) * (yqt(t) - yat) - ht * (xqt(t) - xat)]}{(x3t(t)^2 + ht^2)}$$

t3t(t) = angle(ct3t(t),st3t(t))
xc3t(t) = l3t * cos (t3t(t) + si3t)
yc3t(t) = l3t * sin (t3t(t) + si3t)
rc3xt(t) = xqt + xc3t(t)
rc3yt(t) = yqt + yc3t(t)
xg4t(t) = g4t * cos (t3t(t) + bt4t)
yg4t(t) = g4t * sin (t3t(t) + b4t)
xg3t(t) = g3t * cos (t3t(t) + bt3t)
yg3t(t) = g3t * sin (t3t(t) + bt3t)

$$om3t(t) = \frac{-vaxt(t) * s3t(t) + vayt(t) * c3t(t)}{x3t(t)}$$

vslt(t) = −om3t(t) * ht − vaxt(t) * c3t(t) − vayt(t) * s3t(t)

$$alf3t = \frac{-aatx(t) * s3t(t) + aayt(t) * c3t(t) + ht * om3t(t)^2 + 2 * om3t(t) * vslt(t)}{x3t(t)}$$

aslt(t) = − alf3t(t)*ht − aaxt(t)*c3t(t) − aayt(t)*s3t(t) − x3t(t)*om3t(t)²
vc3xt(t) = vaxt(t) − yc3t(t) *om3t(t)
vc3yt(t) = vayt(t) + xc3t(t) *om3t(t)
ac3xt(t) = aaxt(t) − xc3t(t) *om3t(t)² − yc3t(t) * alf3t(t)
ac3yt(t) = aayt(t) − yc3t(t) *om3t(t)² + xc3t(t) * alf3t(t)
ag3xt(t) = aaxt(t) − xg3t(t) *om3t(t)² − yg3t(t) * alf3t(t)
ag3yt(t) = aayt(t) − yg3t(t) *om3t(t)² + xg3t(t) * al3t(t)
ag4xt(t) = − xg4t(t) *om3t(t)² − yg4t(t) * alf3t(t)
ag4yt(t) = − yg4t(t) *om3t(t)² + xg4t(t) * alf3t(t)
X3t(t) = −m3t * ag3xt(t)
Y3t(t) = −m3t * ag3yt(t)
T3t(t) = −Ig3t(t) * alf3t(t)
X4t(t) = −m4t * ag4xt(t)
Y4t(t) = −m4t * ag4yt(t)
T4t(t) = −Ig4t(t) * alf3t(t)
k3t(t) = −T3t(t) − X3t(t) * yg3t(t) + Y3t(t) * xg3t(t) − Xc3t(t) * yc3t(t)
         + Yc3t(t)* xc3t(t)

$$k4t(t) = -T4t(t) - X4t(t) * yg4t(t) + Y4t(t) * xg4t(t)$$

$$F43t(t) = \frac{-k3t(t) - k4t(t)}{x3t(t)}$$

$$X43t(t) = -F43t(t) * s3t(t)$$
$$Y43t(t) = F43t(t) * c3t(t)$$
$$X14t(t) = -X43t(t) - X4t(t)$$
$$Y14t(t) = -Y43t(t) - Y4t(t)$$

$$F14t(t) = \sqrt{X14t(t)^2 + Y14t(t)^2}$$

$$X23t(t) = X43t(t) - X3t(t) - Xc3t(t)$$
$$Y23t(t) = Y43t(t) - Y3t(t) - Yc3t(t)$$

$$F23t(t) = \sqrt{X23t(t)^2 + Y23t(t)^2}$$

$$X12t(t) = X23t(t) - X2t(t)$$
$$Y12t(t) = Y23t(t) - Y2t(t)$$

$$F12t(t) = \sqrt{X12t(t)^2 + Y12t(t)^2}$$

$$Td(t) = X2(t)*yg2(t) - Y2(t)*xg2(t) - X23t(t) * ya(t) + Y23t(t) * xa(t) - T2$$
$$SFxt(t) = -X14t(t) - X12t(t)$$
$$SFyt(t) = -Y14t(t) - Y12t(t)$$

$$SFt(t) = \sqrt{SFxt(t)^2 + SFyt(t)^2}$$

PROBLEMS

Static Force Analysis

7.1   For the double-slider mechanism shown in Figure P7.1, find the ratio between the forces P and Q.

$$OA = 30 \text{ cm}, \quad AB = AC = 100 \text{ cm}$$

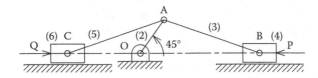

FIGURE P7.1

7.2   For the mechanism shown in Figure P7.2, find the driving torque in terms of the resisting force P.

$$OA = 60 \text{ mm}, \quad AB = 280 \text{ mm}, \quad QB = 120 \text{ mm}, \quad OQ = 300 \text{ mm},$$

$$BC = 300 \text{ mm}$$

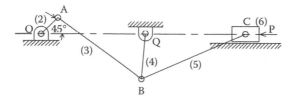

FIGURE P7.2

7.3 Figure P7.3 shows an outline of the Zoller double-piston engine. Find the resisting torque at crank OA due to the driving forces F_1 and F_2 at the two pistons.

$$OA = 40 \text{ mm}, \ AB = 120 \text{ mm}, \ AC = 30 \text{ mm}, \ \text{angle } ACB = 90°,$$
$$CD = 120 \text{ mm}$$

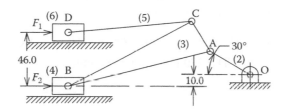

FIGURE P7.3

7.4 For the mechanism shown in Figure P7.4, find the relation between the torques applied at links OA and QD.

$$OA = 60 \text{ mm}, \ AB = 140 \text{ mm}, \ CD = 100 \text{ mm}, \ QD = 120 \text{ mm}$$

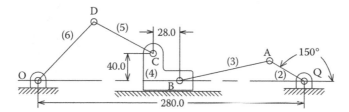

FIGURE P7.4

7.5 For the mechanism shown in Figure P7.5, find the relation between the torques applied at links OA and QD.

$$OA = 80 \text{ mm}, \ AC = CB = 120, \ OQ = 400 \text{ mm}, \ QD = 120 \text{ mm},$$
$$DC = 260 \text{ mm}$$

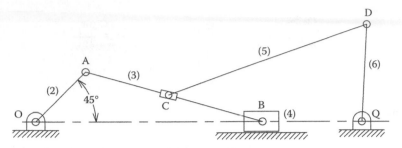

FIGURE P7.5

7.6   For Problem P7.5, if block (4) is subjected to a horizontal force of 200 N to the left and link QD is subjected to an external moment of 2400 N·cm counterclockwise, find the torque applied to link OA.

7.7   For the mechanism shown in Figure P7.7, find the relation between the torques applied at links OA and OD.

$$OA = 60 \text{ mm}, \quad AB = 230 \text{ mm}, \quad QB = QC = 1350 \text{ mm}, \quad BC = 100 \text{ mm},$$
$$CD = 270 \text{ mm}, \quad OD = 180 \text{ mm}$$

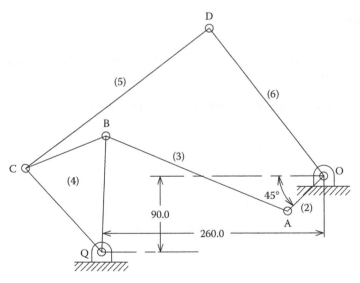

FIGURE P7.7

7.8   For the press machine shown in Figure P7.8, the pressing force $P = 500$ N. Find the driving torque on the crank OA.

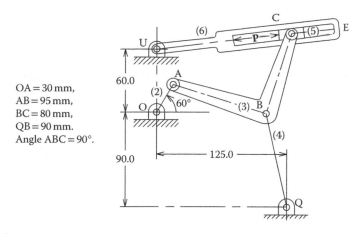

FIGURE P7.8

7.9 Find the torque on crank OA due to the external forces $F_4 = 500$ N on slider (4) and $F_8 = 600$ N on slider (8) (Figure P7.9).

OA = 75 mm, AB = 150 mm, BC = 100 mm, AC = 70 mm,
CD = 60 mm, QD = 150 mm, DE = 75 mm, EH = 150 mm

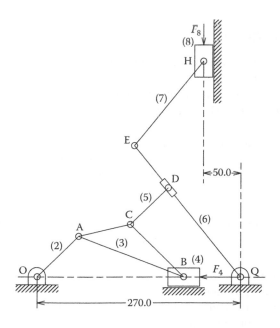

FIGURE P7.9

7.10 Find the torque on crank OA due to the external forces $F_4 = 500$ N on slider (4) and $F_8 = 600$ N on slider (8) (Figure P7.10).

$$OA = 60 \text{ mm, } AB = 240 \text{ mm, } AC = 80 \text{ mm, } CD = 240 \text{ mm,}$$
$$DE = 140 \text{ mm}$$

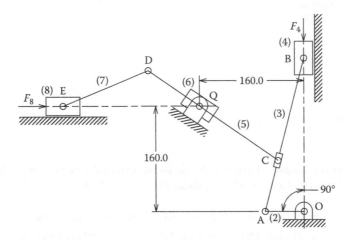

FIGURE P7.10

FORCE ANALYSIS IN GEARS

7.11 A pair of spur gears transmits 20 kW with a speed ratio of 2.5. The speed of the pinion is 600 rpm. The diameter of the pinion is 100 mm. The tooth profile is involute with a pressure angle 20°. Determine the loads.

7.12 Repeat Problem 7.11 replacing the spur gears with helical gears with a helix angle 30°.

7.13 The gear train shown in Figure P7.13 is used to transmit 500 kW at an input speed at gear 1 of 1750 rpm. Determine the forces on the gears, the torque delivered to the worm wheel 8, and the efficiency of the worm drive.

$$N_1 = 18, N_2 = 27, N_3 = 20, N_4 = 41, N_5 = 18, N_6 = 38,$$
$$N_7 = 2RH, N_8 = 24$$

For the worm,
- The pitch p is 45 mm.
- The pitch diameter of the worm D_W is 150 mm.
- The pressure angle φ is 20°.
- The coefficient of friction μ is 0.05.

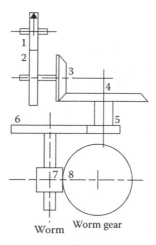

FIGURE P7.13

7.14 For the gear train shown in Figure P7.14, the motor that is connected to worm 1 rotates at 1750 rpm. A wire rope is connected to the drum and is used for lifting loads up to 1000 kg. The drum diameter is 1 m. Determine the forces on the gears, the torque delivered to the worm wheel 8, and the efficiency of the worm drive.

$$N_1 = 3, N_2 = 90, N_3 = 24, N_4 = 72, N_5 = 15, N_6 = 40,$$

$$N_7 = 3, N_8 = 48$$

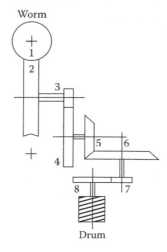

FIGURE P7.14

For the worm,
- The pitch p is 25 mm.
- The pitch diameter of the worm D_W is 120 mm.
- The pressure angle φ is 20°.
- The coefficient of friction μ is 0.05.

7.15 For the gear trains shown in Figure P7.15a and b, the input shaft delivers 20 kW at 1000 rpm. Find the loads on the gear teeth.

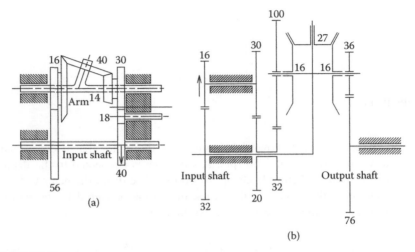

FIGURE P7.15

7.16 In Figure P7.16, the power delivered to the arm is 10 kW at 300 rpm in the counterclockwise direction. Find the teeth loads and the braking torque for the following cases:
a. Gear 1 is fixed.
b. Gear 4 is fixed.

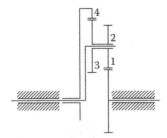

FIGURE P7.16

7.17 The train shown in Figure P7.17 is actually an automotive epicyclic gearbox where the ring gears 3 and 6 can be locked independently by means of band brakes. Let $N_1 = N_4 = 23$ teeth and $N_2 = N_5 = 22$ teeth (accordingly $N_3 = N_6 = 67$ teeth). The input power is 200 hp at 3000 rpm clockwise. Find the teeth loads and the braking torque for both cases.

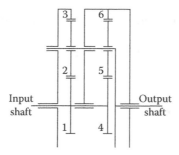

FIGURE P7.17

FRICTION FORCE ANALYSIS

7.18 For the sake of analysis, use the following data for Problems P7.1 through P7.10 to make static force analysis:
- The radius of the friction circle in all turning joints is 10% of the length of cranks OA.
- The friction angle in all sliding joints is 10°.
- The angular speeds of all cranks OA are 10 rad/s clockwise.

DYNAMIC FORCE ANALYSIS

7.19 For the mechanism shown in Figure P7.19, the forces applied on the links are shown in the figure. Find the forces transmitted by the links and the driving torque on crank OA.

$$OA = 80 \text{ mm}, \ OE_2 = 40 \text{ mm}, \ AC = CB = 120 \text{ mm}, \ OQ = 400 \text{ mm},$$
$$QD = 120 \text{ mm}, \ QE_6 = 60 \text{ mm}, \ DC = 260 \text{ mm}, \ CE_5 = 60 \text{ mm}$$

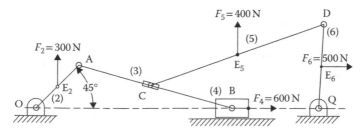

FIGURE P7.19

7.20 For the mechanism shown in Figure P7.20, find the reaction forces between links. Also, find the driving torque on crank OA. Dimensions are in centimeters.

$OA = 25$ cm, $OG_2 = 12.5$ cm, $BG_3 = 25$ cm, ABC is one link,

angle $ABC = 90°$ $F_2 = 200$ N, $T_2 = 10$ N·m, $F_3 = 500$ N, $T_3 = 50$ N·m,

$$F_4 = 100 \text{ N}, T_4 = 3 \text{ N·m}$$

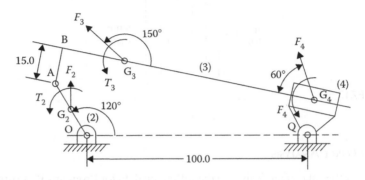

FIGURE P7.20

7.21 Solve Problems 7.19 and 7.20 analytically.
7.22 The input crank of the four-bar mechanism of Figure P7.22 rotates at a constant speed of 500 rad/s clockwise. Use the four-bar chain analysis to determine the following over a complete cycle and plot the results against the crank angle:
   a. The reaction forces between the links
   b. The driving torque on crank OA
   c. The shaking force

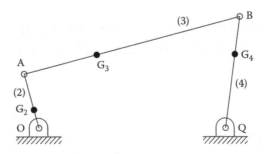

FIGURE P7.22

$OA = 65$ mm, $OG_2 = 25$ mm, $AB = 300$ mm, $AG_3 = 100$ mm,

$QB = 150$ mm, $QG_4 = 100$ mm, $OQ = 250$ mm

$$m_2 = 2.25 \text{ kg}, m_3 = 4.5 \text{ kg}, m_4 = 5.75 \text{ kg}$$

$$I_2 = 0.001 \text{ kg} \cdot \text{m}^2, I_3 = 0.027 \text{ kg} \cdot \text{m}^2, I_4 = 0.15 \text{ kg} \cdot \text{m}^2$$

7.23 The slider mechanism shown in Figure P7.23 is used in a compressor. The radius of gyration for the crank is 8 cm, and for the connecting rod is 12 cm. The gas force on piston P = 4000 N is constant during the compression stroke (piston is moving left) and is equal to zero during the suction stroke. The crank rotates with a uniform angular velocity of 100 rad/s counterclockwise. Use the engine chain analysis to determine the following over a complete cycle and plot the results against the crank angle:
a. The reaction forces between the links
b. The driving torque on crank OA
c. The shaking force

$$OA = 50 \text{ mm}, \quad OG_2 = 40 \text{ mm}, \quad AB = 200 \text{ mm}, \quad AG_3 = 60 \text{ mm},$$

$$m_2 = 5 \text{ kg}, \quad m_3 = 4.0 \text{ kg}, m_4 = 2.0 \text{ kg}$$

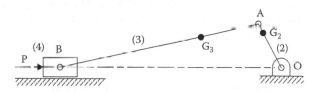

FIGURE P7.23

7.24 The mechanism is partially balanced by placing a counterweight attached to the crank in the opposite direction at a radius of 40 mm. The value of the counter weight affects the balancing of the mechanism. To understand its effect, repeat Problem 7.23, using balancing masses of 5, 6, 7, and 8 kg.

7.25 For the mechanism shown in Figure P7.25, the crank OA rotates at a constant speed of 200 rad/s clockwise. Use the chain analysis to determine the following over a complete cycle and plot the results against the crank angle:
a. The reaction forces between the links
b. The driving torque on crank OA
c. The shaking force

$$OQ = 30 \text{ mm}, \quad OA = 85.0 \text{ mm}, \quad OG_2 = 42.5 \text{ mm}, \quad AB = 80.0 \text{ mm},$$

$$AG_3 = 40.0 \text{ mm}, \quad QB = 125 \text{ mm}, \quad QG_4 = 62.5 \text{ mm}, \quad BC = 130.0 \text{ mm},$$

$$BG_5 = 65.0 \text{ mm},$$

$$m_2 = 2.5 \text{ kg}, m_3 = 3.0 \text{ kg}, m_4 = 2.0 \text{ kg}, m_5 = 3.5 \text{ kg}, m_4 = 6.0 \text{ kg},$$

$$I_2 = 0.001 \text{ kg} \cdot \text{m}^2, I_3 = 0.027 \text{ kg} \cdot \text{m}^2, I_4 = 0.15 \text{ kg} \cdot \text{m}^2,$$

$$I_4 = 0.15 \text{ kg} \cdot \text{m}^2$$

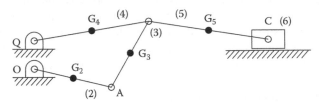

FIGURE P7.25

7.26 Figure P7.26 shows a Whitworth quick-return motion mechanism, which is used in shaping machines. Crank OA rotates at a uniform angular velocity of 30 rad/s clockwise. During the cutting stroke, the ram is subjected to a constant resisting load of 1000 N. The resisting force is zero during the return stroke. Plot the reaction forces, the driving torque, and the shaking force over one cycle.

$$OQ = 100 \text{ mm}, \ OA = 200.0 \text{ mm}, OG_2 = 100.0 \text{ mm}, \ QB = 100.0 \text{ mm},$$
$$QG_4 = 140.0 \text{ mm}, \ BC = 350.0 \text{ mm}, \ BG_3 = 175.0 \text{ mm},$$

$$m_2 = 5.0 \text{ kg}, m_3 = 8.0 \text{ kg}, m_4 = 15.0 \text{ kg}, m_5 = 15.0 \text{ kg}, m_6 = 50.0 \text{ kg}$$

$$I_2 = 0.01 \text{ kg} \cdot \text{m}^2, I_3 = 0.027 \text{ kg} \cdot \text{m}^2, I_4 = 0.15 \text{ kg} \cdot \text{m}^2, I_5 = 0.15 \text{ kg} \cdot \text{m}^2$$

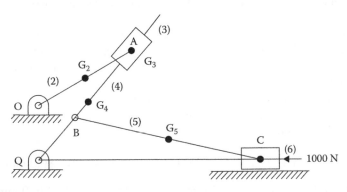

FIGURE P7.26

FLYWHEEL

7.27 The torque output diagram of a single-cylinder, four-stroke engine is shown in Figure P7.27. Determine:
   a. The average output torque
   b. The power in kilowatts delivered by the engine if its speed is 3000 rpm
   c. The speed diagram and then locate the crank angles where the engine speed is a maximum and a minimum during the cycle
   d. The energy that causes the maximum speed variation
   e. The mass moment of inertia of a flywheel to keep the speed variation within 3%
   f. The maximum values of the angular acceleration and angular deceleration

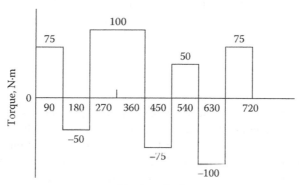

FIGURE P7.27

7.28 Repeat Problem 7.27 if the engine consists of four cylinders. The ignition timing between the successive cylinders is 180°.

7.29 The torque diagram of a single-cylinder, two-stroke engine is approximated by a triangle as shown in Figure P7.29. The speed of the engine is 3000 rpm. Determine the moment of inertia of the flywheel such that the total speed variation is 6%.

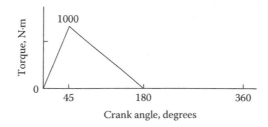

FIGURE P7.29

7.30 Repeat Problem 7.27 if the engine consists of two cylinders. The torque of each cylinder is repeated every 180°.

7.31 Repeat Problem 7.27 if the engine consists of three cylinders. The torque of each cylinder is repeated every 120°.

7.32 A 2.0-hp motor is used to drive a machine at a mean speed of 300 rpm. The resisting torque of the machine is shown in Figure P7.32 while the driving torque is constant. Determine the number of cycles per minute and the inertia of the flywheel to keep the speed variation within 3%.

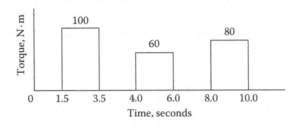

FIGURE P7.32

7.33 A press operated by a Scotch yoke is used to punch holes in steel plates with a thickness of 10 mm. The punching starts when the crank is 60° from the lowest position. The force required for punching starts with 20,000 N and ends with zero. Determine the power of the driving motor if the speed is 150 rpm. Also, determine the inertia of the flywheel to keep the speed total variation within 3%.

7.34 The driving torque in a mechanical system is constant while the resisting torque is given by

$$T_r = 100 \sin^2 \theta \ \text{N} \cdot \text{m}$$

Determine the inertia of the flywheel such that the total speed variation is within 3%. The maximum speed is 15 rad/s.

7.35 Figure P7.35 shows the driving and resisting torques for a machine; both contain half sine curves over one 180°. If the speed of the machine is 300 rpm and the maximum speed variation is 3%, calculate the power and the inertia of the flywheel.

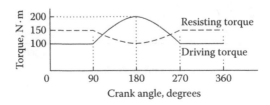

FIGURE P7.35

7.36 Figure P7.36 shows the driving and resisting torques for a machine; both
contain half sine curves as shown. If the speed of the machine is 600
rpm and the maximum speed variation is 2%, calculate the power and
the inertia of the flywheel.

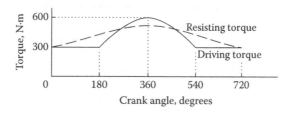

FIGURE P7.36

7.37 The resisting torque of a double-acting pump consists of two sine curves
with peaks at 90° and 270° as shown in Figure P7.37. The cycle is 360°.
The driving torque is uniform. If the speed of the pump is 600 rpm and
the maximum speed variation is 2%, calculate the power of the driving
motor and the inertia of the flywheel.

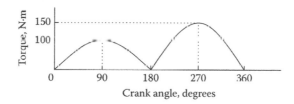

FIGURE P7.37

7.38 Repeat Problem 7.37 if the pump is driven by a DC motor and its power
is increased to be twice, and then, three times the required power.
7.39 Suppose that the pump of Problem 7.37 has a flywheel of a mass of 2 kg
at a radius of gyration of 20 cm. Determine:
a.   The total speed variation
b.   The maximum acceleration and the maximum deceleration
c.   Make practical suggestions to how to reduce the maximum acceleration by 50%
7.40 In a mechanical system, the driving and resisting torques are given by

$$T_d = 150 + 35\sin\theta - 100\cos 2\theta$$

$$T_r = K35\sin^2\theta$$

Calculate the power of the driving motor and the inertia of the flywheel if the speed is 600 rpm and the total speed variation is 3%.

Cam Dynamics

7.41 In a cam with a reciprocating roller follower, the base circle diameter is 80 mm, the roller diameter is 40 mm, the amount of offset is 15 mm (positive for the rise), the lift is 30 mm, and the rise and return angles are 120° each. The motion of the follower is described by

Rise	Return
1. Simple harmonic	Simple harmonic
2. Parabolic	Parabolic
3. Cycloid	Cycloid

The mass of the follower system is 1 kg, the external load on the follower is 50 N, and the initial deflection on the spring is 10 mm. For the three types of motions, determine the spring stiffness. Plot the driving torque and the normal force over one complete cam rotation.

8 Balancing

8.1 INTRODUCTION

In Chapter 7, we found that inertia forces are created in machines due to the masses and the acceleration of the members. These inertia forces in turn cause shaking forces in the machines. The shaking forces, in most machines, cause vibrations, which have harmful effects. In a close look at any mechanism, for instance, the engine mechanism, we see that it consists of a rotating member (the crank), floating member (the connecting rod), and a reciprocating member (the piston). Generally speaking, the inertia forces due to the rotating members can be completely eliminated by adding countermasses. Complete elimination of the inertia forces due to the floating and reciprocating parts requires expensive solutions. However, it is possible to reduce their effects by partial balancing, which is acceptable in most applications.

8.2 BALANCING OF ROTATING PARTS

In many applications, the machines consist only from rotating parts, for example, turbine rotors, centrifugal pumps, transmission shafts with gears, and pulleys. These parts are manufactured by different processes, which apparently guarantee balancing. However, under the most optimized circumstances, there is no guarantee that these parts are completely balanced due to the following reasons:

1. *Blow holes in castings*: It may be present within the material and cannot be detected by normal visual inspection.
2. *Eccentricity*: Exists whenever the geometric centerline of a part does not coincide with its rotating centerline.
3. *Addition of keys and key ways*: These are elements used to fix hubs on shafts.
4. *Distortion*: It may be the cause of stress relief in rotors fabricated by welding. Also, parts shaped by pressing, drawing, or extruding are liable to have distortion. Change of temperature causes thermal distortion.
5. *Clearance tolerances*: It is caused due to the accumulation of tolerances during the assembly.
6. *Corrosion and wear*: Caused in rotors used in environments, which subjects them to abrasion, corrosion, or wear-like fans, blowers, compressors, and pumps.
7. Deposit buildup.

Due to these reasons, a balancing process is required to eliminate the vibrations due to the inherent unbalance.

8.2.1 Static Balance

The term "static" refers to the state of rest. Static balancing process means that the sum of the inertia forces in the rotating parts is zero without the consideration of the axial location of the forces. That is, neglecting the effect of the moments of the inertia forces. This is the situation for balancing of rotating masses, which are, or nearly, in the same plane. It is essentially a two-dimensional problem. Some examples of common devices that meet this criterion, and thus can successfully be statically balanced, are a single gear or pulley on a shaft, a thin flywheel, an airplane propeller, and an individual turbine blade wheel (but not the entire turbine). An automobile tire can be critically statically balanced since it has an axial dimension.

8.2.1.1 Balancing of a Single Mass

Consider a thin disk of mass m mounted on a shaft that is rotating with an angular velocity ω. The center of gravity of the disk is located at a distance r from the geometrical center (Figure 8.1). Due to the eccentricity of the center of gravity, a centrifugal force $\mathbf{F}$ with magnitude F is created such that

$$F = mr\omega^2$$

This force causes a shaking force at the bearings of the shaft causing vibration. To reduce this effect, we place a mass m_b at a distance r_b from the center of gravity and opposite to it. This countermass develops a force $\mathbf{F}_b$ with magnitude F_b such that

$$F_b = m_b r_b \omega^2$$

To completely eliminate the shaking force,

$$\mathbf{F}_b = \mathbf{F}$$

$$m_b r_b \omega^2 = mr\omega^2$$

Therefore,

$$m_b r_b = mr \tag{8.1}$$

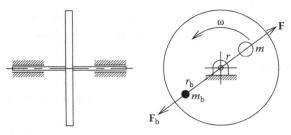

FIGURE 8.1   Balancing of a single rotating mass.

Usually, we place the counter balancing mass m_b on the disk on a suitable radius. It is important that $m_b r_b$ is equal to mr and is placed in a radial position opposite to the center of gravity.

8.2.1.2 Balancing of Several Masses in a Single Plane

When several masses are attached to a thin disk at different locations, each mass imposes an inertia force. The resultants of these forces are obtained either graphically or analytically and then treated as a single mass as done in Section 8.2.1.1.

EXAMPLE 8.1

Figure 8.2a shows a thin disk with three masses attached to it. Find the balancing mass.

$m_1 = 400$ g, $r_1 = 20$ cm, $m_2 = 200$ g, $r_2 = 30$ cm, $m_3 = 100$ g, $r_3 = 30$ cm.

SOLUTION

(a) Graphical method

All the forces are proportional to ω^2. Thus, the force polygon (Figure 8.2b) is drawn in terms of mr.

$$m_1 r_1 = 8 \text{ kg} \cdot \text{cm}$$

$$m_2 r_2 = 6 \text{ kg} \cdot \text{cm}$$

$$m_3 r_3 = 3 \text{ kg} \cdot \text{cm}$$

From the polygon,

$$m_e r_e = 9.24 \text{ kg} \cdot \text{cm at an angle } 54° \text{ from } m_1 r_1$$

Therefore, $m_b r_b = 9.24$ kg·cm. The balancing mass m_b is placed at an angle 234° from m_1 at a suitable radius.

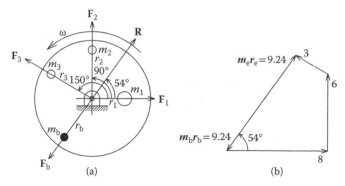

(a)   (b)

FIGURE 8.2 Balancing of several masses in a single plane. (a) Angular position of the force (b) force polygon.

(b) Analytical method

The analytical method is based on resolving the vectors to horizontal and vertical components, determining the horizontal and vertical components of the resultant, and then determining the value of the resultant and its angle. For the unbalance masses,

$$\left(m_e r_e\right)_x = 8 + 3 \times \cos 150° = 5.4 \text{ kg} \cdot \text{cm}$$

$$\left(m_e r_e\right)_y = 6 + 3 \times \sin 150° = 7.5 \text{ kg} \cdot \text{cm}$$

$$m_e r_e = 9.243 \text{ kg} \cdot \text{cm}$$

$$\text{Angle of } \left(m_e r_e\right) = \tan^{-1} \frac{7.5}{5.4} = 54.236°$$

8.2.2 Dynamic Balance

In most applications, the unbalance masses are not located in one plane as was demonstrated in Section 8.2.1. Usually, they are located at axial distances over the shaft. Static balance ensures that the sum of all centrifugal forces is zero. However, this action does not eliminate the transmitted forces to the supports due to the presence of moments. In this case, complete balance is achieved by placing masses to counteract these moments. These masses are located some distance apart. In other words, balancing is performed by placing balancing masses in two different planes. This is called *dynamic balance*. The process of dynamic balance is illustrated in Sections 8.2.2.1 and 8.2.2.2.

8.2.2.1 Two Equal and Opposite Masses in Two Planes

Figure 8.3 shows two equal and opposite and equal unbalance masses each of magnitude *mr* with distance *a* apart. The shaft rotates with an angular speed ω. The masses are attached to a shaft that is supported by two bearings LS and RS at a distance *L* apart. The forces transmitted to bearings are obtained by considering the moments about each bearing. It is given by,

$$F_{SL} = F_{SR} = \frac{a}{L} mr\omega^2$$

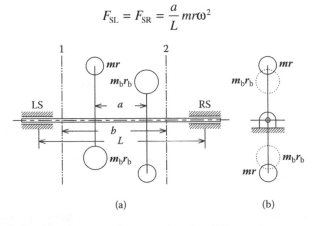

(a)   (b)

FIGURE 8.3   Balancing of two equal masses placed at different planes.

But both forces are in opposite directions.

To create a complete balance for the system, we have two choices,

1. To place counterbalance masses $m_b r_b$, each is equal to mr in a direction opposite to the original unbalance.
2. To place counterbalance masses $m_1 r_1$ and $m_2 r_2$ at suitable planes 1 and 2 (Figure 8.3), if it is difficult to use the first solution. These planes are known as balancing planes.

$$m_1 r_1 = m_2 r_2 = \frac{a}{b} mr$$

8.2.2.2 Several Masses in Several Planes

The balancing of masses in different planes requires

1. Locating two balancing planes, say planes L and R, at suitable positions on the shaft.
2. Placing balancing masses $m_L r_L$ and $m_R r_R$ in the balancing planes.
3. The magnitude and the angular position of the balancing masses are determined by considering moments about each balancing plane at a time. Then applying,

$$\sum \mathbf{M}_i \Big|_L = 0$$

$$\sum \mathbf{M}_i \Big|_R = 0$$

(8.2)

The moment about each plane includes the moment of the balancing masses. The balancing conditions may be applied either graphically or analytically.

8.2.2.2.1 Graphical Method

The moment vectors in Equation 8.2 are obtained by applying the cross product of the vectors representing the unbalance masses and the distances from the balancing plane. The directions are obtained by using the right-hand rule. This is illustrated in Figure 8.4.

The three unbalance masses $m_1 r_1$, $m_2 r_2$, and $m_3 r_3$ are attached to a shaft at different axial positions and different angular positions. Points L and R are the moment planes. The vectors $\mathbf{a}_1$, $\mathbf{a}_2$, and $\mathbf{a}_3$ are the vectors from point L to the unbalance masses, while the vectors $\mathbf{b}_1$, $\mathbf{b}_2$, and $\mathbf{b}_3$ are the vectors from point R. Vectors $\mathbf{M}_{1L}$, $\mathbf{M}_{2L}$, $\mathbf{M}_{3L}$, $\mathbf{M}_{1R}$, $\mathbf{M}_{2R}$, and $\mathbf{M}_{3R}$ are moment vectors about points L and R. Their directions are obtained using the right-hand rule. If we rotate the moment vectors 90°, say, counterclockwise, we see that the moment vectors of the unbalance masses to the right of the moment points coincide with the unbalance masses while those to the left are in the opposite directions as shown in Figure 8.5. The reverse is true if we rotate the vectors in the opposite direction. Using this concept facilitates the analysis.

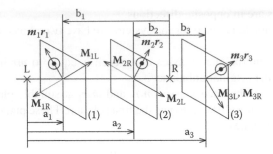

FIGURE 8.4   Vectors representing the unbalance masses.

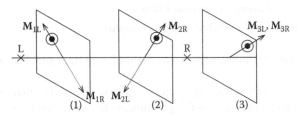

FIGURE 8.5   Representation of vectors for the moments.

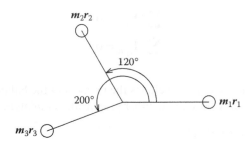

FIGURE 8.6   Angular positions for the masses.

EXAMPLE 8.2

The following data are provided for the system shown in Figure 8.5,

$$m_1 = 1.5 \text{ kg}, \ m_2 = 3.0 \text{ kg}, \ m_3 = 1.5 \ kg.$$

$$r_1 = 50 \text{ mm}, \ r_2 = 20 \text{ mm}, \ r_3 = 40 \text{ mm}.$$

$$a_1 = 200 \text{ mm}, \ a_2 = 600 \text{ mm}, \ a_3 = 1400 \text{ mm}.$$

$$b_1 = 800 \text{ mm}, \ b_2 = 400 \text{ mm}, \ b_3 = 400 \text{ mm}.$$

The angular positions of the unbalance masses relative to $m_1 r_1$ are shown in
Figure 8.6.

TABLE 8.1
Moment about L

Plane	m_i (kg)	r_i (mm)	a_i (mm)	$M_i = mra$	Angle
1	1.5	50	200	15,000	0°
2	3.0	20	600	36,000	120°
3	1.5	40	1,400	84,000	200°
R	m_{bR}	r_{bR}	$s = 1,000$	$m_{bR}r_{bR}s$	θ_{bR}

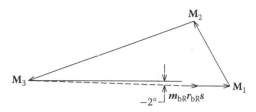

FIGURE 8.7 Vector diagram for moment about L.

Determine the magnitudes and the angular positions of the balancing masses at planes located at L and R. The distance s between the balancing planes is 1000 mm.

SOLUTION

To obtain the magnitudes and the angular positions of the balancing masses at planes $m_L r_L$ and $m_R r_R$, we take moments about points L and R respectively and apply Equation 8.1.

For the moments about L,

$$M_{1L} + M_{2L} + M_{3L} + M_{Lb} = 0$$

where M_{bL} is the moment of the balancing mass at plane R about L. Table 8.1 represents the elements of the preceding equation.

A vector diagram is drawn (Figure 8.7), from which we obtain,

$$m_{bR}r_{bR}s = 82,000 \text{ kg mm}^2$$

$$m_{bR}r_{bR} = 82 \text{ kg mm}$$

The angle of $m_{bR}r_{bR}$ is −2°.

For the moment about R (Table 8.2), the distances for the unbalance masses to the left of R are indicated by a negative sign to remind that the moment vectors are in the opposite direction to unbalance masses. The vector diagram is shown in Figure 8.8.

$$m_{bL}r_{bL}s = -76,300 \text{ kg mm}^2$$

The angle of $-m_{bR}r_{bR}$ is 22°.

$$M_{bR}m_{bR} = 76.3 \text{ kg mm}$$

TABLE 8.2
Moment about R

Plane	m_i (kg)	r_i (mm)	b_i (mm)	$M_i = mra$	Angle
1	1.5	50	−800	−60,000	0°
2	3.0	20	−400	−24,000	120°
3	1.5	40	400	24,000	200°
L	m_{bL}	r_{bL}	$s = -1,000$	$m_{bL}r_{bL}s$	θ_{bL}

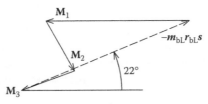

FIGURE 8.8   Victor diagram for moment about R.

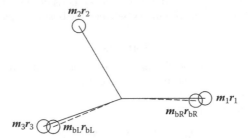

FIGURE 8.9   Balanced system.

The angle of $m_{bR}r_{bR}$ is 202°.
The balanced system is shown in Figure 8.9.

ANALYTICAL METHOD

The procedure for the analytical method is summarized as follows:

- Resolve the moment vectors to horizontal and vertical components.
- Determine the algebraic sum of the components.
- Determine the resultant of the components in magnitude and direction.

The analytical solution is presented in Tables 8.3 and 8.4.
From Table 8.3,

$$15,000 - 18,000 - 78,930 + m_{bR}\,r_{bR}s\cos\theta_R = 0$$

$$31,180 - 28,730 + m_{bR}\,r_{bR}s\sin\theta_R = 0$$

TABLE 8.3

Moment about L for Analytical Method

Plane	m_i	r_i	a_i	θ	mra cos θ	mra sin θ
1	1.5	50	200	0°	15,000	0
2	3.0	20	600	120°	$-18,000 \times 10^4$	31,180
3	1.5	40	1,400	200°	−78,930	−28,730
R	m_{bR}	r_{bR}	1,000	θ_R	$m_{bR}r_{bR}s \cos\theta_R$	$m_{bR}r_{bR}s \sin\theta_R$

TABLE 8.4

Moment about R for Analytical Method

Plane	m_i	r_i	a_i	θ	mra cos θ	mra sin θ
1	1.5	50	−800	0°	15,000	0
2	3.0	20	−400	120°	12,000	−20,780
3	1.5	40	400	200°	−22,550	−8,208
L	m_{bL}	r_{bL}	− 1,000	θ_L	$m_{bR}r_{bR}s \cos\theta_L$	$m_{bR}r_{bR}s \sin\theta_L$

Thus,

$$m_{bR}r_{bR}s \cos\theta_R = 81{,}930$$

$$m_{bR}r_{bR}s \sin\theta_R = -2447$$

Dividing by $s = 1000$,

$$m_{bR}r_{bR} = 81.97 \text{ kg mm}$$

The angle of $m_{bR}r_{bR}$ is −1.711°.
From Table 8.4,

$$15{,}000 + 12{,}000 - 22{,}550 + m_{bL}r_{bL}s \cos\theta_L = 0$$

$$-20{,}780 - 8{,}208 + m_{bL}r_{bL}s \sin\theta_L = 0$$

Thus,

$$m_{bL}r_{bL}s \cos\theta_L = 70{,}550$$

$$m_{bL}r_{bL}s \sin\theta_L = 28{,}990$$

Dividing by $s = -1000$,

$$m_{bL}r_{bL} = 76.27 \text{ kg mm}$$

TABLE 8.5

Sum of the Masses

Plane	m_i	r_i	θ	mra cos θ	mra sin θ
1	1.5	50	0°	75.0	0
2	3.0	20	120°	−30.0	51.96
3	1.5	40	200°	−56.38	−20.52
R	81.97	1	−1.771°	81.93	−2.53
L	m_{bL}	r_{bL}	θ_L	$m_{bL} r_{bL} \cos\theta_L$	$m_{bL} r_{bL} \sin\theta_L$

The angle of $m_{bL} r_{bL}$ is 202.34°.

It should be noticed that we still satisfy the condition of equilibrium if we replace the moment about R represented by Table 8.4 by the summation of the components of the masses as presented in Table 8.5.

$$m_{bL}\, r_{bL} \cos\theta_L = -96.931$$

$$m_{bR}\, r_{bR} \sin\theta = -28.907$$

$$m_{bR}\, r_{bR} = 76.27 \text{ kg mm}$$

The angle of $m_{bR} r_{bR}$ is 202.24°.

8.3   BALANCING OF RECIPROCATING PARTS

8.3.1   EXACT MODEL OF A SINGLE-CYLINDER ENGINE

To get an understanding of the effect of the reciprocating and floating parts of a mechanism, consider the case of a single slider mechanism (Figure 8.10), with the following data:

Speed of crank 4000 rpm
Length of crank $r_2 = 5$ cm
Length of connecting rod $r_3 = 20$ cm
Mass of the crank (considered concentrated at the crank end) $m_2 = 1.0$ kg
Mass of connecting rod $m_3 = 1.5$ kg
Mass moment of inertia of the connecting rod $I_{G3} = 0.007$ kg·m²
Position of the center of gravity of the connecting rod from the crank
    end $g_3 = 5$ cm
The mass of the piston $m_4 = 1.25$ kg

The data was used in the MathCAD program presented in Appendix A.2 in Chapter 7 to obtain the shaking force. A plot for the shaking force due to all inertia forces are shown by the dotted line (Figure 8.11). The shaking force due to the

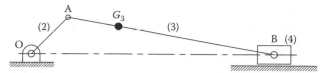

FIGURE 8.10 The engine mechanism.

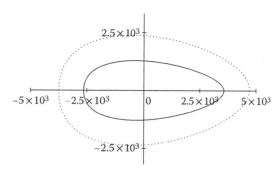

FIGURE 8.11 Polar diagrams for the engine mechanism.

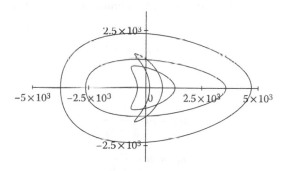

FIGURE 8.12 Balancing of the engine mechanism.

reciprocating link (piston) and the floating link (connecting rod) only (solid line) are shown by the solid line.

The rotating masses can be completely balanced by adding a balancing mass at the opposite side of the crank. Complete balance of the reciprocating and the floating masses can be achieved by a complicated setup. However, partial balance is obtained by increasing the balancing mass by a specific amount to compensate their effect, which is a good compromise. The effect of the added mass on the shaking force is shown in Figure 8.12.

8.3.2 APPROXIMATE MODEL OF A SINGLE-CYLINDER ENGINE

In Section 8.3.1, the exact analysis for balancing a single-cylinder engine was presented by using MathCAD software. However, it is possible to use some approximations that are very effective in simplifying the analysis, especially for the case of multicylinder engines. The approximations are performed in two steps.

8.3.2.1 Equivalent Masses of Links

A rigid link in a plane motion can be replaced by an equivalent system of two concentrated masses, placed on a mass less link, which are kinetically equivalent.

Consider link AB having a mass m, a moment of inertia I, and a center of gravity at G (Figure 8.13a). This link is kinetically equivalent to a massless link AB with mass m_A placed at, say, point A and a mass m_E placed at point E with a distance h_E from the center of gravity G (Figure 8.13b). The two links are kinetically equivalent when the masses on the equivalent link satisfy the following conditions,

1. The sum of the two masses is equal to the mass of the original link.

$$m_A + m_E = m \qquad\qquad (a)$$

2. The position of the center of gravity of the two links is the same.

$$m_A \times g + m_E \times h_E = m \qquad\qquad (b)$$

3. The mass moment of inertia about the center of gravity of the two links is the same.

$$m_A \times g^2 + m_E \times h_E^2 = I \qquad\qquad (c)$$

From Equations a and b,

$$m_A = m \frac{h_E}{g + h_E}$$

$$\qquad\qquad (8.3)$$

$$m_E = m \frac{g}{g + h_E}$$

From Equations c and 8.3,

$$g \times h_E = \frac{I}{m} \qquad\qquad (8.4)$$

The application of the equivalent link yields exact shaking force and shaking moment. The mass m_A is added to the mass of the crank and is considered as rotating masses. To estimate the shaking force and the shaking moment, it is necessary to evaluate the acceleration of point E, which does not simplify the analysis.

(a)   (b)

FIGURE 8.13   Equivalent masses of a link.

The simplification is achieved by placing a mass m_A at point A and the other mass at point B (Figure 8.14).

The values of m_A and m_B are obtained by satisfying conditions (a) and (b). In this case,

$$m_A = m \frac{h_B}{r_3}$$

$$m_E = m \frac{g}{r_3}$$

(8.5)

where r_3 is the length of the connecting rod. This approximation violates the third condition (c), which will make a change in the shaking moment. However, the shaking force is exactly the same since the position of the center of gravity is not altered. The mass m_A is added to the mass of the crank as a rotating mass while the mass m_B is added to the mass of the piston as a reciprocating mass. The total rotating mass m_r is given by,

$$m_r = m_2 + m_A$$

The total reciprocating mass m_c is given by,

$$m_c = m_4 + m_B$$

8.3.2.2 Shaking Force of the Piston

Referring to Figure 8.10, the distance x of the piston from point O when the crank rotates an angle θ is given by,

$$x = r_2 \cos \theta + \sqrt{r_3^2 - \left(r_2 \sin \theta \right)^2}$$

$$= r_2 \cos \theta + r_3 \sqrt{1 - \left(\frac{r_2}{r_3} \sin \theta \right)^2}$$

Expanding the radical we get,

$$x = r_2 \cos \theta + r_3 \left[1 - \frac{1}{2} \left(\frac{r_2}{r_3} \sin \theta \right)^2 - \frac{1}{8} \left(\frac{r_2}{r_3} \sin \theta \right)^4 + \frac{1}{16} \left(\frac{r_2}{r_3} \sin \theta \right)^6 - \dots \right]$$

Usually, the ratio between r_2 and r_3 is less than 1/3. This means that the value of the term raised to power 4 is less than 1.5%. Thus, this term together with all terms

FIGURE 8.14 Approximate masses for the connection.

containing higher power can be neglected. Therefore, the value of x can be approximated, without appreciable loss of accuracy, to,

$$x \approx r_2 \cos\theta + r_3 \left[1 - \frac{1}{2}\left(\frac{r_2}{r_3}\sin\theta\right)^2 - \frac{1}{8}\left(\frac{r_2}{r_3}\sin\theta\right)^4 \right]$$

$$= r_3 + r_2 \left[\cos(\theta) - \frac{1}{2}\frac{r_2}{r_3}\sin^2\theta \right]$$

(8.6)

The velocity V_4 and the acceleration A_4 of the piston can be obtained by differentiating Equation 8.6 once or twice with respect to time. Thus,

$$V_4 = -\omega\, r_2\left(\sin\theta + \frac{r_2}{2r_3}\sin 2\theta\right)$$

$$A_4 = -\omega^2\, r_2\left(\cos\theta + \frac{r_2}{r_3}\cos 2\theta\right)$$

(8.7)

The shaking force due to the reciprocating masses is given by,

$$SF = m_c\,\omega^2\, r_2\left(\cos\theta + \frac{r_2}{r_3}\cos 2\theta\right)$$

(8.8)

The shaking force consist of two parts, namely, the primary force F_p and the secondary force F_s, which are given by,

$$F_p = m_c\omega^2 r_2\cos\theta$$

(8.9)

$$F_s = \frac{m_c}{4}(2\omega)^2\, r_2\,\lambda\cos 2\theta$$

(8.10)

where λ is the ratio between the lengths of the crank and the connecting rod.

$$\lambda = \frac{r_2}{r_3}$$

8.3.3 Direct and Reverse Crank Representation

The shaking force due to the reciprocating mass is along the line of action of the piston. The primary and the secondary forces can be represented by rotating masses as shown in Figure 8.15. The primary force is represented by two masses, each with magnitude of $m_c/2$, placed on a crank of length r_2, and rotates in opposite directions with an angular velocity ω; they are called direct D and reverse R (Figure 8.15a). Each mass produces a centrifugal force F_{p1} such that

$$F_{p1} = \frac{m_c}{2}\omega^2 r_2$$

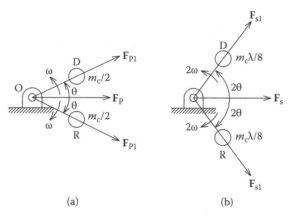

FIGURE 8.15   (a) Direct and reverse cranks for the primary force (b) direct and reverse cranks for the secondary force.

The resultant of the two forces is equal to the primary force. The secondary force is represented by two masses, each with magnitude of $m_c\lambda/8$, placed on a crank (D and R) of length r_2, and rotates in opposite direction with an angular velocity 2ω (Figure 8.15a). Each mass produces a centrifugal force F_{s1} such that

$$F_{s1} = \frac{m_c}{8} (2\omega)^2\, r_2$$

The concept of direct and reverse cranks permits the use the analysis of the rotating masses, which is simple. Furthermore, it is possible to achieve complete balance for the slider mechanism as will be discussed in the forthcoming sections.

8.3.4   BALANCING OF A SINGLE-CYLINDER ENGINE

As we pointed out, the shaking force causes vibrations especially in high-speed engines. Thus, it is appropriate to eliminate or, at least, reduce the shaking force. This is achieved by one of the following methods.

8.3.4.1   Dummy Cylinders

The balancing system consists of two dummy cylinders placed at both sides of the original cylinder (Figure 8.16). The crank of the dummy cylinders is at 180° of the original crank. The rotating masses and reciprocating masses of the balancing cylinders are exactly the same of those of the original cylinder. This system offers a complete balance for the engine although it is very expensive.

8.3.4.2   Direct and Reverse Balancing Masses

It is possible to obtain complete balance by counteracting the primary and the secondary forces by using the concept of the direct and reverse cranks.

The balancing setup is shown in Figure 8.17. It consists of a pair of gears, P_1 and P_2, for balancing the primary force and another pair S_1 and S_2 for balancing the

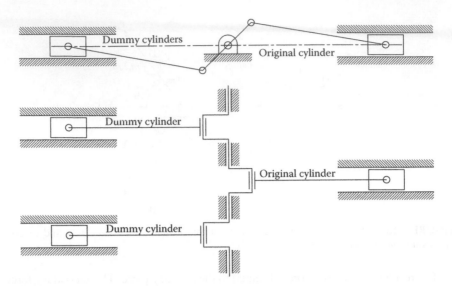

FIGURE 8.16   Complete balancing of a single cylinder engine.

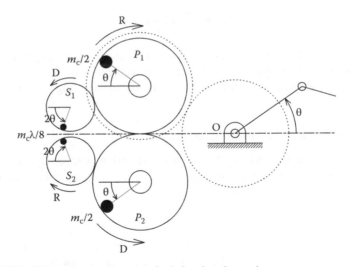

FIGURE 8.17   Using direct and reverse for balancing the engine.

secondary force. Gears S_1 and S_2 are meshed with gears P_1 and P_2 with a speed rise of 2. The mass attached to P_2, with magnitude $m_c/2$, is placed opposite to the crank. The mass on P_1 is the same as that of P_1 but placed in a mirror image position. The mass attached to S_1, with magnitude $m_c\lambda/8$, is placed opposite to twice the crank angle. The mass on S_2 is the same as that of S_1 but placed in a mirror image position. The balancing of the rotating mass is done by placing a balance mass at the opposite end of the crank.

8.3.4.3 Partial Balancing

The complete balancing of the single-cylinder engine could be very costly. However, it is possible to reduce the shaking force by choosing a suitable countermass at the other end of the crank. This process is called partial balance. The vector representing the shaking force is given by,

$$SF = F_r + F_p + F_s + F_b \tag{8.11}$$

where F_r is the inertia force of the rotating parts, F_p and F_s are the primary and the secondary forces of the reciprocating mass and are given by Equations 8.9 and 8.10, and F_b is the force due to the balancing mass m_b.

$$F_r = m_r\,\omega^2\,r_2\,e^{i\theta}$$

$$F_p = m_c\,\omega^2\,r_2\,\cos\theta$$

$$F_c = m_c\,\omega^2\,r_2\,\lambda\cos 2\theta$$

$$F_b = m_b\,\omega^2\,r_2\,e^{i\theta}$$

We can get an understanding of the resulting shaking force by plotting Equation 8.11 over a complete cycle of the crank. This is called the polar diagram.

The diagram is constructed by drawing three concentric circles. The first circle is with a radius equal to $m_r\omega^2 r_2$, the second with a radius $(m_r\omega^2 r_2 + m_c\omega^2 r_2)$, and the third with a radius $(m_r\omega^2 r_2 + m_c\omega^2 r_2 + m_c\omega^2 r_2\lambda)$ (Figure 8.18). Consider the engine mechanism presented in Section 8.2.

According to the data, the total rotating mass m_r is given by,

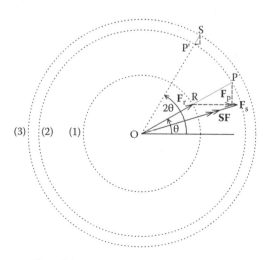

FIGURE 8.18 Construction of the polar diagram.

$$m_r = m_2 + m_3 \frac{r_3 - g_3}{r_3} = 2.125 \, \text{kg}$$

The total reciprocating mass m_c is given by,

$$m_c = m_4 + m_3 \frac{g_3}{r_3} = 1625 \, \text{kg}$$

$$m_r \, \omega^2 \, r_2 = 1.864 \times 10^4 \, N, m_c \, \omega^2 r_2 = 1.426 \times 10^4 \, N$$

$$m_c \omega^2 r_2 \, \lambda = 0.356 \times 10^4 \, N$$

The construction of the polar diagram is shown in Figure 8.18 and is outlined as follows,

1. Draw circle (1) with of radius equal to $F_r = 1.864 \times 10^4$ with a suitable scale. This circle bounds the inertia force due to the rotating masses.
2. Draw circle (2) with radius equal to $F_r + F_p = 3.29 \times 10^4$ with the same scale. Circles (1) and (2) bound the amplitude of the primary force.
3. Draw circle (3) with radius equal to $F_r + F_p + F_s = 3.646 \times 10^4$ with the same scale. Circles (2) and (3) bound the amplitude of the secondary force.
4. For any crank angle θ, draw line OR to represent $\mathbf{F_r}$.
5. From point R, draw line RP. The horizontal projection of this line represents F_p.
6. Draw line P'S with an angle 2θ between circles (2) and (3). The horizontal projection of this line represents F_s. This projection is added to F_p.
7. The shaking force is the resultant of the three forces.

Partial balance of the engine mechanism is made by placing a balancing mass with magnitude equal to the rotating masse plus two-thirds of the reciprocating mass. The resultant polar diagram for the unbalanced (dotted line) and the partially balanced system is shown in Figure 8.19.

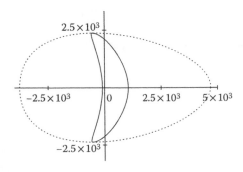

FIGURE 8.19 Polar diagram.

8.3.5 BALANCING OF A TWO-CYLINDER V ENGINE

A two-cylinder V engine consists of two cylinder banks operating one crank through two connecting rods (Figure 8.20). In general, the angle between the cylinder centerlines is 2α. Usually, the rotating masses can be easily balanced. Thus, we shall be concerned with the study of the effect of the reciprocating masses from now on.

Usually, the reciprocating mass of both cylinders is the same. Let us measure the angle of rotation of the crank from the midline of the cylinders, which is vertical according to Figure 8.20. The reciprocating forces F_1 and F_2 are along the centerline of cylinders (1) and (2) respectively and are given by,

$$F_1 = F_{p1} + F_{s1}$$

$$F_2 = F_{p2} + F_{s2}$$

where

$$F_{p1} = m_c\omega^2 r_2 \cos(\theta + \alpha)$$

$$F_{p2} = m_c\omega^2 r_2 \cos(\theta - \alpha)$$

$$F_{s1} = m_c \omega^2 r_2 \lambda \cos 2(\theta + \alpha)$$

$$F_{s2} = m_c \omega^2 r_2 \lambda \cos 2(\theta - \alpha)$$

To obtain the resultant, we resolve the forces into the horizontal and vertical directions. For the primary forces,

$$X_p = m_c \omega^2 r_2 [\cos(\theta + \alpha)\sin\alpha - \cos(\theta - \alpha)\sin\alpha]$$
$$= -2 m_c \omega^2 r_2 \sin^2(\alpha)\sin(\theta)$$
$$Y_p = m_c \omega^2 r_2 [\cos(\theta + \alpha)\cos\alpha + \cos(\theta - \alpha)\cos\alpha]$$
$$= 2 m_c \omega^2 r_2 \cos^2(\alpha)\cos(\theta)$$

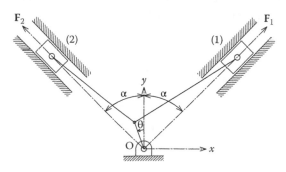

FIGURE 8.20 Forces in a two-cylider V-engine.

The resultant primary force is given by,

$$F_p = 2m_c \omega^2 \; r_2 \sqrt{\cos^4 \alpha \cos^2 \theta + \sin^4 \alpha \sin^2 \theta}$$

For the secondary forces,

$$X_s = m_c\omega^2 r_2 \, \lambda \, [\cos 2(\theta + \alpha)\sin \alpha - \cos 2(\theta - \alpha)\sin \alpha]$$
$$= -2m_c\omega^2 r_2 \lambda \sin (\alpha) \sin (2\alpha) \sin (2\theta)$$
$$Y_s = m_c\omega^2 r_2 \, \lambda \, [\cos 2(\theta + \alpha)\cos \alpha + \cos 2(\theta - \alpha)\cos \alpha]$$
$$= 2m_c\omega^2 r_2 \, \lambda \cos (\alpha)\cos (2\alpha)\cos (2\theta)$$

The resultant secondary force is given by,

$$F_s = 2m_c\omega^2 r_2 \, \lambda \sqrt{\cos^2 \alpha \cos^2 2\alpha \cos^2 \theta + \sin^2 \alpha \sin^2 2\alpha \sin^2 \theta}$$

The value of α affects the values of the primary and the secondary forces. The most appropriate value of α is 45°. In this case,

$$X_p = -m_c\omega^2 r_2 \sin (\theta)$$
$$Y_p = m_c\omega^2 r_2 \cos (\theta)$$
$$F_p = m_c\omega^2 r_2$$

This means that the primary force is equivalent to a rotating mass equal to m_c. For the secondary force,

$$X_s = -\sqrt{2} \; m_c \, \omega^2 \; r_2 \, \lambda \sin (2\theta)$$

$$Y_s = 0$$

The resultant secondary force is equal to X_s. It can be represented by two masses, each of magnitude $\dfrac{\sqrt{2}}{8} m_c r_2 \lambda$, and are rotating in opposite directions with a speed of 2ω. The mechanism can be completely balanced by placing balancing mass in the opposite direction for each mass (Figure 8.21).

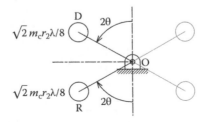

FIGURE 8.21   Balancing of a two-cylinder V-engine.

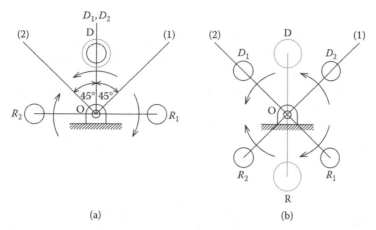

FIGURE 8.22 Direct and reverse crank representation of the V–engine. (a) Primary forces (b) secondary forces.

8.3.5.1 Analysis Using Direct and Reverse Cranks

Using the concept of the direct and reverse cranks simplifies the analysis to a great extent. The original crank is placed at some position. However, there is no loss of generality to place the crank in a position, which further simplifies the analysis. It could be placed at $\theta = 0$, α, or $-\alpha$.

For the primary forces, the direct and reverse masses for each cylinder have a value of $m_c/2$. The direct cranks are always located on the original crank. The reverse cranks are located at the mirror image of the original crank from the centerline of each cylinder (Figure 8.22a).

Since the angle between the cylinders is 45°, the reverse cranks are opposite to each other and their effect is cancelled. Therefore, the effect of the primary forces is equivalent to a rotating mass with magnitude m_c located at the original crank.

For the secondary forces, the direct and reverse masses for each cylinder have a value of $m_c r_2 \lambda/8$. The direct crank of cylinder (1), when the angle of the original crank is zero, is 90° ahead its centerline. That is, it coincides with the centerline of cylinder (2). The reverse crank is at angle −90° (Figure 8.22b). For cylinder (2), the direct crank is at −90° from the centerline, that is, coincides with the centerline of cylinder (1) while the reverse crank is ahead by 90°. The resultant of the direct masses is equal to a mass of magnitude of $\dfrac{\sqrt{2}}{8} m_c r_2 \lambda$. The resultant of the reverse masses is equal to mass of magnitude of $\dfrac{\sqrt{2}}{8} m_c r_2 \lambda$. Both resultant masses rotate in opposite directions with a speed of 2ω.

8.3.6 BALANCING OF RADIAL ENGINES

Radial engines are used in aircraft engines where a group of cylinders are arranged in a radial position with the crank shaft (Figure 8.23). The number of cylinders is usually an odd number, which offers a good balancing characteristic as will be described.

Figure 8.23 shows a five-cylinder radial engine equally distributed around the crank. The angle between the centerlines of the adjacent cylinders is $\alpha = 72°$. The balancing analysis is best performed by using the concept of direct and reverse cranks. For the sake of simplicity, we consider that the original crank coincides with the centerline of cylinder (1).

For the primary forces, the direct cranks of all cylinders coincide with the original crank (Figure 8.24a). The reverse crank of each cylinder is located at the mirror image of the direct crank with respect to its centerline (Figure 8.24b).

It is clear that the resultant of all the reverse masses is zero, while the resultant of the direct masses is a single mass of magnitude 2.5 m_c located at the original crank. Therefore, the primary forces can be completely balanced by placing a balancing mass at the crank.

For the secondary forces, the direct crank of each cylinder is placed at an angle equal to twice the angle between the original crank and its centerline (Figure 8.25a). The reverse crank of each cylinder is located at the mirror image of the direct crank with respect to its centerline (Figure 8.25b). It is clear that the resultants of both direct and reverse masses are zero. Therefore, the secondary forces are completely

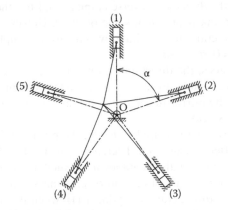

FIGURE 8.23 Five-cylinder radial engine.

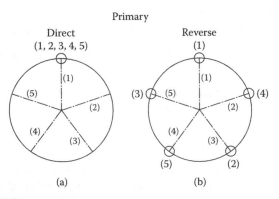

FIGURE 8.24 (a) Direct and (b) reverse of primary forces.

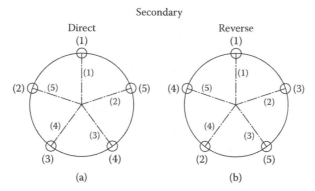

FIGURE 8.25   (a) Direct and (b) reverse of secondary forces.

balanced without placing balancing masses. This situation applies for any radial engine with odd number of cylinders. This explains why radial engines offer the best balancing solution.

8.3.7 IN-LINE ENGINES

Figure 8.26 shows the outline of n cylinders placed in parallel and are connected to one crank shaft with axial distances $a_2, a_3, \ldots, a_n$ from cylinder number (1). Usually, all cylinders have the same data in regard of the lengths of cranks, connecting rods, and the masses. In general, let $R_1, R_2, \ldots, R_n$ be the lengths of cranks, $L_1, L_2, \ldots, L_n$ be the lengths of the connecting rods, and $m_1, m_2, \ldots, m_n$ be the reciprocating masses of the cylinders. The crank of cylinder number i is ahead of the crank of cylinder number 1 by an angle φ_i.

Each cylinder develops a shaking force. Besides, the shaking forces produce a shaking moment. All these effects result in vibration to the engine. The shaking force is given by,

$$SF = \sum_{i=1}^{n} F_i$$

The shaking force of each cylinder consists of a primary force F_{pi} and a secondary force F_{si}.

$$F_i = F_{pi} + F_{si}$$
$$F_{pi} = m_i R_i \omega^2 \cos(\theta + \varphi_i)$$
$$F_{si} = m_i R_i \gamma_i \omega^2 \cos 2(\theta + \varphi_i)$$

The terms on the right-hand side of F_{pi} and F_{si} can be expanded to take the following forms:

$$F_{pi} = m_i R_i \omega^2 \left(\cos\theta \cos\varphi_i - \sin\theta \sin\varphi_i\right)$$
$$F_{si} = m_i R_i \gamma_i \omega^2 \left(\cos 2\theta \cos 2\varphi_i - \sin 2\theta \sin 2\varphi_i\right)$$

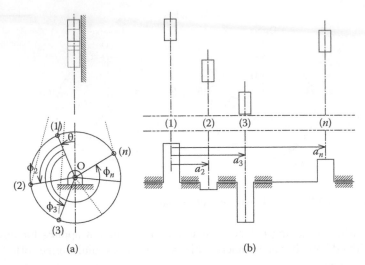

FIGURE 8.26   In-line engine.

Therefore, the components of the shaking force are given by,

$$F_{\mathrm{p}} = \omega^2 \left(\cos\theta \sum_{i=1}^{n} m_i\, R_i \cos\varphi_i - \sin\theta \sum_{i=1}^{n} m_i\, R_i \sin\varphi_i \right) \qquad (8.12)$$

$$F_{\mathrm{s}} = \omega^2 \left(\cos 2\theta \sum_{i=1}^{n} m_i\, R_i \cos 2\varphi_i - \sin 2\theta \sum_{i=1}^{n} m_i\, R_i \sin 2\varphi_i \right) \qquad (8.13)$$

The shaking moment can be obtained by considering the moment about the centerline of cylinder number (1). Therefore,

$$M_{\mathrm{p}} = \omega^2 \left(\cos\theta \sum_{i=1}^{n} m_i\, R_i\, a_i \cos\varphi_i - \sin\theta \sum_{i=1}^{n} m_i\, R_i\, a_i \sin\varphi_i \right) \qquad (8.14)$$

$$M_{\mathrm{s}} = \omega^2 \left(\cos 2\theta \sum_{i=1}^{n} m_i\, R_i\, a_i \cos 2\varphi_i - \sin 2\theta \sum_{i=1}^{n} m_i\, R_i\, a_i \sin 2\varphi_i \right) \qquad (8.15)$$

Usually, the masses, crank radii, and the lengths of connecting rod of all cylinders are same. The conditions for complete balance of multicylinder in-line engine are as follows:

For the primary forces,

$$\sum_{i=1}^{n} \cos\varphi_i \qquad (8.15\mathrm{a})$$

$$\sum_{i=1}^{n} \sin\varphi_i \qquad (8.15\mathrm{b})$$

For the secondary forces,

$$\sum_{i=1}^{n} \cos 2\varphi_i \tag{8.16a}$$

$$\sum_{i=1}^{n} \sin 2\varphi_i \tag{8.16b}$$

For the primary moment,

$$\sum_{i=1}^{n} a_i \cos \varphi_i \tag{8.17a}$$

$$\sum_{i=1}^{n} a_i \sin 2\varphi_i \tag{8.17b}$$

For the secondary moment,

$$\sum_{i=1}^{n} a_i \cos 2\varphi_i \tag{8.18a}$$

$$\sum_{i=1}^{n} a_i \sin 2\varphi_i \tag{8.18b}$$

8.3.8 Applications of In-Line Engines

8.3.8.1 Four-Cylinder Four-Stroke Engine

To check the balancing, we have to establish the crank configuration. The cycle for four-stroke engines is 720°. The firing of the consequent cylinders should be at equal intervals. Thus, the angle between the cranks should be 180° (720° divided by 4). For the axial configuration of the cranks, we have two possibilities. The first is to place the crank of cylinder (2) at 180° from (1), that of (3) at 180° from (1), and that of (4) at 0° from (1) (Figure 8.27).

The firing order of this arrangement is determined from the top crank. At this position, cylinder (1) is firing. After 180°, either (2) or (3) is firing. After another

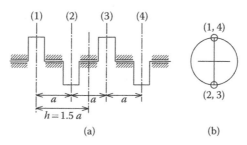

(a)   (b)

FIGURE 8.27 Four-cylinders four-stroke engine, first configuration.

180°, cylinder (4) is firing. After another 180°, either (2) or (3) is firing. The firing orders are indicated by the following chart,

$$1 \Rightarrow 3 \Rightarrow 4 \Rightarrow 2$$

$$1 \Rightarrow 2 \Rightarrow 4 \Rightarrow 3$$

The balancing condition is determined by applying Equations 8.15 through 8.18. The analysis is simplified by using a tabular form (Table 8.6).

From the table, it is seen that the primary force and the primary moment are completely balanced. The secondary force and the secondary moment are given by,

$$F_s = 4 \, m_c \omega^2 r_2 \, \lambda \cos(2\theta)$$

$$M_s = 6 \, a \, m_c \omega^2 r_2 \, \lambda \cos(2\theta)$$

The secondary force and the secondary moment can be replaced by a single force of magnitude F_s placed at a distance h from the centerline of cylinder (1),

$$h = \frac{M_s}{F_s} = 1.5 \, a$$

The secondary force and the secondary moment can be completely balanced by placing two equal masses, each of magnitude $m_c/2$ with radius r_2, at a distance $1.5a$ from the centerline of cylinder (1) rotating in opposite directions with speed of 2ω as shown in Figure 8.28.

TABLE 8.6

Anlaysis of the First Configuration of a Four-Cylinder Engine

Cylcle No.	φ_i	F_p		$2\varphi_i$	F_s		a_i	M_p		M_s	
		$\cos \varphi_i$	$\sin \varphi_i$		$\cos 2\varphi_i$	$\sin 2\varphi_i$		$a \cos \varphi_i$	$a \sin \varphi_i$	$a \cos 2\varphi_i$	$a \sin 2\varphi_i$
1	0	1	0	0	1	0	0	0	0	0	0
2	180	−1	0	0	1	0	a	−a	0	a	0
3	180	−1	0	0	1	0	$2a$	−$2a$	0	$2a$	0
4	0	1	0	0	1	0	$3a$	$3a$	0	$3a$	0
Sum	m	0	0	m	4	0	m	0	0	$6a$	0

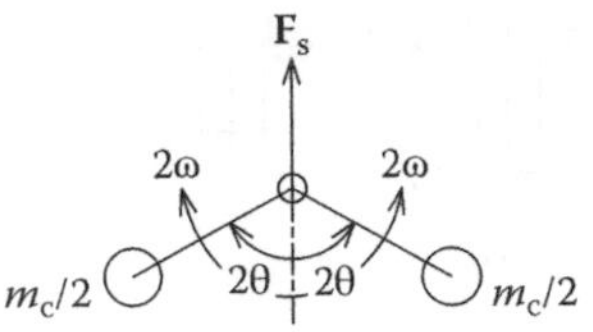

FIGURE 8.28 Balancing of the first configuration of a four-cylinder engine.

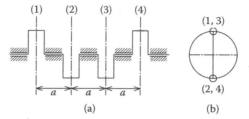

FIGURE 8.29 Four-cylinders four-stroke engine, second configuration.

TABLE 8.7
Analysis of Second Configuration of a Four-Cylinder Engine

Cylcle No.		F_p			F_s			M_p		M_s	
	φ_i	$\cos\varphi_i$	$\sin\varphi_i$	$2\varphi_i$	$\cos 2\varphi_i$	$\sin 2\varphi_i$	a_i	$a\cos\varphi_i$	$a\sin\varphi_i$	$a\cos 2\varphi_i$	$a\sin 2\varphi_i$
1	0	1	0	0	1	0	0	0	0	0	0
2	180	−1	0	0	1	0	a	−a	0	a	0
3	0	1	0	0	1	0	2a	2a	0	2a	0
4	180	−1	0	0	1	0	3a	−3a	0	3a	0
Sum	m	0	0	m	4	0	m	−2a	0	6a	0

The second possibility of the crank arrangement is shown in Figure 8.29.

Table 8.7 is used for analyzing the balancing condition of this arrangement.

It is clear that the primary force, the secondary force, and the secondary moment are the same as in the previous arrangement. The primary moment in the first arrangement is balanced while it is not balanced in this one.

8.3.8.2 Four-Cylinder Two-Stroke Engine

The cycle for four-stroke engines is 360°. Thus, the angle between the cranks should be 90°. The crank arrangement is shown in Figure 8.30.

Table 8.8 is used for analyzing the balancing of this configuration.

From the table, we see that the primary forces, the secondary force, and the secondary moment are balanced. The resultant primary moment is,

$$M_p = \sqrt{10}\, m_c\omega^2 r_2 a$$

The primary moment is represented by two equal and opposite masses placed at a distance d apart, each of magnitude of $\dfrac{\sqrt{10}\,a}{d}\, m_c$. The right mass makes an angle 162° with the crank of cylinder (1).

8.3.8.3 Six-Cylinder Four-Stroke Engine

The angle between the cranks should be 120° (720° divided by 6). The most appropriate axial arrangement of the cranks is shown in Figure 8.31.

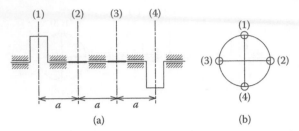

FIGURE 8.30 Four-cylinder two-stroke engine.

TABLE 8.8
Analysis of a Four-Cylinder Two-Stroke Engine

Cylce No.	φ_i	F_p $\cos \varphi_i$	$\sin \varphi_i$	$2\varphi_i$	F_s $\cos 2\varphi_i$	$\sin 2\varphi_i$	a_i	M_p $a \cos \varphi_i$	$a \sin \varphi_i$	M_s $a \cos 2\varphi_i$	$a \sin 2\varphi_i$
1	0	1	0	0	1	0	0	0	0	0	0
2	270	0	−1	180	−1	0	a	0	$-a$	$-a$	0
3	90	0	1	180	−1	0	$2a$	0	$2a$	$-2a$	0
4	180	−1	0	0	1	0	$3a$	$-3a$	0	$3a$	0
Sum	m	0	0	m	0	0	m	$-3a$	a	0	0

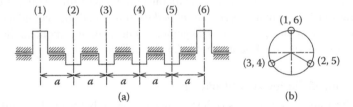

FIGURE 8.31 Six-cylinder engine.

The best firing order is to separate the adjacent cylinders from consequent firing to ensure smooth distribution of the torque. There are several firing orders that can be used. The most appropriate is,

$$1 \Rightarrow 5 \Rightarrow 3 \Rightarrow 6 \Rightarrow 2 \Rightarrow 4$$

Studying the balancing condition for the engine is performed by constructing a table as in the previous cases. Two additional rows are added as shown. Taking cylinder (1) as a reference, the angle between (2) and (1) is 240°, between (3) and (1) is 120°, between (4) and (1) is 120°, between (5) and (1) is 240°, and between (2) and (1) is 0°. Table 8.9 is used for studying the balancing of this configuration.

It is clear that all the inertia effects are zero and this type of engine is completely balanced.

TABLE 8.9

Analysis of a Six-Cylinder Two-Stroke Engine

Cyclce No.	φ_i	$\cos \varphi_i$	$\sin \varphi_i$	$2\varphi_i$	$\cos 2\varphi_i$	$\sin 2\varphi_i$	a_i	$a \cos \varphi_i$	$a \sin \varphi_i$	$a \cos 2\varphi_i$	$a \sin 2\varphi_i$
		F_p			F_s			M_p		M_s	
1	0	1	0	0	1	0	0	0	0	0	0
2	240	−.5	−.87	120	−.5	.87	a	−.5a	−.87a	−.5a	.5a
3	120	−.5	.87	240	−.5	−.87	$2a$	−a	1.74a	−a	−1.74a
4	120	−.5	.87	240	−.5	−.87	$3a$	−1.5a	2.6a	−1.5a	−2.6a
5	240	−.5	−.87	120	−.5	.87	$4a$	−2a	−3.46a	−2a	3.46a
6	0	1	0	0	1	0	$5a$	$5a$	0	$5a$	0
Sum	m	0	0	m	0	0	m	0	0	0	0

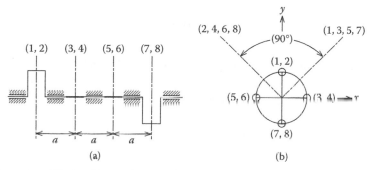

FIGURE 8.32 V–eight engine.

8.3.8.4 V-Eight Four-Stroke Engine

V8 engines are used in automobiles requiring high power like in the case of race cars. Using eight in-line cylinders requires large axial space. To overcome the space problem, four cylinder banks are used. Each bank consists of two radial cylinders with a radial angle of 90° as described in Section 8.3.8.2.

The firing interval is 90°. Thus, the angle between the cranks is 90°. The arrangement of the banks and the cranks is shown in Figure 8.32.

The results obtained in Section 8.3.8.2 can be used in the analysis for this engine; the component of the primary force is given by,

$$X_p = -m_c \omega^2 r_2 \sin(\theta)$$

$$Y_p = m_c \omega^2 r_2 \cos(\theta)$$

Thus, for cylinders (1) and (2),

$$X_{p12} = -m_c \omega^2 r_2 \sin(\theta)$$

$$Y_{p12} = m_c \omega^2 r_2 \cos(\theta)$$

The components of the primary force of (3) and (4) can be deduced by adding 270° to θ. Thus,

$$X_{p34} = -m_c\omega^2 r_2 \sin(\theta + 270) = m_c\omega^2 r_2 \cos(\theta)$$

$$Y_{p34} = m_c\omega^2 r_2 \cos(\theta + 270) = m_c\omega^2 r_2 \sin(\theta)$$

Similarly, for (5) and (6), we add 90° to θ, and for (7) and (8), we add 180° to θ. Thus,

$$X_{p56} = -m_c\omega^2 r_2 \sin(\theta + 90) = -m_c\omega^2 r_2 \cos(\theta)$$

$$Y_{p56} = m_c\omega^2 r_2 \cos(\theta + 90) = -m_c\omega^2 r_2 \sin(\theta)$$

$$X_{p78} = -m_c\omega^2 r_2 \sin(\theta + 180) = m_c\omega^2 r_2 \sin(\theta)$$

$$Y_{p78} = m_c\omega^2 r_2 \cos(\theta + 180) = -m_c\omega^2 r_2 \cos(\theta)$$

The components of the resultant primary forces are given by,

$$X_p = X_{p12} + X_{p34} + X_{p56} + X_{p78} = 0$$

$$Y_p = Y_{p12} + Y_{p34} + Y_{p56} + Y_{p78} = 0$$

This means that the primary forces are balanced. For the secondary forces, we see that

$$X_s = -\sqrt{2}m_c\omega^2 r_2 \lambda \sin(2\theta)$$

$$Y_s = 0$$

The vertical components of all cylinders are zero. Thus,

$$X_{s12} = -\sqrt{2}m_c\omega^2 r_2\lambda \sin(2\theta)$$

$$X_{s34} = -\sqrt{2}m_c\omega^2 r_2\lambda \sin(2\theta + 540) = \sqrt{2}m_c\omega^2 r_2\lambda \sin(2\theta)$$

$$X_{s56} = -\sqrt{2}m_c\omega^2 r_2\lambda \sin(2\theta + 180) = \sqrt{2}m_c\omega^2 r_2\lambda \sin(2\theta)$$

$$X_{s78} = -\sqrt{2}m_c\omega^2 r_2\lambda \sin(2\theta + 360) = -\sqrt{2}m_c\omega^2 r_2\lambda \sin(2\theta)$$

The resultant of the secondary forces is given by,

$$X_s = X_{s12} + X_{s34} + X_{s56} + X_{s78} = 0$$

This means that the secondary forces are balanced. The components of the primary moments are obtained by taking the moment of the forces about the centerline of cylinders (1) and (2). For simplicity, consider the value of θ to be zero. Thus, the horizontal components of the moment are given by,

$$M_{p12}^x = 0 \times X_{p12} = 0$$

$$M_{p34}^x = a \times X_{p34} = m_c \omega^2 r_2 a$$

$$M_{p56}^x = 2a \times X_{p56} = -2m_c \omega^2 r_2 a$$

$$M_{p78}^x = 3a \times X_{p56} = 0$$

Hence,

$$M_p^x = -m_c \omega^2 r_2 a$$

The vertical components of the moment are given by,

$$M_{p12}^y = 0 \times Y_{p12} = 0$$

$$M_{p34}^y = a \times Y_{p34} = 0$$

$$M_{p56}^y = 2a \times Y_{p56} = 0$$

$$M_{p78}^y = 3a \times Y_{p78} = -3 m_c \omega^2 r_2 a$$

Hence,

$$M_p^y = -3m_c \omega^2 r_2 a$$

The resultant primary moment is,

$$M_p = \sqrt{10}\, m_c \omega^2 r_2 a$$

The primary moment is represented by two equal and opposite masses placed at a distance d apart, each of magnitude of $\dfrac{\sqrt{10}\,a}{d} m_c$. The right mass makes an angle $252°$ with the x-axis. It makes an angle of $162°$ with the crank of cylinder (1). The moments of the secondary forces are given by,

$$M_{s12} = 0 \times X_{s12} = 0$$

$$M_{s12} = a \times X_{s34} = \sqrt{2}\, a\, m_c \omega^2 r_2 \lambda \sin(2\theta)$$

$$M_{s56} = 2a \times X_{s56} = 2\sqrt{2}\, a\, m_c \omega^2 r_2 \lambda \sin(2\theta)$$

$$M_{s78} = 3a \times X_{s34} = -3\sqrt{2}\, a\, m_c \omega^2 r_2 \lambda \sin(2\theta)$$

The sum of the secondary moments is zero. This means that the secondary moment is balanced.

8.4   IN-PLACE BALANCING

Most faults in machines can be corrected by changing the defected parts and proper fitting. It is only the unbalance that requires balancing process. Rotating elements can be balanced using balancing machines before assembly. Balancing machines are equipped with means to measure the amount of unbalance and to locate the position where to fix the balancing masses. Heavy components and parts that undergo unbalance during operation can be balanced in place. This process is called *in-place* balancing. The process is based on measuring the vibration signal due to the unbalance. For balancing machines, the correspondence between the amount of unbalance and its location and the vibration signal is calibrated. This correspondence is to be determined for in-place balancing. Instruments such as phase meter and vibration meter or analyzer are needed. The balancing process is described from Section 8.4.1 to the end of the chapter.

8.4.1   SINGLE-PLANE BALANCING (STATIC BALANCING)

Single-plane balancing applies to machine elements in the form of thin disks such as fans, flywheels, pulleys, gears, grinding wheels, and similar components. The setup of single-plane balancing is shown in Figure 8.33. It consists of a pickup, analyzer, and a stroboscope for measuring the phase angle.

The pickup is fixed on the bearing and is connected to the analyzer. Also, the stroboscope is connected to the analyzer. Reference marks are placed on both the rotor and the bearing. To understand the basis of the balancing process, suppose that the disk is completely balanced. Place a mass on the disk at a certain position and

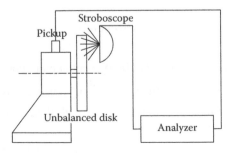

FIGURE 8.33   Set up for measuring the vibration signal.

mark this position. Measure the amplitude and phase angle of the vibration signal, which are represented by V and φ respectively. If the amount of the mass is double without changing the position, we find that the amplitude of vibration is doubled and the phase angle reading is the same. On the other hand, if we keep the mass the same and move its angular position (with the same radius) an angle β in a direction *opposite to the direction* of rotation, we find that the vibration amplitude is the same while *the phase angle reading increases with an amount* β. Of course, if we change the position of the mass in the direction of rotation, the reading decreases. Accordingly, we conclude that the amplitude of the vibration signal is proportional to the amount of the unbalance and the phase angle changes with the change of the position of the unbalance. The balancing process of the disk is described as follows:

1. Measure the amplitude and the phase angle of the vibration signal due to the original unbalance V_o and φ_o respectively.
2. Place a trial mass of known quantity m_t at a certain position; mark this position (Figure 8.34). Measure the amplitude and the phase angle of the vibration signal V and φ respectively. This signal is due to the combined effect of the original unbalance and the trial mass.

The analysis may be done either graphically or analytically. The radii of all masses are unified. In this case, the unbalance quantities are in terms of the masses.

8.4.1.1 Graphical Method

1. Draw a vector $\mathbf{V}_o$ of length V_o and an angle φ_o (Figure 8.35).
   Draw a vector $\mathbf{V}$ with length V and an angle φ. This vector is the resultant effect of the original unbalance and the trial mass.
2. The effect of the trial mass is obtained by subtracting $\mathbf{V}_o$ from $\mathbf{V}$.

$$\mathbf{V}_t = \mathbf{V} - \mathbf{V}_o$$

This is a vector of length V_t and with an angle φ_t. From this vector, we can obtain the amount of the original unbalance m_o and its location relative to the position of the trial mass.

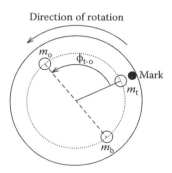

FIGURE 8.34   Locating the position of the balancing mass.

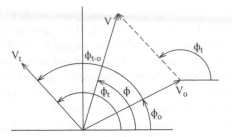

FIGURE 8.35   Graphical determination of balancing mass.

$$m_{o} = m_{t} \frac{V_{o}}{V_{t}}$$

The position of the original unbalance makes an angle $\varphi_{t} -_{o} = \varphi_{t} - \varphi_{o}$ in the direction opposite to the direction of rotation (notice that, in this demonstration, φ_{t} is larger than φ_{o}, which means that m_{t} is behind m_{o} according to the direction of rotation) (Figure 8.34).

8.4.1.2   Analytical Method

A vector in the complex form is given by,

$$\mathbf{V} = V e^{i\varphi}$$
$$= V \cos\varphi + i\, V \sin\varphi$$
$$= a + i\, b$$

Since the vibration signal in the second step is the combined effect of the original unbalance and the trial mass,

$$\mathbf{V} = \mathbf{V}_{o} + \mathbf{V}_{t}$$

Thus,

$$\mathbf{V}_{t} = \mathbf{V} - \mathbf{V}_{o}$$

$$= (a - a_{o}) + i(b - b_{o})$$

Hence,

$$V_{t} = \sqrt{(a - a_{o})^{2} + (b - b_{o})^{2}}$$

$$\varphi_{t} = \tan^{-1} \frac{b - b_{o}}{a - a_{o}}$$

The exact value of φ_{t} is determined from the signs of the expressions $(a - a_{o})$ and $(b - b_{o})$.

For complete balance, we change the amount of the trial mass by a factor c and change its position by an angle β opposite to the direction of rotation. Changing the amount of the mass and its position changes the amplitude of its vibration signal by a factor c and changes the phase angle by an angle β. Thus, the vibration signal due to the balancing mass m_b is given by,

$$\begin{aligned} \mathbf{V}_b &= c\, V_t\, e^{i(\varphi_t+\beta)} \\ &= c\, e^{i\beta}\, V_t\, e^{i\varphi_t} \end{aligned}$$

or,

$$\mathbf{V}_b = \mathbf{C}\mathbf{V}_t$$

where

$$\mathbf{C} = ce^{i\beta}$$

To determine the values of c and β, we apply the condition of balancing, which is the vibration signal of the original unbalance and the balancing mass is zero. Thus,

$$\mathbf{V}_o + \mathbf{V}_b = 0$$

or,

$$\mathbf{V}_o + \mathbf{C}\mathbf{V}_t = 0$$

$$\mathbf{C} = -\frac{\mathbf{V}_o}{\mathbf{V}_t}$$

$$= \frac{V_o}{V_t}e^{i(\varphi_o-\varphi_t+\pi)}$$

Therefore,

$$c = \frac{V_o}{V_t} \tag{8.19}$$

$$\beta = \varphi_o - \varphi_t + \pi \tag{8.20}$$

For complete balance, we fix a balancing mass $m_b = \dfrac{V_o}{V_t}m_t$ placed at an angle equal to $\varphi_o - \varphi_t + \pi$ opposite to the direction of rotation from the mark of the trial mass.

EXAMPLE 8.3

The amplitude and phase angle due to the original unbalance of a grinding wheel are 50 mm/s and 40° respectively. A trial mass of 100 g placed at a marked place makes the amplitude and phase angle to be 90 mm/s and 150° respectively. Find the magnitude and the position of the balancing mass.

SOLUTION

The data given is
$V = 90$ mm/s, $\varphi = 150°$, $V_o = 50$ mm/s, $\varphi_o = 40°$. Then,

$$a = V \cos\varphi = -77.942$$

$$b = V \sin\varphi = 45$$

$$a_o = V_o \cos\varphi_o = 38.302$$

$$b_o = V_o \sin\varphi_o = 32.139$$

$$a_t = a - a_o = -116.244$$

$$b_t = b - b_o = 6.163$$

From the above result, the effect of a 100 g trial mass at the specified position gives a vibration signal of magnitude $V_t = 116.407$ mm/s and a phase angle $\varphi_t = 177°$. The magnitude of the balancing mass is,

$$m_b = m_t \frac{V_o}{V_t} = 43\,\text{g}$$

The position of the balancing mass from the position of the trial mass is, $\varphi_b = \varphi_o - \varphi_t + \pi = 43°$ opposite to the direction of rotation.

8.4.2 STATIC BALANCING WITHOUT PHASE MEASUREMENTS

1. Measure the amplitude of the vibration signal due to the original unbalance V_o.
2. Place a trial mass m_t at some position and mark it. Measure the vibration amplitude V_1.
3. Remove the trial mass and place it 180° from the first position. Measure the vibration amplitude V_2.

We consider that the preceding readings represent the magnitudes of vectors $\mathbf{V}_o$, $\mathbf{V}_1$, and $\mathbf{V}_2$. These vectors are such that

$$\mathbf{V}_1 = \mathbf{V}_o + \mathbf{V}_t$$

$$V_2 = V_o - V_t$$

Writing the vectors in the complex form,

$$V_1 e^{i\varphi_1} = V_o e^{i\varphi_o} + V_t e^{i\varphi_t}$$

$$V_2 e^{i\varphi_2} = V_o e^{i\varphi_o} - V_t e^{i\varphi_t}$$

Multiplying each side by its conjugate,

$$V_1 e^{i\varphi_1} V_1 e^{-i\varphi_1} = (V_o e^{i\varphi_o} + V_t e^{i\varphi_t})(V_o e^{-i\varphi_o} + V_t e^{-i\varphi_t})$$

$$V_2 e^{i\varphi_2} V_2 e^{-i\varphi_1} = (V_o e^{i\varphi_o} - V_t e^{i\varphi_t})(V_o e^{-i\varphi_o} - V_t e^{-i\varphi_t})$$

Expanding the above equations

$$V_1^2 = V_o^2 + V_t^2 + 2V_o V_t \cos(\varphi_t - \varphi_o)$$

$$V_2^2 = V_o^2 + V_t^2 - 2V_o V_t \cos(\varphi_t - \varphi_o)$$

Solving these equations, we get,

$$V_t = \sqrt{\frac{V_1^2 + V_2^2 - 2V_o^2}{2}}$$

$$\varphi_t - \varphi_o = \pm\cos^{-1}\frac{V_1^2 - V_2^2}{4V_o V_t}$$

The amount of balancing mass and its location is given by,

$$m_b = m_t \frac{V_o}{V_t}$$

$$\varphi_b = \varphi_o - \varphi_t + \pi$$

Since there are two solutions for the angle, the exact value can be determined by trial, that is, placing the corrective mass at one of the positions. If balancing is not achieved, we put it at the other position.

EXAMPLE 8.4

It is possible to carry out the balancing process by doubling the amount of the trial mass in its position instead of replacing it in the opposite direction. The amplitude of vibration due to the original unbalance of a grinding wheel is 50 mm/s. A trial mass of 50 g placed at a marked place makes the amplitude 80 mm/s. Doubling the amount of the trial mass makes the amplitude 120 mm/s. Find the magnitude and the position of the balancing mass.

SOLUTION

When the amount of the trial mass is doubled in its position, the amplitude of vibration V_t due to the trial mass is doubled and the phase angle is the same. The vector equations are given by,

$$V_1\, e^{i\varphi_1} = V_o\, e^{i\varphi_o} + V_t\, e^{i\varphi_t}$$

$$V_2\, e^{i\varphi_2} = V_o\, e^{i\varphi_o} + 2V_t\, e^{i\varphi_t}$$

Multiplying each complex quantity by its conjugate leads to

$$V_1^2 = V_o^2 + V_t^2 + 2V_o V_t \cos(\varphi_t - \varphi_o)$$

$$V_2^2 = V_o^2 + 4V_t^2 + 4V_o V_t \cos(\varphi_t - \varphi_o)$$

Solving these equations gives

$$V_t = \sqrt{\frac{V_o^2 + V_2^2 - 2V_1^2}{2}}$$

$$\varphi_t - \varphi_o = \pm\cos^{-1}\frac{V_1^2 - V_o^2 - V_t^2}{2V_o V_t}$$

Substituting with the given data,

$$V_t = 45.277 \text{ mm/s}$$

$$m_b = 45.277 \text{ g}$$

$$\varphi_t - \varphi_o = \pm 65.86°$$

$$\varphi_o - \varphi_t = \pm 65.86°$$

$$\varphi_b = \varphi_o - \varphi_t + 180$$

$$= 114.14° \text{ or } 245.86° \text{ opposite to the direction of rotation}$$

8.4.3 TWO-PLANE BALANCING (DYNAMIC BALANCING)

In most cases, unbalance problem occurs in machines with long rotors such as turbines, compressors, or long shafts carrying several disks. In such cases, balancing must be performed in two planes. Such machines must be equipped with special disks specially mounted for balancing.

The setup for the balancing process (Figure 8.36) consists of a long rotor mounted on a shaft that is supported by two bearings. The vibration signals are picked from

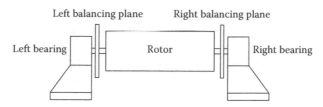

FIGURE 8.36   Set up rig for rotor balancing.

the left and right bearings. The amplitudes and the phase angles of the vibration signals are displayed by an analyzer.

The balancing process is outlined as follows:

1. Measure the amplitude and phase angle of the vibration signal due to the original unbalance at the left and right bearings. They are denoted by V_0^L, V_0^R, φ_0^L, and φ_0^R.
2. Place a trial mass m_{TL} at the left balancing plane. Measure the amplitude and the phase angle of the vibration signal due to the combined effect of the original unbalance and m_{TL} at the left and right bearings. They are denoted by V_1^L, V_1^R, φ_1^L, and φ_1^R.
3. Remove the trial mass at the left plane. Place a trial mass m_{TR} at the right balancing plane. Measure the amplitude and the phase angle of the vibration signal due to the combined effect of the original unbalance and m_{TR} at the left and right bearings. They are denoted by V_2^L, V_2^R, φ_2^L, and φ_2^R.

The preceding readings form six vectors. The components of these vectors are

$$a_0^L = V_0^L \cos\varphi_0^L,\ b_0^L = V_0^L \sin\varphi_0^L,\ a_0^R = V_0^R \cos\varphi_0^R,\ b_0^R = V_0^R \sin\varphi_0^R \qquad (1)$$

$$a_1^L = V_1^L \cos\varphi_1^L,\ b_1^L = V_1^L \sin\varphi_1^L,\ a_1^R = V_1^R \cos\varphi_1^R,\ b_1^R = V_1^R \sin\varphi_1^R \qquad (2)$$

$$a_2^L = V_2^L \cos\varphi_2^L,\ b_2^L = V_2^L \sin\varphi_2^L,\ a_2^R = V_2^R \cos\varphi_2^R,\ b_2^R = V_2^R \sin\varphi_2^R \qquad (3)$$

The effect of the trial masses on the left and the right bearings is obtained by subtracting row (1) from rows (2) and (3).

$$a_{TL}^L = a_1^L - a_0^L,\ b_{TL}^L = b_1^L - b_0^L$$

$$a_{TL}^R = a_1^R - a_0^R,\ b_{TL}^R = b_1^R - b_0^R$$

$$a_{TR}^L = a_2^L - a_0^L,\ b_{TR}^L = b_2^L - b_0^L$$

$$a_{TR}^R = a_2^R - a_0^R,\ b_{TR}^R = b_2^R - b_0^R$$

The preceding quantities are the components of vectors $\mathbf{V}_{TL}^L$, $\mathbf{V}_{TL}^R$, $\mathbf{V}_{TR}^L$, and $\mathbf{V}_{TR}^R$ with magnitudes and phase angles given by,

$$V_{TL}^L = \sqrt{(a_{TL}^L)^2 + (b_{TL}^L)^2}, \quad \varphi_{TL}^L = \tan^{-1} \frac{b_{TL}^L}{a_{TL}^L}$$

$$V_{TL}^R = \sqrt{(a_{TL}^R)^2 + (b_{TL}^R)^2}, \quad \varphi_{TL}^R = \tan^{-1} \frac{b_{TL}^R}{a_{TL}^R}$$

$$V_{TR}^L = \sqrt{(a_{TR}^L)^2 + (b_{TR}^L)^2}, \quad \varphi_{TR}^L = \tan^{-1} \frac{b_{TR}^L}{a_{TR}^L}$$

$$V_{TR}^R = \sqrt{(a_{TR}^R)^2 + (b_{TR}^R)^2}, \quad \varphi_{TR}^R = \tan^{-1} \frac{b_{TR}^R}{a_{TR}^R}$$

For complete balance, we can change the left trial mass m_{TL} by a factor c_L and change its position by an angle β_L and change the right trial mass m_{TR} by a factor c_R and change its position by an angle β_R. These quantities form the correcting vectors, which are given by,

$$\mathbf{C}_L = c_L \, e^{i\beta_L}$$

$$\mathbf{C}_R = c_R \, e^{i\beta_R}$$

The conditions for complete balance are such that the amplitudes of the vibration signal due to the original unbalance plus the effect of the corrected masses at the left and right bearings are zero.

$$\mathbf{V}_o^L + \mathbf{C}_L \, \mathbf{V}_{TL}^L + \mathbf{C}_R \, \mathbf{V}_{TR}^L = 0$$

$$\mathbf{V}_o^R + \mathbf{C}_L \, \mathbf{V}_{TL}^R + \mathbf{C}_R \, \mathbf{V}_{TR}^R = 0$$

Solving these equations for $\mathbf{C}_L$ and $\mathbf{C}_R$,

$$\mathbf{C}_L = \frac{\mathbf{V}_o^R \, \mathbf{V}_{TR}^L - \mathbf{V}_o^L \, \mathbf{V}_{TR}^R}{\mathbf{V}_{TL}^L \, \mathbf{V}_{TR}^R - \mathbf{V}_{TL}^R \, \mathbf{V}_{TR}^L} \tag{8.21}$$

$$\mathbf{C}_R = \frac{\mathbf{V}_o^L \, \mathbf{V}_{TL}^R - \mathbf{V}_o^R \, \mathbf{V}_{TL}^L}{\mathbf{V}_{TL}^L \, \mathbf{V}_{TR}^R - \mathbf{V}_{TL}^R \, \mathbf{V}_{TR}^L} \tag{8.22}$$

The magnitudes and the angles of these complex numbers are the corrective quantities for the balancing masses.

EXAMPLE 8.5

In balancing a turbine, the following data is provided:

- $V_o^L = 85$ mm/s, $V_o^R = 65$ mm/s, $\varphi_o^L = 60°$, $\varphi_o^R = 205°$.
- A trial mass of 50 g placed at the left plane makes the signal.

$$V_1^L = 60 \text{ mm/s}, V_1^R = 45 \text{ mm/s}, \varphi_1^L = 125°, \varphi_1^R = 230°.$$

- A trial mass of 60 g placed at the right plane makes the signal.

$$V_1^L = 60 \text{ mm/s}, V_1^R = 105 \text{ mm/s}, \varphi_1^L = 35°, \varphi_1^R = 160°.$$

Find the amount and locations of the balancing masses.

SOLUTION

Tables 8.8 and 8.10 are used to write the components of the vectors. To find the corrective vectors, we apply Equations 8.21 and 8.22

$$C_L = \frac{V_0^R V_{TR}^L - V_0^L V_{TR}^R}{V_{TL}^L V_{TR}^R - V_{TL}^R V_{TR}^L}$$

This value gives

$$c_L = 1.006$$

$$\beta_L = 55.5°$$

Similarly,

$$C_R = \frac{V_0^L V_{TL}^R - V_0^R V_{TL}^L}{V_{TL}^L V_{TR}^R - V_{TL}^R V_{TR}^L}$$

$$C_R = \frac{(42.5 + i73.6)(30.0 - i7.0) - (-58.9 + i27.5)(-76.9 + i24.5)}{(-76.9 - i24.5)(-39.8 + i63.4) - (30 - i7.0)(6.6 - i39.2)}$$

$$= -0.181 - i0.454$$

This value gives

$$c_R = 0.489$$

$$\beta_R = 248.3°$$

TABLE 8.10
Analysis for EXAMPLE 8.5

Vector	V (mm/s)	φ (degree)	a	b
V_0^L	85	60	42.5	73.6
V_0^R	65	205	−58.9	−27.5
V_1^L	60	125	−34.4	49.1
V_1^R	45	240	−22.5	−39.0
V_2^L	60	45	42.5	42.4
V_2^R	105	180	−105.0	0.0
V_{TL}^L	56.64	−122.0	−30.0	−48.0
V_{TL}^R	46.79	1.3	46.8	1.0
V_{TR}^L	71.52	−53.6	42.4	−57.6
V_{TR}^R	53.63	131.8	−35.7	40.0

The preceding results show that for complete balance, a mass of 50 g is to be fixed on the left plane, placed at 55.5° from the trial mass in a direction opposite to the direction of rotation, and a mass of 29.34 g is to be fixed on the right plane, placed at 248.3° from the trial mass in a direction opposite to the direction of rotation.

PROBLEMS

ROTATING MASSES

8.1 A thin rotor carries three unbalance masses: $m_1r_1 = 100$ kg cm with angle $0°$, $m_2r_2 = 90$ kg cm with angle $210°$, $m_3r_3 = 150$ kg cm with angle $150°$. The rotor is to be balanced by adding a balancing mass at a radius of 20 cm. Find the magnitude and the angular location of the balancing mass graphically and analytically.

8.2 A thin rotor carries three unbalance masses: $m_1r_1 = 80$ kg cm, $m_2r_2 = 100$ kg cm, $m_3r_3 = 150$ kg cm. Find the relative angular position of the three masses such that the rotator is balanced.

8.3 Two masses of 8 and 16 kg rotate in the same plane at radii 1.5 and 2.25 cm respectively. The radii of these masses are 60° apart. Find the position of the third mass of magnitude 12 kg in the same plane, which can produce complete dynamic balance of the system.

8.4 Two equal and opposite masses, each of magnitude 100 kg cm, are mounted on a shaft as shown in Figure P8.4. If the shaft rotates at 900 rpm, find the reactions at the bearings.

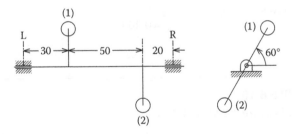

FIGURE P8.4

8.5 Two masses are mounted at right angle on a shaft as shown in Figure P8.5. The masses are $m_1r_1 = 50$ kg cm, $m_2r_2 = 80$ kg cm. If the shaft rotates at 900 rpm, find the reactions at the bearings.

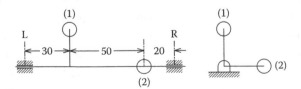

FIGURE P8.5

8.6 For the shaft shown in Figure P8.6, determine the bearing reactions at L and R due to the unbalance masses indicated if the rotor speed is 1500 rpm. $m_1 r_1 = 100$ kg cm, $m_2 r_2 = 50$ kg cm, $m_3 r_3 = 150$ kg cm.

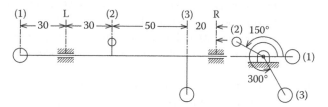

FIGURE P8.6

8.7 Three pulleys are out of balance. The amount of unbalance is 120, 180, and 160 kg mm. The pulleys are keyed to the shaft at planes (1), (2), and (3) respectively as shown in Figure P8.7. Find the following:
a. The relative angular position of the three pulleys if the resultant of the unbalanced forces is zero.
b. The out-of-balance moment when the shaft rotates at 600 rpm.
c. The dynamic load on each bearing (A and B).

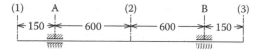

FIGURE P8.7

8.8 A shaft carries two masses as shown in Figure P8.8. The amount of unbalance are $m_1 r_1 = 30$ kg cm, $m_2 r_2 = 20$ kg cm. Find the magnitudes and the angular positions of the balancing masses to be located at planes (L) and (R) so that the system is in complete dynamic balance. The balancing masses are located with radii of 5 cm.

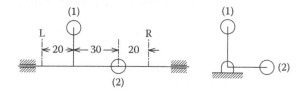

FIGURE P8.8

8.9 A shaft carries four masses at planes (1), (2), (3), and (4). The planes are equally spaced from each other with an axial distance of 240 mm. The amount of unbalance of the masses are $m_2 r_2 = 60$ kg cm, $m_3 r_3 = 40$ kg cm, and $m_4 r_4 = 24$ kg cm. The radius of the mass at plane (1) is 4 cm. Find the magnitude of m_1 and the relative angular position of the masses so that the shaft is in complete balance.

8.10 Four disks A, B, C, and D are attached to a uniformly rotating shaft, spaced at equal intervals along the shaft, and have masses of 7.5, 12.5, 7, and 6 kg respectively. The mass centers of the disks are 4, 3, 5, and 8 mm from the axis of rotation respectively. An additional mass M is to be attached to disk D at an effective radius of 60 mm from the axis of rotation. Find the minimum value of the mass M and the relative angular positions of the mass centers of the masses to ensure complete dynamic balance of the rotating shaft.

8.11 The shaft shown in Figure P8.11 is supported by two bearings A and B. Three pulleys are attached to the shaft in planes C, D, and E. The pulleys have masses of 20, 50, and 48 kg respectively. The mass centers radii are 12.5, 15.5, and 15 mm respectively. The pulleys have been arranged so that the resultant force is zero.

Determine the dynamic forces produced on the bearings when the shaft rotates at 300 rpm.

If two masses are placed in planes C and E at a radius of 80 mm to balance the system completely, find the two masses and their relative angular settings.

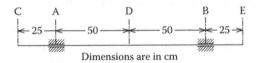

Dimensions are in cm

FIGURE P8.11

RECIPROCATING MASSES

8.12 A V engine has two identical cylinders. The length of each connecting rod is 350 mm, and the reciprocating mass of each cylinder is 12 kg. The crank radius is 75 mm.
   a. Find the V angle, which results in a minimum primary force.
   b. Calculate the primary and the secondary forces when the engine runs at 500 rpm.

8.13 In an opposed double-cylinder radial engine (Figure P8.13), the reciprocating mass of the cylinders is 3 kg. The length of the crank is 5 cm, the length of the connecting rods is 30 cm, and the speed of the crank shaft is 1500 rpm. Determine the magnitude of the shaking force.

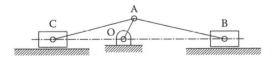

FIGURE P8.13

8.14 Repeat Problem 8.13 when the engine has two opposite cranks as shown in Figure P8.14.

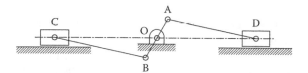

FIGURE P8.14

8.15 If the cranks of the engine of Problem 2.13 are placed with a phase angle as shown in Figure P8.15, determine the shaking force.

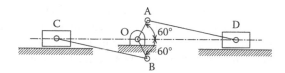

FIGURE P8.15

8.16 For the three-cylinder radial engine shown in Figure P8.16, determine the shaking force. The value of $m_c r_2 \omega^2$ is 5000 N and the ratio of length of the crank to the connecting rod λ is 0.25. Find the resultant shaking force.

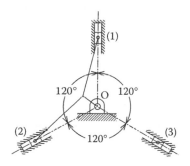

FIGURE P8.16

8.17 A two-stroke in-line two-cylinder engine runs at 1500 rpm. Each piston has a stroke of 100 mm. The reciprocating mass of each cylinder is 2 kg. The ratio of length of the crank to the connecting rod λ is 0.25. The distance between the centerlines of the cylinders is 100 mm. Find the resultant primary force, the resultant secondary shaking force, the primary shaking moment, and the secondary shaking moment.

8.18 If the cranks of the engine of Problem 2.16 are at right angles, find the resultant primary force, the secondary shaking force, the primary shaking moment, and the secondary shaking moment.

8.19 A four-stroke in-line three-cylinder engine runs at 1500 rpm. Each piston has a stroke of 100 mm. The reciprocating mass of each cylinder is 2 kg.

The ratio of length of the crank to the connecting rod λ is 0.25. The distance between the centerlines of the cylinders is 100 mm. Determine the firing order, the resultant primary force, the resultant secondary force, the primary shaking moment, and the secondary shaking moment.

8.20 A four-stroke in-line six-cylinder engine runs at 3600 rpm and each piston has a stroke of 100 mm. Each piston weights 1.5 kg and the connecting rods are 200 mm long each. If the total reciprocating mass per cylinder is 2 kg, what are the unbalance forces and moments (primary and secondary)? The firing order is 1–3–5–6–4–2 and distance between the centerline of the cylinder is 100 mm.

8.21 A two-stroke in-line engine has eight identical cylinders spaced at equal intervals of 150 mm. The mass of the reciprocating part per cylinder is 2 kg, the piston has a stroke of 100 mm, and the connecting rods are 180 mm long. The firing order is 1–3–2–4–8–6–7–5. Determine the magnitude and direction of the resultant shaking force or couple if the engine runs at 600 rpm.

8.22 A two-stroke four-cylinder in-line engine has a firing order of 1–4–3–2. The crank radius is 150 mm, the connecting rod is 475 mm, and the reciprocating mass per cylinder is 22.5 kg. The engine drives two cylinders (A and B) (Figure P8.22), which are mounted in-line with the engine as shown in the figure. The cranks of A and B make angles of 135° and 315° with the crank of cylinder (1). For cylinders A and B, the crank radius is 190 mm, the connecting rod is 375 mm, and the reciprocating mass per cylinder is 15 kg. Determine the unbalanced forces and moments when the engine rotates at 400 rpm.

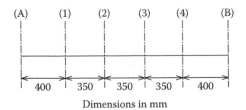

FIGURE P8.22

8.23 A four-cylinder engine has the outer cranks set at 120° to each other, and their reciprocating masses are 360 kg. The distances between the planes of rotation of adjacent cranks are 0.45, 0.75, and 0.6 m.
   a. If the engine is to be in a complete primary balance, find the value of reciprocating masses and the angular position for each of the inner cylinders.
   b. If the length of each crank is 0.30 m, the length of the connecting rod is 1.2 m, and the speed of rotation is 240 rpm, find the maximum secondary unbalance force.

8.24 Figure P8.24 shows a two-cylinder V–90° in-line engine. Let $mr\omega^2 = 1$, $\lambda = 0.25$, and a = 10 mm. Find the resultant primary force, the resultant secondary force, the resultant shaking force in magnitude and direction, and the distance a_R of the shaking force from the centerline of cylinder (1) for $\theta_1 = 60°$.

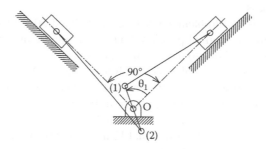

FIGURE P8.24

8.25 For the four-cylinder opposed engine shown in Figure P8.25, derive in terms of θ_1 the primary force F_p, the secondary force F_s, shaking force S, the distance a_R of S from the plane of cylinder (1). Evaluate S and a_R for $\theta_1 = 90°$, assuming that $mr\omega^2$ and the distance between sets of cylinders are unity, $\lambda = 0.25$. For what angle or angles θ_1, if any, will the resultant primary force be zero?

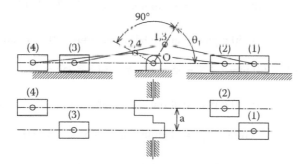

FIGURE P8.25

IN-PLACE BALANCING

8.26 A grinding wheel has an unbalance of unknown magnitude and position. When a trial mass of 50 g–cm is placed on the wheel at 0° with respect to the x-axis, the resulting unbalance is on a line making 90°. When an additional trial mass of 20 g–cm is placed at 30°, the resulting unbalance is on a line making 60°. Find the magnitude and direction of the original unbalance.

8.27 In static balancing, a mass of 20 g gave a signal of 20 mm/s with an angle of 30°. When the same mass was placed at the opposite direction, the signal became 20 mm/s with an angle of 150°. Find the amount of the balancing mass and its relative position relative to the original position.

8.28 A disk has an unbalance, which gives a vibration signal with amplitude of 10 mm/s and the phase angle 30°. A trial mass of 20 g makes the amplitude 20 mm/s and the phase angle 90°. Find the amount and the location of the balancing mass.

8.29 If the trial mass of Problem 7.5 is 25 g and is fixed 30° in the direction of rotation from the original position, find the amplitude and phase angle of the vibration signal due to the new trial mass.

8.30 A disk has an original unbalance with a signal of 30 mm/s and a phase angle of 60°. A trial mass of 20 g made the signal 25 mm/s and 90°. The balancing mass was placed by mistake in the opposite direction. Find the amplitude and the phase angle of the signal.

8.31 A trial mass of 20 g makes the signal 12 mm/s and 0°. When the trial mass is placed in the opposite direction, the signal becomes 12 mm/s and 90°. Find the amount and position of the original unbalance.

8.32 The original unbalance of a disk gives a signal with amplitude of 25 mm/s. A trial mass of 50 g made the signal 50 mm/s. With the trial mass placed in its position, another mass of 75 g placed in the opposite direction to the first mass makes the signal 50 mm/s. Find the amount of the balancing mass and its position from the first trial mass.

8.33 In balancing a turbine, the following data is provided:

$$V_o^L = 100 \text{ mm/s}, V_o^R = 80 \text{ mm/s}, \varphi_o^L = 90°, \varphi_o^R = 210°.$$

A trial mass of 50 g placed at the left plane makes the signal

$$V_1^L = 60 \text{ mm/s}, V_1^R = 45 \text{ mm/s}, \varphi_1^L = 120°, \varphi_1^R = 240°.$$

A trial mass of 50 g placed at the right plane makes the signal

$$V_1^L = 60 \text{ mm/s}, V_1^R = 105 \text{ mm/s}, \varphi_1^L = 45°, \varphi_1^R = 180°.$$

Find the amount and locations of the balancing masses.

8.34 In a balancing process, the original unbalance gave a signal of 10 mm/s, 30° at the left bearing and 20 mm/s, 60° at the right bearing. A 20-g trial mass at the left bearing made the signals 15 mm/s, 0°, and 15 mm/s, 90°, at the left and right bearings respectively. A 30-g trial mass at the right bearing made the signals 5 mm/s, 120°, and 25 mm/s, 45°, at the left and right bearings respectively. Find the amount and positions of the balancing masses.

Bibliography

Alger, J. R. M., and C. V. Hays. *Creative Synthesis in Design*, Prentice-Hall, Upper Saddle River, NJ, 1964.

Ambekar, A. G. *Mechanism and Machine Theory*, Prentice-Hall of India, New Delhi, 2007.

American Gear Manufacturers Association. 'AGMA 933—B03. *"Basic Gear Geometry,"* March 13, 2003.'

Angles, J. *Computational Kinematics*, Kluwer Academic, Boston, MA, 1993.

Ballaney, P. L. *Theory of Machines*, Khanna Publishers, New Delhi, 1979.

Bansal, R. K., and J. S. Brar. *A Textbook of Theory of Machines (In S.I. Units)*, Firewall Media, New Delhi, January 1, 2004.

Bevan, T. *Theory of Machines*, 3rd ed., CBS Publisher, New Delhi, 2004.

Biggs, J. S. *Mechanisms*, McGraw-Hill Company, New York, 1955.

Billings, J. H. *Applied Kinematics for Students and Mechanical Designers*, D. Van Nostrand Company, Inc., New York, 1943.

Bose, S. K. *Theory of Machines*, Allied Publisher Pvt. Ltd., New Delhi, 2004.

Carson, W. L. *Teaching Unit on Complex Numbers as Applied to Linkage Modeling, in Monograph on Mechanical Design*, McGraw-Hill Book Company, New York, 1977.

Chen, F. Y. *Mechanics and Design of Cam Mechanisms*, Pergamon Press, New York, 1982.

Chironis, N. P. *Mechanisms, Linkages, and Mechanical Controls*, McGraw-Hill, New York, 1965.

Colbourne, J. R. *The Geometry of Involute Gears*, Springer, New York, 1987

Doina, P., C. Marco, and M. Husty, *New Trends in Mechanism Science, Analysis and Design*, Springer Science+Business Media B. V., New York, 2010.

ENTC 463, 'Mechanical Design Applications II, *Worm and Worm Gear*.' Texas A&M University.

Erdman, A. G. Three and Four Precision Point Kinematic Synthesis of Planar Linkages, *Mechanisms and Machine Theory*, Vol. 16, 227–245, 1981.

Erdman, A. G. *Modern Kinematics*, John Wiley, New York, 1993.

Erdman, A. G., and G. N. Sandors. *Mechanism Design: Analysis and Synthesis*, Vol. 1, 3rd ed., Prentice-Hall International, Inc., Upper Saddle River, NJ, 1997.

Gladwel, G. M. I. *Cam Synthesis*, Kluwer Academic Publishers, Boston, MA, 1993.

Grashof, F. *Theory of Machines*, BiblioBazaar, September 7, 2010.

Green, W. G. *Theory of Machines*, Blackie, London, 1962.

Green, W. G. *Theory of Machines*, Blackie, London, 1964.

Guillet, G. L., and A. H. Church, *Kinematics of Machines*, John Wiley, New York, 1950.

Gupta, B. V. R. *Theory of Machines, Kinematics and Dynamics*, International Publishing House Pvt. Ltd., New Delhi, 2011.

Gupta, K. C. A General Theory of Synthesizing Four Bar Function Generators with Transmission Angle Control, *Journal of Applied Mechanics*, Vol. 45, no. 2, June 1968.

Hagen, D., A. G. Erdman, D. Harvey, and J. Tacheny. "Rapid Algorithm for Kinematic and Dynamic Analysis of Planar Rigid Links with Revolute Joints," ASME Paper No. 78-DET-64, 1976.

Hain, K. *Applied Kinematics*, 2nd ed., McGraw-Hill Book Company, New York, 1967.

Hall, A. S. *Kinematics and Linkage Design*, Balt Publishers, West Lafayette, IN, 1961.

Hartenberg, R. S., and J. Denavit. *Kinematics Synthesis of Linkages*, McGraw-Hill Book Company, New York, 1964.

Haug, E. J. *Computer Aided Kinematics and Dynamics of Mechanical Systems*, Allyn and Bacon, Boston, MA, 1989.

Heywood, J. B., *Internal Combustion Engine Fundamentals*, McGraw-Hill, New York, 1988.

Hill, B. D. *Machine Dynamics*, Deakin University, Melbourne, Australia, 1997.

Hinkle, R. T. *Design of Machines*, Prentice-Hall, Inc., Englewood Cliffs, NJ, 1957.

Hinkle, R. T. *Kinematics of Machines*, 2nd ed., Prentice-Hall, Englewood Cliffs, NJ, 1960.

Hornes, J. A., and G. L. Nelson. *Analysis of the Four-Bar Linkage*, The Technology Press of MIT and John Wiley, New York, 1951.

Howard. P. J. *Theory of Machines*, Macdonald & Co., London, 1966.

Kapelevich, A. *Geometry and Design of Involute Spur Gears with Asymmetric Teeth*, Elsevier Science Ltd., MO, 1999.

Khurmi, R. S., and J. K. Gupta. *Theory of Machines*, S. Chand & Company Ltd., New Delhi, 1997.

Kinzel, G. L. *Mechanism Design and Synthesis*, American Society of Mechanical Engineers, January 1, 1992.

Kohavi, Z., and A. Paz. *Theory of Machines and Computations*, Academic Press, New York, 1971.

Koloc, Z., and M. Václavík. *Cam Mechanisms*, Elsevier, Amsterdam, 1993.

Latvin, F. L. *Gear Geometry and Applied Theory*, Cambridge University Press, 2004.

Low, B. B. *Theory of Machines*, Longmans, Green, London, 1954.

Mabie, H. H., and C. F. Reinholtz. *Kinematics and Dynamics of Machinery*, 4th ed., John Wiley, New York, 1978.

Marghitu, D. B. *Kinematic Chains and Machine Component Design*, Elsevier Academic Press, Burlington, VT, 2005.

Marghitu, D. B., and J. C. Malcolm. *Analytical Elements of Mechanisms*, Cambridge University Press, 2001.

Martin, J. H. *Kinematics and Dynamics of Machinery*, Waveland Press, Inc., Long Grove, Illinois, 2002.

McKay, R. F. *The Theory of Machines*, E. Arnold & Co., London, 1938.

Molian, S. *The Design of Cam Mechanisms and Linkages*, American Elsevier Pub. Co., New York, 1968.

Molian, S. *Mechanism Design: An Introductory Index*, Cambridge University Press, 1982.

Mostafa, M. Axially Moving Cams with Translating Followers, *Journal of Engineering and Science*, Vol. 3, no. 2, 1977a. College of Engineering, University of Riyadh, Saudi Arabia.

Mostafa, M. The Exact Cam Contour, *The Bulletin of the Faculty of Engineering*, Vol. 10, 1971a. Alexandria University.

Mostafa, M. A Graphical Method for Determining the Velocity and Acceleration of Plane Mechanisms. *The Bulletin of the Faculty of Engineering*, Vol. 10, Alexandria University, 1971b.

Mostafa, M. The Minimum Base Circle Radius of Plane Cams, *Journal of Engineering and Science*, Vol. 3, no. 2, 1977b. College of Engineering, University of Riyadh, Saudi Arabia.

Mostafa, M. The Minimum Cam Size, *The Bulletin of the Faculty of Engineering*, Vol. 10, 1971c. Alexandria University.

Mostafa, M. "The Relative Acceleration of Sliding Links," Research Report Presented to the Committee for Promoting Associate Professors, December 1972.

Mostafa, M. The Relative Angular Velocity of Links, *Journal of Engineering for Industry, Transactions of ASME*, Vol. 95, 1973.

Mostafa, M. "Specified Contour Space Cams with Oscillating Followers," *Lahore 1st Mechanical Engineering Conference*, Pakistan, 1975.

Mostafa, M. "Synthesis of a Four-Bar Linkage for a Limited Variation in the Velocity Ratio," Presented at IFToMM Conference, Montreal, Canada, 1979. Also, *Journal of Engineering and Science*, Vol. 6, no. 2, 1977c. College of Engineering, University of Riyadh, Saudi Arabia.

Mostafa, M., T. Awad, and M. Trabia. A Computer Method for Kinematics and Dynamic Analysis of Plane Mechanisms, *The Bulletin of the Faculty of Engineering*, Vol. 12, 1983a. Alexandria University.

Mostafa, M., T. Awad, and M. Trabia. Optimum Dynamic Performance Using Nonlinear Programming, *The Bulletin of the Faculty of Engineering*, Vol. 12, 1983b. Alexandria University.

Mostafa, M., E. Badawy, H. Elhares, and N. Moharem. *Mechanics of Machinery,* Vol. 1, *Kinematics*, The General Egyptian Book Organization, Alexandria Branch, Egypt, Nov 1973.

Mostafa, M., and S. El-Shakiry. Roller Chain Mechanisms. Part I—General Concept, *Alexandria Engineering Journal*, Vol. 30, no. 2, 1991. Alexandria University.

Mostafa, M., A. El-Sherif, and M. Aziz. "Linkage Adder Mechanism," 1st International Symposium on Design and Synthesis, Tokyo, Japan, 1984.

Norton, R. L. *An Introduction to Synthesis and Analysis of Mechanisms and Machines*, 2nd ed., McGraw-Hill Inc., New York, 1999.

Phillips, J. *Freedom in Machinery*, Cambridge University Press, 1990.

Prakashan, N. *Theory of Machines and Mechanisms-II*, H. G. Phacatkar, 2008.

Rangwala, A. S. *Reciprocating Machinery Dynamics*, New Age International Ltd., New Delhi, 2006.

Rattan, S. S. *Theory of Machines*, Tata McGraw-Hill Education, New Delhi, 2009.

Ravi, V. *Theory of Machines, Kinematics*, PHI Learning Private Ltd., New Delhi, 2011.

Rua, J. S. *Theory of Machines Through Solved Problems*, New Age International (P) Ltd., New Delhi, 1996.

Rothbart, H. A. *Cam Design Handbook*, McGraw-Hill Prof Med/Tech, New York, 2004.

Rudman, J. *Auto Mechanics*, National Learning Corporation, October 1, New York, 2005.

Sadhu, S. *Theory of Machines*, 2nd ed., Dorling Kidersley (India) Pvt. Ltd., 2006.

Sandors, G. N. *Advanced Mechanism Design: Analysis and Synthesis*, Vol. 2, Prentice-Hall, Upper Saddle River, NJ, 1984.

Sandors, G. N. *Kinematics of Mechanisms*, 2nd ed., McGraw-Hill Book Company, New York, 1985.

Sclater, N. *Mechanisms and Mechanical Devices*, 5th ed., McGraw-Hill Book Company, New York, 1991.

Sharma, C. S. *Theory of Mechanisms and Machines*, Prentice-Hall of India, 2006.

Shigley, J. E. *Kinematic Analysis of Mechanisms*, McGraw-Hill Book Company, New York, 1969.

Shigley, J. E. *Theory of Machines and Mechanisms*, 2nd ed., McGraw-Hill Education, Pvt., New York, 1981.

Shigley, J. E. *Theory of Machines, Parts 1-2*, McGraw-Hill, New York, 1961.

Sneck, H. J. *Machine Dynamics*, Prentice Hall, NJ, 1991.

Soh, N. P. *The Principles of Design*, Oxford University Press, New York, 1990.

Taylor, C. F. *The Internal Combustion Engines in Theory and Practice*, MIT Press, Cambridge, MA, 1960.

Taylor, C. W. *Widening Horizons in Creativity*, John Wiley, New York, 1964.

Wison, C. E., and J. P. Sadler. *Kinematics and Dynamics of Machinery*, Pearson Education, Upper Saddle River, NJ, 2003.

Yan, H. S. *Creative Design of Mechanical Devices*, Springer, Singapore, 1998.

Index